D0966352

INSECT BIOCHEMISTRY
AND FUNCTION

Insect Biochemistry and Function

Edited by

D. J. CANDY

Lecturer in Biochemistry
University of Birmingham

and

B. A. KILBY

Reader in Biochemistry
University of Leeds

CHAPMAN AND HALL
London

A HALSTED PRESS BOOK

JOHN WILEY & SONS
New York

First published 1975
by Chapman and Hall
11 New Fetter Lane, London EC4P 4EE

© *1975 Chapman and Hall Ltd.*

Typeset by Preface Limited, Salisbury, Wilts

Printed in Great Britain by
Whitstable Litho

Library of Congress Cataloging in Publication Data

Candy, David John, 1936–
    Insect biochemistry and function.

    "A Halsted Press book."
    Includes bibliographical references.
    1. Insects——Physiology. 2. Biological chemistry.
I. Kilby, B. A., joint editor. II. Title
[DNLM: 1. Biochemistry. 2. Insects. QX500 C219i]
QL495.C26   1975   595.7'01'92   74-26569
ISBN 0-470-13347-3

Contributors

E. Bailey   Dept: of Biochemistry, The University, Sheffield, Yorkshire.

D. G. Cochran Dept: of Entomology, Virginia Polytechnic, Institute and State University, Virginia, U.S.A.

G. G. Lunt   Dept: of Biochemistry, School of Biological Sciences, University of Bath, Somerset.

B. Sacktor   Gerontology Research Center, National Institute of Child Health and Human Development, National Institute of Health, Baltimore, U.S.A.

Contents

	Page
Preface	xi
Chapter 1. Biochemistry of Insect Flight. B. SACKTOR	
Part 1 – Utilization of Fuels by Muscle	1
Introduction	3
Pathways for the Utilization of Carbohydrates, Amino acids and Fats in Flight Muscle	4
Physiological aspects of the bioenergetics of flight	4
Glycogenolysis	5
Utilization of trehalose	20
Glycolysis	22
Glycolytic-mitochondrial interactions	29
Auxiliary glycolytic reactions	33
Oxidation of the end-products of glycolysis and the permeability of mitochondria to metabolic intermediates	35
Utilization of amino acids	40
Oxidation of fat	45
The Control of Flight Muscle Mitochondrial Metabolism	49
Morphological and biochemical organization of flight muscle mitochondria	50
Oxidative phosphorylation	56
Control of the mitochondrial oxidation of α-glycerol-P	61
Control of the mitochondrial oxidation of proline	63
Control of the mitochondrial oxidation of pyruvate and the citric acid cycle	65

Mitochondrial ion movements and the special role of Ca²⁺
in the regulation of flight muscle metabolism   70
The oxidation-reduction state of the respiratory
components in mitochondria and the control of
metabolism   76
References   81

Chapter 2. Biochemistry of Insect Flight. E. BAILEY
Part 2 – Fuel Supply   89
Introduction   91
The Nature of Respiratory Fuels   92
Stores of Respiratory Fuels   93
Lipids and their Metabolism   94
  Tissue content of lipids   96
  Absorption of dietary lipids and incorporation into fat
    body lipids   107
  Lipid degradation by fat body   110
  Biosynthesis of lipids   117
  Lipid release from the fat body   137
Carbohydrates and their Metabolism   146
  Tissue content of carbohydrates   146
  Utilization of carbohydrate reserves during flight 148
  Metabolism of sugars other than glucose   150
  Glycogen biosynthesis   152
  Trehalose biosynthesis   154
  Carbohydrate synthesis from non-carbohydrate
    precursors   157
  Glycogen breakdown   159
  Hormonal control of carbohydrate metabolism   161
The Supply of Amino Acid Fuels for Flight   164
Conclusion   166
References   167

Chapter 3. Excretion in Insects. D. G. COCHRAN   177
Introduction   179
Excretion Associated with the Malpighian Tubule-Rectum
System   181
  Ions, water and active transport   182
  Uric acid   191
  Ammonia   215
  Urea   223
  Amino acids   228

Tryptophan derivatives 232
Pteridines 243
Other nitrogenous products 252
Carbohydrates 254
Miscellaneous excretory products 255
Xenobiotics 256
Minor Excretory Systems 258
Labial glands 259
Pericardial and other cells 260
Utriculi majores 261
The gut 262
Concluding Remarks 264
References 265
Addendum 282

Chapter 4. Synaptic Transmission in Insects. G. G. LUNT 283
Introduction 285
General Outline of the Insect Nervous System 285
Chemical Transmitters in the Central Nervous System 288
Acetylcholine 288
Acetylcholine receptors 289
Glutamic acid and γ-amino butyric acid (GABA) 290
Biogenic amines 293
Chemical Transmitters at the Neuromuscular Junction 294
Glutamic acid 294
Glutamate receptor 297
Enzymes associated with the glutamate system 298
γ-Amino butyric acid (GABA) 300
Biogenic amines 301
Summary 301
References 302
Addendum 305

Index 307

Preface

There has been a considerable upsurge in interest in insect bio-chemistry and physiology in recent years and this has been reflected in a notable expansion in the number of original papers in this field. Whereas insect physiology has tended to receive ample attention from reviewers, the same has not always been true for the more biochemical aspects of insect research. This book is a venture to help redress the balance. No attempt has been made to cover all aspects of insect biochemistry, but rather a few topics have been selected which seemed to us to merit a review at the present time.

One reason for this increased interest in insect biochemistry is perhaps the growing realization that insects can be very useful organisms to act as model systems for the experimental study of general biochemical principles. One remembers, for instance, that Keilin's perceptive observations on the flight muscles of living bees and wax moths led to his discovery of the cytochromes. The fundamental unity of biochemistry has long been accepted as a dogma by the faithful and the insect kingdom provides no exception to it. The main biochemical processes in insects are being revealed as essentially the same as in other life forms but, as so often found in comparative biochemistry, there are interesting variations on the central theme. Often by choosing an appropriate insect species, one can find an example of a particular biochemical mechanism which has been developed or modified to a considerable extent and the study of such a system may be most informative about general biochemical mechanisms. It is this correlation between biochemical specialization and physiological function that can make the subject so fascinating. A good example of such a process is insect flight, which clearly illustrates the numerous biochemical adaptations that can occur concomitant with a specific function. A discussion of this subject has been given particular attention in the first half of this book.

Another reason for the increased activity in insect biochemistry is an economic one. The production of insecticides is now a major industry and one in which constant research for new products is essential. Insects soon develop resistant strains under selection pressure from the use of a particular insecticide and the development of new ones is frequently required. One would hope that an increased knowledge of insect biochemistry and physiology would facilitate a rational approach to the development of new lethal agents which interfere with the normal life processes in insects. The aim, however, must be selective toxicity in which the agent is insect-specific while man, other animals and if possible, beneficial insects are unaffected and unharmed. This selectivity would not be expected if the insecticide interferes with some fundamental process common to most organisms as happens, for instance, with the use of cyanide to inhibit cytochrome oxidase. Rather, the ultimate target of the insecticide must be some biochemical or physiological adaption or modification which is essential for the insect but absent or unimportant in the species which one does not wish to harm. A detailed knowledge of biochemical differences between insects and other life forms could have great potential and importance and we hope that this book may perhaps serve to stimulate further interest and research in these areas, and eventually assist in the rational design of selective insecticides.

D. J. C.

October 1974   B. A. K.

1 Biochemistry of Insect Flight
Part 1 *Utilization of fuels by muscle*
B. SACKTOR

INTRODUCTION

The intensity and precise control of metabolic processes in the flight muscle of insects during flight have engaged the attention of biochemists and physiologists for many years. The over-all level of metabolism in the working muscle may be estimated from either the depletion of the animals' depots of fuel or the respiratory exchange. Values as high as 2400 cal/g muscle/h for the bee during prolonged periods of continuous flight have been reported (Weis-Fogh, 1952). This value is 30- to 50-fold those for leg and heart muscle of man at maximum activity. With respect to the utilization of energy reserves, it has been found that the locust during flight consumes fat at a rate of 4.1 mg per h (Beenakkers, 1965). The cost of flight in insects becomes fully evident from a comparison of the rate of oxygen uptake during flight with that of the same insect at rest. On initiation of flight, some blowflies consume oxygen at a rate of about 3000 μl/min/g, elevating their basal rates approximately 100-fold (Davis and Fraenkel, 1940). This is the most intense respiration known in biology and also the most controlled. By contrast, the respiration of humming birds in flight, although much greater per unit weight than that recorded for any other vertebrate, is subject to a control of only 5-fold (Pearson, 1950). From these few selected examples, and others reported elsewhere (Sacktor, 1965, 1970), it is apparent that insect flight muscle is the tissue of choice for the study of the control of catabolism and biological oxidations, and many of the results obtained with insects have a significance and a relevance which transcend the boundaries between classes.

It is proposed in the sections that follow to summarize the more significant early findings and to detail recent contributions to our knowledge of bioenergetic systems in insect flight muscle. Emphasis will be placed on the principal pathways for the utilization of carbohydrates, fats, and some amino acids in this highly specialized tissue; the functional description of the unique mitochondria of flight muscle and the role of these mitochondria in biological oxidations; and the control and integration of the different metabolic processes. For historical perspectives and more complete accounts of the earlier studies on oxidative processes the reader is referred to previous reviews (Sacktor, 1965, 1970; Hansford and Sacktor, 1971), and for descriptions of glycolysis and intermediary metabolism to Wyatt (1967), Friedman (1970), and Sacktor (1970). A discussion of the storage, metabolism, and transport of energy reserves in the fat body and other tissues of the insect are to be found in the review by Dr Bailey (pp. 91–176).

PATHWAYS FOR THE UTILIZATION OF CARBOHYDRATES, AMINO ACIDS, AND FATS IN FLIGHT MUSCLE

Physiological aspects of the bioenergetics of flight

The rates of oxygen uptake by flying insects are indicative of the rates of oxidative processes taking place in the working muscles. During flight the respiratory rates are often increased 20- and may be as high as 100-times those found in non-flying insects. In the example of the blowfly, *Lucilia sericata,* already cited, Davis and Fraenkel (1940) reported an average consumption of oxygen during flight of 1625 $\mu l/min/g$, and at rest, (but not basal), of 33–50 $\mu l/min/g$, indicating a 60-fold change on flight as a minimum. Such large increases in respiration on initiation of flight are not restricted to the *Diptera* and *Hymenoptera,* which have the asynchronous type of excitation-contraction coupling, and are characterized by high frequency of movements of their wings. Essentially identical increases in oxygen uptake between individuals at rest and during flight have been observed in *Orthoptera* and *Lepidoptera,* which have the synchronous type of excitation-contraction coupling and, in general, have relatively slow rates of wingbeat. For instance, in a variety of moth species, Zebe (1954) reported oxygen uptakes of from 7–12 $\mu l/min/g$ at rest, which increased to values of 700–1660 $\mu l/min/g$ during flight, – an increment of over 100-times in some cases. Similar increases at the start of flight have been observed with a variety of species. (Sacktor, 1965).

In spite of such enormous respiratory rates, the observations that fruit flies (Chadwick, 1953) and locusts (Krogh and Weis-Fogh, 1951) can maintain flight for hours while accruing no, (or only a small) oxygen debt, indicates that the metabolic processes are not limited by the availability of oxygen. In insects, haemoglobin and myoglobin are absent, and air is conveyed directly to the flight muscle through an elaborate conduit of tracheae which invade the fibres and are in close juxtaposition with each mitochondrion. The minute distances between the tracheoles and the mitochondria suggest that diffusion suffices to transport at least part of the extra oxygen utilized during flight. In fact, Weis-Fogh (1964, 1967) calculated that in small insects (e.g., flies) diffusion of respiratory gases is sufficient to account for the entire transport between the spiracles and the end of the tracheoles, even at the highest rate of metabolism. In larger insects, such as dragon flies, locusts, and wasps, the primary tracheole supply must be strongly ventilated, while diffusion is sufficient in the remaining part of the air tubes.

The nature of the metabolic fuel reserve that is utilized during flight of various insects was discussed previously (Sacktor, 1965). In general, *Diptera* and *Hymenoptera* have a RQ equal to unity and carbohydrates are the main, if not the exclusive, substrate. In other insects, including many *Lepidoptera* and *Orthoptera,* RQ values of 0.73 are found and fats are depleted, even though some species (moths) were gorged with glucose (Zebe, 1954). Locusts, roaches, and aphids may use both carbohydrates and fat. Glycogen and trehalose are used during initial periods of flight but as flight continues the RQ decreases and fat becomes the principal fuel, being able to sustain flight for hours. More recently, Van Handel and Nayar (1972) have seriously questioned the exclusive use of fats in *Lepidoptera*; in fact, they clearly demonstrated the direct use of carbohydrates during the flight of the moth, *Spodoptera frugiperda.* The use of carbohydrates during flight of some moths would also be indicated by the findings of Gussin and Wyatt (1965) and Stevenson (1968a) that *Hyalophora cecropia* and *Prodenia eridania* flight muscle homogenates oxidize glycogen and sugars and the enzyme trehalase is moderately active in these preparations. Perhaps there is a distinction between *Lepidoptera* that feed and those that do not feed as adults, since the activities of some glycolytic enzymes of flight muscles of *Pieris* and *Agrotis,* which feed, are much greater than those of *Philosamia* and *Actias,* which do not feed (Beenakkers, 1969). Accordingly, the flight muscles of the moths and butterflies that feed on nectar may utilize sugars as well as fats whereas the muscles of those *Lepidoptera* that utilize reserves derived from larval feeding rely primarily on lipids. The perfusion technique for flight muscle, developed by Candy (1970), may be useful to test whether or not a potential substrate can be utilized directly by flight muscle. The role of amino acids as substrates for flight was considered earlier (Sacktor, 1961, 1965). The rapid utilization of proline on the initiation of flight of the blowfly, and its significance (Sacktor and Wormser-Shavit, 1966) will be dealt with later. A unique example of the utilization of proline as an energy-furnishing reserve is found in the tsetse fly (Bursell, 1963; 1966).

Glycogenolysis

The large deposits of glycogen in flight muscle as well as the depletion of these reserves during flight indicate that in many insects, especially *Diptera,* glycogen provides a major vehicle for storage of flight energy which is rapidly mobilized to meet the metabolic requirements of the muscle. For example, in the blowfly, *Phormia regina,* fed *ad lib.,* glycogen comprises 10—15 mg/g wet wt. of the thorax (Childress *et al.,* 1970). During flight, approximately 2.5 µmoles (as equivalents of glucose)/g wet wt. are utilized each minute until the polysaccharide is depleted, after about 10—15 min (Sacktor and Wormser-Shavit, 1966). In an uninterrupted forced flight, lasting an hour or more, different loci of carbohydrate stores, including those in the fat body and gut, are used (Clegg and Evans, 1961; Sacktor, 1965). In prolonged flights of the blowfly, Sacktor and Wormser-Shavit (1966) found that during the first 5 min the concentration of glycogen in the fat body does not significantly decrease but after about 15 min the depletion of this depot is large and becomes larger with flights of longer duration. Thus, during flight, glycogen is mobilized from its two principal storage loci at independent rates, glycogen in muscle being used before that in fat body.

Histochemical examination of the flight muscle of *Drosophila melanogaster* by light microscopy indicates glycogen deposits along the surface of the muscle masses, around their points of insertion, and between smaller fibre bundles and individual fibrils (Wigglesworth, 1949). Electron microscopic examination of *P. regina* flight muscle fibres *in situ* reveals that glycogen is located mostly in the interfibrillar sarcoplasm, with scattered deposits on the myofibrils (Childress *et al.,* 1970). In locust flight muscle, Seiss and Pette (1960) showed histochemically that the polysaccharide is aggregated principally in the isotropic zone. Fibres with a high concentration of glycogen occasionally show an additional deposition in the M-band. Similar myofibrillar distributions are seen in the flight muscle of the blowfly, *P. regina,* (Sacktor and Shimada, 1972) and blackfly, *Simulium vittatum* (Liu and Davies, 1971). As illustrated in Fig. 1.1a, glycogen exists in the form of rosettes or *alpha* particles which vary in size and may reach a diameter of 0.25 micron. The individual components comprising the rosettes, the *beta* particles, are not clearly delineated. However, the polydisperse nature of native glycogen is apparent from electron micrographs of pure glycogen (Fig. 1.1b) isolated by mild buffer extraction (Childress *et al.,* 1970). Incubation of the flight muscle tissue with amylase, following fixation results in the disappearance of the glycogen rosettes, leaving only an amorphous network of electron dense material in the

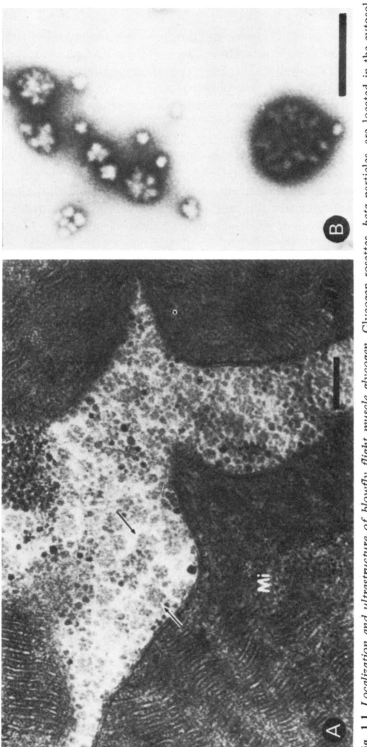

Fig. 1.1 *Localization and ultrastructure of blowfly flight muscle glycogen.* Glycogen rosettes, *beta* particles, are located in the cytosol between the mitochondria and myofibrils. (A) Glycogen rosettes (arrows) are prominent between mitochondria (Mi). X 75 000, bar represents 0.2 μm. (B) Pure glycogen, isolated by mild procedures, in the form of *beta* particles is shown after negative staining with phosphotungstic acid. X 76 590, bar represents 0.2 μm. (From Childress *et al.*, 1970).

interfibrillar region. Sections of the muscle examined by light microscopy following this treatment are no longer PAS positive.

Sedimentation analyses of pure native glycogen from blowfly flight muscle show that it is polydisperse. Over 50 per cent of the glycogen particles have sedimentation coefficients in excess of 300 corresponding to molecular weights as high as 50–100 million (Childress *et al.,* 1970). Chemical characterization of the isolated native glycogen indicates that the outer chains are quite short, only about 25 per cent of the total glucosyl residues being released by phosphorylase alone. Treatment of native glycogen with hot alkali (typical conditions for the extraction of glycogen from tissues), increases the release of glucosyl residues by phosphorylase, to 31 per cent. This reflects the availability of additional terminal residues, which previously were resistent to enzymatic attack, perhaps because of steric hindrance. On incubation with phosphorylase *b*, amylo-1, 6-glucosidase, AMP and inorganic phosphate (P_i), purified native glycogen is degraded completely to glucose-1-P (90.5 per cent) and glucose (9.5 per cent). Less than 0.1 per cent is protein.

The significance of using native glycogen in kinetic studies of glycogenolysis is evident from the findings that flight muscle phosphorylase *a* has a lower affinity for native flight muscle glycogen than for the same substrate treated with alkali or for preparations of glycogen from other species. The apparent K_m values are 0.09 and 0.29 mM for the alkali-treated and native glycogens, respectively, in the presence of saturating levels of P_i and AMP. A much greater difference between K_m values for the two substrates is observed at low AMP levels. In addition, phosphorylase *a* has differing affinities for P_i depending on which glycogen is present as co-substrate. With saturating levels of AMP and glycogen, the apparent K_m values for P_i are 4.5 and 9.5 mM, for alkali-treated and native glycogens respectively. Values for maximum velocity with native substrate are approximately 50 per cent of the values obtained with KOH-treated or commercial glycogens. Differences between native glycogen and glycogen artificially modified either by isolation procedures or by commercial preparation from other organisms, emphasizes the importance of using the natural substrate in kinetic measurements used in predicting parameters involved in the control of glycogenolysis *in vivo.*

Glycogen phosphorylase, in both *a* and *b* forms, from flight muscle of the blowfly, *P. regina,* has been purified to a high degree of homogeneity, and characterized (Childress and Sacktor, 1970). Glycogen phosphorylase catalyses the reaction:

$$\text{Glycogen}_n + P_i \rightleftharpoons \text{Glycogen}_{n-1} + \text{Glucose-1-P}$$

It has been calculated that the phosphorylases comprise approximately 1.5 per cent of the total muscle protein and have a potential activity of 9.6 μmol/min/g wet wt. of thorax, a value more than adequate to account for the rate of glycogenolysis during flight. Estimations of the molecular weight of phosphorylase *b* by sedimentation equilibrium, gel filtration, and calculation from sedimentation and diffusion coefficients indicate a value of approximately 100 000. Flight muscle phosphorylase *a*, purified as the *a* form or converted from *b* to *a* with phosphorylase *b* kinase and ATP, has the same sedimentation coefficient (S_{20} = 7.4) and molecular weight as does the *b* form of the enzyme. The amino acid composition of flight muscle phosphorylase *b* has been determined (Childress and Sacktor, 1970), and compared with analyses of human, rabbit, and frog muscle phosphorylases (Sacktor, 1970). Although the overall compositions are similar, striking differences between the flight muscle enzyme and the vertebrate muscle enzymes are found in the contents of half-cystine, arginine, and lysine residues. So far, no characterization of phosphorylase from the flight muscle of other insect species has been carried out, although enzymic activities have been noted in the muscles of the roach, *Blaberus discoidalis,* (Wyatt, 1967), the locust, *Locusta migratoria,* (Goldsworthy, 1970) and the silkmoth, *Hyalophora cecropia* (Wiens and Gilbert, 1967).

The kinetic properties of the purified phosphorylases *a* and *b*, from blowfly flight muscle have been examined (Childress and Sacktor, 1970), and the interactions between the cosubstrates, glycogen and P_i, with phosphorylase *a* are shown in Fig. 1.2. These double, reciprocal plots result in a series of straight lines which intersect in the fourth quadrant, suggesting a bimolecular, sequential mechanism in which increasing levels of either substrate enhance the binding of the other. The values for apparent K_m for P_i range from 47 mM at 0.15 mM glycogen to 7.3 mM at 1.7 mM glycogen; the values for apparent K_m for glycogen range from 6.4 mM at 1.6 mM P_i to 0.72 mM at 40.3 mM P_i. In contrast to phosphorylase *a*, double reciprocal plots of velocity against substrate concentration for phosphorylase *b* are non-linear. However, as in the case of phosphorylase *a*, increasing levels of one substrate enhance the binding of the other.

Although it is not required for activity, AMP stimulates phosphorylase *a* 2- and 3-fold at saturating levels of substrates and

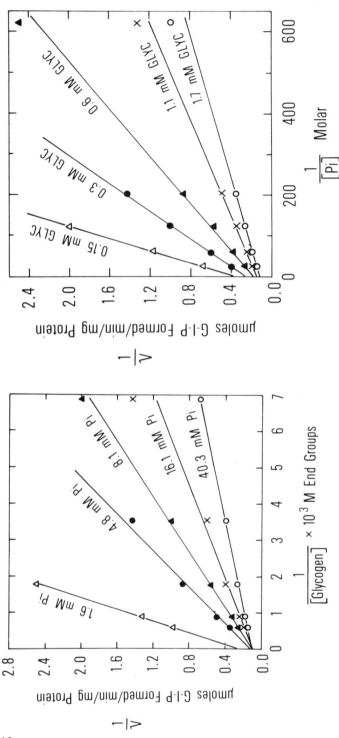

Fig. 1.2 *Kinetic properties of blowfly flight muscle phosphorylase a.* Reciprocals of initial velocity of phosphorylase *a* activity are shown as a function of reciprocals of concentration of substrate, at several fixed levels of cosubstrate. The AMP concentration was 1.6 mM. The values of apparent K_m for P_i were, respectively, 47.6, 23.7, 16.3, 11.2, and 7.3 mM in order of the lowest to the highest glycogen concentration. The values of apparent K_m for glycogen were, respectively, 6.4, 4.5, 2.6, 1.7 and 0.7 mM in order of the lowest to the highest P_i concentration. (From Childress and Sacktor 1970.)

10-fold or higher at low substrate levels. The apparent K_m for AMP is 0.6 μM at saturating levels of P_i and glycogen. Lowering the level of either substrate decreases the affinity of the enzyme for the activator. Moreover, AMP increases the affinity of phosphorylase *a* for both substrates. For example, the apparent K_m for P_i is lowered from 100 mM to 9 mM in the presence of low and high concentrations of AMP, respectively. Unlike the *a* form of the enzyme, phosphorylase *b* has an absolute requirement for AMP. Furthermore, levels of AMP 100-fold greater than those which stimulate phosphorylase *a* are needed to stimulate phosphorylase *b*. Increasing amounts of AMP lower the apparent K_m values for both P_i and glycogen. As illustrated in Fig. 1.3, the apparent K_m values for P_i range from 100 mM at the lowest, to 5 mM at the highest level of AMP. Double, reciprocal plots of velocity and AMP concentration reveal that increasing levels of glycogen or P_i lowers the affinity of phosphorylase *b* for AMP. It has been estimated from the data in Fig. 1.3 that the apparent K_m values for AMP decreases from 2 mM to 0.2 mM as the P_i concentration increases from 4.8 mM to 40 mM.

Childress and Sacktor (1970) have found that ATP is a potent inhibitor of phosphorylase *b* but not of phosphorylase *a*. The K_i value is approximately 2 mM (Fig. 1.3). At high AMP concentrations the inhibition by ATP is less, suggesting competitive inhibition with respect to AMP. This view is strengthened by the observations that ATP increases the apparent K_m values for both glycogen and P_i (Fig. 1.3), as if AMP has been displaced from its binding site on the enzyme.

These detailed kinetic data for blowfly flight muscle phosphorylases, coupled with a knowledge of the concentrations of the substrates, activator and inhibitor in the muscle during rest and flight (Sacktor and Wormser-Shavit, 1966; Sacktor and Hurlbut, 1966), permit estimates of the phosphorylase activities under simulated conditions *in vivo*. As shown in Table 1.1, the level of glycogen in the muscle is sufficient to saturate the enzymes, except after 10 min or more of flight when the muscle glycogen reserve is near depletion. The concentration of P_i in the muscle at rest is 7.0 mM, increasing to 7.5 mM during flight. On the other hand, the concentration of AMP in the muscle is increased sharply from 0.1 mM to 0.3 mM in the rest to flight transition. However, examination of the apparent K_m values for P_i and AMP reveals that each ligand has a marked effect upon the affinity of the enzyme for the other in these concentration ranges. For phosphorylase *b*, the apparent K_m for Pi at 0.1 mM AMP is about 100 mM, very much above the level found in the muscle *in*

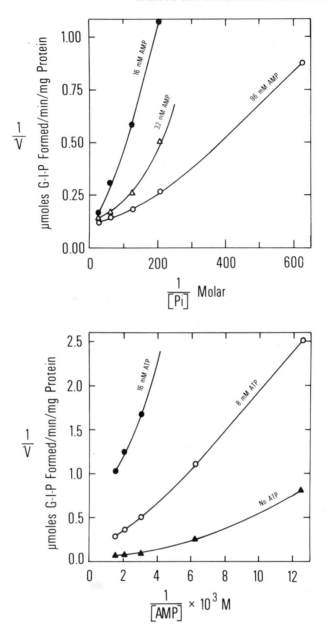

Fig. 1.3. *Kinetic properties of blowfly flight muscle phosphorylase b. Upper left:* Double reciprocal plots of initial velocity of phosphorylase *b* as a function of P_i concentration at several levels of AMP. The concentration of glycogen is 2 mM. *Upper right*: Double reciprocal plots of initial velocity of the enzyme as a function of AMP concentration at several levels of P_i. Glycogen concentration is 2 mM. *Lower left*: Double reciprocal plots showing the effect of ATP on the

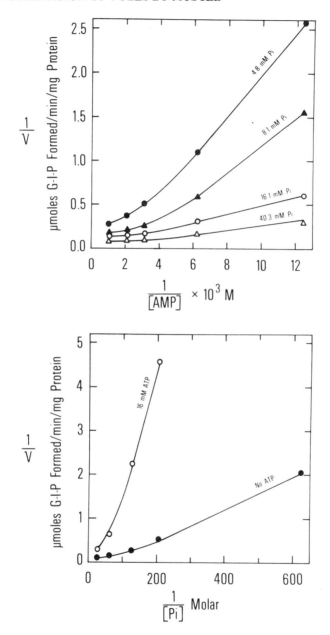

initial activity of phosphorylase *b* as a function of AMP concentration. Substrate concentrations are 16 mM P_i and 2 mM glycogen. *Lower right*: Double reciprocal plots showing the effect of ATP on the initial activity of phosphorylase *b* as a function of P_i concentration. The concentrations of AMP and glycogen are 0.32 mM and 2 mM, respectively. (From Childress and Sacktor, 1970.)

Table 1.1

Comparison of metabolite levels and apparent K_m values for flight muscle phosphorylases under conditions in vivo

Metabolite*	Level at rest in vivo	Level during flight in vivo	Apparent K_m value[†]		Potential activity[‡] at simulated conditions of Rest		Flight	
			b	a	b	a	b	a
	mM		mM					
Glycogen	6–9	6 → 0.75§	0.6	1.7				
P_i	7.0	7.5	100	8				
AMP	0.1	0.3	1.0	0.001	nil	50	nil	50
ATP	7.0	6.5	2(K_i)					

*Metabolite levels are reported as μmol/g wet wt. of thorax. These are probably minimum values for the metabolite concentrations in the muscle. If one assumes that the flight muscle represents half of the wet weight of the thorax and the chitinous exoskeleton accounts for the other half of the thoracic weight but contributes nothing to the metabolite content, maximal values double those given are obtained. The true values are in between the two extremes but closer to those reported in the table. The differences between use of the minimal or maximal values in calculating the potential activity at simulated conditions are essentially insignificant.

†The apparent K_m values were determined from the kinetic data with the concentrations of co-substrate and activator, as found *in vivo*, at rest.

‡Activities are expressed as the percentage of the potential phosphorylase activity with saturating conditions of substrate and AMP, without ATP.

§The level of glycogen in the muscle *in vivo* decreased steadily during flight to a value of 0.75 mM after 10 min of flight. It remained steady at this exhausted level throughout the remainder of the flight. A glycogen concentration of 1 percent corresponds to 5.9 mM end groups.

(From Childress and Sacktor 1970.)

vivo. The apparent K_m of phosphorylase *b* for AMP at 8 mM P_i is about 1.0 mM, about 10-fold that in the muscle at rest. These observations indicate that the activity of the *b* form would be limited to only a trace of its potential activity at substrate saturation. In addition, the strong inhibitory effect of ATP on phosphorylase *b*, at the low concentration of AMP in the tissue, would decrease the activity of phosphorylase *b* to almost zero.

In contrast, for phosphorylase *a*, the concentration of AMP in the muscle is 100-fold the apparent K_m of the nucleotide. Furthermore, ATP does not inhibit the *a* form of the enzyme. The apparent K_m of phosphorylase *a* for P_i, with conditions simulating those *in vivo*, is

8 mM, a value approximating the concentration in the muscle. Thus, about 50 per cent of the potential activity of phosphorylase *a* would be found and glycogenolysis by phosphorylase *a* would be moderately responsive to changes in the level of P_i.

Based on the specific activity for flight muscle phosphorylase of 9.6 units per g wet wt. of thorax, assayed with native glycogen at 30°C (Childress *et al.*, 1970), and the fact that the potential activity of phosphorylase is the same whether it exists in the *a* or the *b* form, it can be estimated from the kinetic data that at least 50 per cent of the total phosphorylase must be in the *a* form in order to account for the rate of glycogen breakdown that occurs during flight, 2.4 μmol of glucosyl residues per min per g wet wt. of thorax (Sacktor and Wormser-Shavit, 1966). It is suggested, therefore, that to satisfy this rate of glycogenolysis, phosphorylase *b* must be converted to phosphorylase *a* on initiation of flight. This view is supported by direct measurements of the two forms of phosphorylase in the muscle, during rest and during flight (Childress and Sacktor, 1970). In 'resting' blowflies 18 per cent of the phosphorylase is in the *a* form. Initiation of flight induces an immediate increase in the relative amount of the enzyme in the *a* form, to reach a maximum of about 70 per cent at 15 s of flight. This maximum is maintained during a flight of 10 min. The total amount of phosphorylase present (*a* plus *b*) does not change during flight. This level of phosphorylase *a* is adequate to account for the observed rate of glycogenolysis during flight. From the kinetic data obtained under *in vitro* conditions, a 70 per cent level of phosphorylase *a* can catalyze a rate of glycogenolysis of 4.3 μmol glucosyl residues /min/g wet wt. of thorax as compared to the rate of 2.4 actually measured. If, in 'resting' blowflies, the value of 18 per cent of the total phosphorylase in the *a* form is too high because of technical difficulties in extracting the phosphorylases in their state *in situ*, and if these factors increase the amount of phosphorylase *a* in the flown fly to the same extent, the agreement between the predicted rate of glycogenolysis from studies *in vitro* and that in the flying insect would be even more exact.

The conversion of phosphorylase *b* to phosphorylase *a* is catalyzed by the enzyme, phosphorylase *b* kinase. The mechanisms by which phosphorylase *b* kinase becomes activated are, therefore, of significance to the regulation of flight muscle metabolism. The kinase is localized with phosphorylase *a* phosphatase, phosphorylases *a* and *b*, and glycogen in the post-mitochondrial supernatant of the blowfly flight muscle and is readily isolated. Hansford and Sacktor (1970a) have found that Ca^{2+}, at physiological concentrations, activates phosphorylase *b* kinase. Stimulation of the kinase is evident at concentrations of Ca^{2+} as low as 10^{-8} M, with maximal enhancement at about 10^{-6} M. Mammalian skeletal muscle phosphorylase *b* kinase is also activated by Ca^{2+} (Ozawa *et al.*, 1967; Heilmeyer *et al.*, 1970). However, comparative studies show that the responses of the enzymes from the two tissues to Ca^{2+} are different (Sacktor *et al.*, 1974). As illustrated in Fig. 1.4, rabbit, skeletal muscle kinase has virtually no activity in the absence of Ca^{2+}, and exhibits a large increase in activity over a narrow range of Ca^{2+} concentrations. The phosphorylase kinase from blowfly flight muscle has appreciable

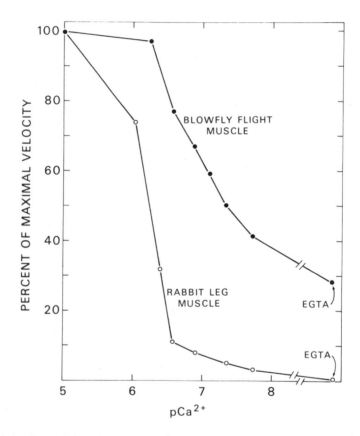

Fig. 1.4. *A comparison of the effect of different concentrations of Ca^{2+} on the activities of phosphorylase b kinases from blowfly flight muscle and rabbit leg muscle.* The incubation mixture contained 100 mM tris acetate buffer, pH 7.1, 10 mM magnesium acetate, 5 mM ATP, and 1 unit/ml of flight muscle or leg muscle phosphorylase *b*. Each incubation contained 2 mM EGTA, and an amount of $CaCl_2$ sufficient to give the levels of Ca^{2+} indicated. The points EGTA on the plots correspond to no added $CaCl_2$. (Sacktor *et al.*, 1974.)

activity in the absence of exogenous Ca^{2+}, and has a smaller increase, only 3-fold, over a wide range of Ca^{2+} concentration. The concentrations of Ca^{2+} required for half-activation are 10^{-7} M and 10^{-6} M for blowfly and rabbit enzymes, respectively.

The activation of phosphorylase *b* kinase by Ca^{2+} provides support for the concept that in muscle the conversion of phosphorylase *b*, to *a* represents an essential component in the mechanism coupling contraction to glycogenolysis. The low Ca^{2+} sensitivity of the kinase in blowfly flight muscle may be of additional physiological importance. Flight muscle of the blowfly show contraction-relaxation cycles at a constant Ca^{2+} concentration (Jewell and Ruegg, 1966), and Ca^{2+} concentration can be varied considerably without loss of oscillatory work (Pringle and Tregear, 1969). Moreover, blowfly muscle has very little sarcoplasmic reticulum (Smith, 1966) and it appears that rapid segregation of Ca^{2+} within membrane-bounded vesicles is not required for oscillatory behaviour of the asynchronous flight muscle. Therefore, a high Ca^{2+}-sensitivity for phosphorylase *b* kinase would be disadvantageous. On the other hand, muscles having a well developed sarcoplasmic reticulum system and possessing a relatively slow, synchronous pattern of contraction-relaxation, are able to segregate Ca^{2+} more effectively. Thus, phosphorylase *b* kinase activity in muscles of this type can be fully controlled by changes in Ca^{2+} concentration. It is probably more than coincidental that the distinction between asynchronous and synchronous types of muscle in the Ca^{2+}-sensitivities of their phosphorylase *b* kinases has a precise counterpart in the distinction between asynchronous and synchronous muscles in the Ca^{2+}-sensitivities of their actomyosin ATPases (Maruyama *et al.*, 1968).

In the blowfly, *P. regina*, flight muscle phosphorylase *b* kinase is also stimulated several-fold by high concentrations of P_i (Hansford and Sacktor, 1970a; Sacktor *et al.*, 1974). The apparent K_m is about 20 mM. Maximal activation of the insect kinase requires Mg^{2+} in addition to P_i. The apparent K_m for Mg^{2+} is 3 mM. This enhancement in activity seems to involve the conversion of the kinase from a 'non-active' to an 'active' form. It differs, however, from the conversion of the 'non-activated' to the 'activated' form of the enzyme from rabbit skeletal muscle, as described by Krebs *et al.* (1964) and Huston and Krebs (1968), in that the blowfly kinase activated by P_i has an immeasurably high K_m for phosphorylase *b* at neutral pH (Hansford and Sacktor, 1970a). The insect enzyme shows no appreciable change in the relative activities at pH 6.8 and 8.2 (Sacktor *et al.*, 1974), and Ca^{2+}-dependent proteolysis is prevented by the presence of EGTA or EDTA. Further contrasting the mammalian phosphorylase *b* kinase from the blowfly enzyme, is the observation that the activation of the insect muscle kinase does not require ATP, nor is the activation dependent on cyclic AMP (Sacktor *et al.*, 1974). Thus, no evidence is found to suggest the presence in blowfly flight muscle of a protein kinase that phosphorylates and activates phosphorylase *b* kinase. Although corpus cardiacum hormone may induce glycogenolysis in the fat body of insects (Bowers and Friedman, 1963), the slow (2 h) response of the fat body enzyme to the hormone plus the apparent absence of cyclic AMP-dependent protein kinase in blowfly flight muscle makes a hormonal mechanism for the regulation of glycogenolysis in muscle highly questionable.

Conversion of phosphorylase *a* to phosphorylase *b* is catalysed by the enzyme, phosphorylase *a* phosphatase. The phosphatase is very active in flight muscle of the blowfly, *P. regina,* and its activity is inhibited by NaF (Childress and Sacktor, 1970), as is true for the enzyme in mammalian tissues. However, no information is available as to the regulation of this enzyme in the flight process. Active and inactive forms of the phosphatase in several mammalian tissues have been suggested (Merlevede and Riley, 1966). Studies on whether analogous forms are present in insect tissues have not been reported.

Little is known of the mechanisms by which glycogen is synthesized in flight muscle, despite the importance of the utilization of the polysaccharide during muscular work. The evidence indicates that in insect muscle, as in both insect fat body and mammalian tissue, synthesis of the α-1, 4-glucosidic linkage in glycogen is catalyzed by the enzyme UDP-glucose-glycogen transglycosylase (glycogen synthetase):

$$\text{UDP-glucose} + \text{Glycogen}_n \longrightarrow \text{UDP} + \text{Glycogen}_{n+1}$$

The UDP-glucose is generated by UDP-glucose pyrophosphorylase by the reaction:

$$\text{Glucose-1-P} + \text{UTP} \rightleftharpoons \text{UDP-glucose} + \text{PP}_i$$

Trivelloni (1960) has shown that extracts of thoracic muscles of the locust, *Schistocerca cancellata,* incorporate ^{14}C-glucose from UDP-^{14}C-glucose into glycogen with release of UDP. Histochemical observations on locust flight muscle, show that little glycogen is formed by the transglycosylase reaction (Hess and Pearse, 1961). They suggest that a reversal of the phosphorylase reaction is more likely to be the principal mechanism, although this claim has yet to be verified as quantitation by the histochemical procedure is always difficult (Sacktor, 1970).

Control of glycogen synthesis in flight muscle has not been studied. In other insect tissues the synthetase is activated by glucose-6-P. In the fat body of larvae of the silkmoth, *Hyalophora cecropia,* the K_a for glucose-6-P is 0.6 mM and the K_m for UDP-glucose is 1.6 mM (Murphy and Wyatt, 1965). Glucose-6-P activates without significantly changing the K_m for UDP-glucose, an effect which resembles that found with the D-form of the enzyme in mammalian tissues (Rosell-Perez and Larner, 1964). In honey-bee larvae, the K_a is slightly less than that in the silkmoth, but glucose-6-P decreases the K_m of the enzyme for UDP-glucose about 10-fold, 2.1×10^{-3} M to 3.3×10^{-4} M (Vardanis, 1967). Thus, the enzyme from the bee has properties of both the D- and L-forms of the mammalian enzyme.

In mammalian tissues, glycogen synthetase, like glycogen phophorylase, exists in interconvertible phosphorylated and de-phosphorylated forms, except that the less active, D-form, is phosphorylated and the more active, L-form, is dephosphorylated (Larner and Villar-Palasi, 1971). Compared to the dephosphorylated form, the phosphorylated form is more sensitive to inhibition by ATP and this inhibition is not as readily reversed by glucose-6-P (Piras *et al.,* 1968). In fact, in the presence of physiological concentrations of ATP the phosphorylated form is essentially inactive. Again analogous to modifications of phosphorylase, phos-phorylation of the synthetase is mediated by a kinase and a phosphatase converts the phosphorylated synthetase to its dephos-phorylated form. Significantly, the protein kinase that phos-phorylates glycogen synthetase is the same enzyme that phosphorylates phosphorylase *b* kinase. Thus, processes that stimu-late protein kinase enhance the phosphorylation of glycogen synthetase and minimize the synthesis of glycogen, also bring about the concomitant phosphorylation of phosphorylase *b* kinase, so stimulating glycogen breakdown. Conversely, the actions of phosphatases in dephosphorylating the phosphorylated forms of glycogen synthetase and phosphorylase *b* kinase and phosphorylase *a* increase the synthesis of glycogen and coincidently decrease glyco-genolysis.

As noted above, however, there is no evidence to suggest that blowfly flight muscle possesses a protein kinase. Rather, phos-phorylase *b* kinase seems to be activated primarily by Ca^{2+} and P_i. If this is true, then glycogen synthetase in this muscle may not be subject to control by a phosphorylation-dephosphorylation mechanism. Obviously, the question remains to be resolved.

Utilization of trehalose

In addition to glycogen, the disaccharide trehalose (1-α-D-gluco-pyranosyl-α-D-glucopyranoside) supports flight muscle activity. Trehalose is the principal blood sugar in many species of insects (Wyatt, 1967) and it is also found in muscle. The concentrations of the sugar in these loci are markedly reduced after flight of locusts (Bücher and Klingenberg, 1958), cockroaches (Polacek and Kubista, 1960), blowflies (Clegg and Evans, 1961; Sacktor and Wormser-Shavit, 1966), and houseflies (Srivastava and Rockstein, 1969). A correlation has been observed between the concentration of trehalose in the blood and the frequency of wingbeat in *P. regina* (Clegg and Evans, 1961). Sacktor and Wormser-Shavit (1966) have measured sequential changes in the concentration of trehalose on initiation of flight and during the steady state of sustained flight of the blowfly. The concentration of disaccharide falls precipitously, by approximately 1 μmol/g wet wt. of thorax during the first 5 s, and continues to decrease rapidly for about 30 s. The level decreases progressively but at a lesser rate during the remainder of flight. This discontinuity in the rate at which trehalose is utilized indicates two kinetically different pools of the sugar. The pool that is metabolized at a greater rate and becomes exhausted within 30 s is considered to be the trehalose in the muscle. The other pool, that is depleted gradually during sustained flight has kinetics of utilization resembling closely those of the decrease in the concentration of trehalose in the thoracic blood during flight.

The enzyme trehalase hydrolyses trehalose into two glucose moieties:

$$\text{Trehalose} + \text{H}_2\text{O} \longrightarrow 2 \text{ Glucose}$$

Trehalase was first demonstrated in housefly, *Musca domestica,* flight muscle (Sacktor, 1955) and the activity of the enzyme in a host of insect species has been ably tabulated by Wyatt (1967). In *P. regina* flight muscle homogenates, trehalase has a specific activity of 0.2 μmol of glucose formed/min/mg protein (Reed and Sacktor, 1971). This value is sufficient to account for the rate of trehalose utilization by the blowfly on initiation of flight (Sacktor and Wormser-Shavit, 1966).

In general, trehalase is found in insects in multiple forms, and these may be found in different tissues of a single insect species as well as in the same tissue. A soluble form is represented by the enzyme in the ʼintestine and blood. The other type, associated with particulate fractions, is found in muscle. However, it is now clear that about 25 per cent of the total trehalase activity of muscle is in a soluble form in the blowfly (Hansen, 1966; Reed and Sacktor, 1971), silkmoth (Gussin and Wyatt, 1965), and cockroach (Gilby *et al.*, 1967). The soluble form found in *P. regina* muscle is electrophoretically distinct from the soluble enzyme in the blood of this species (Friedman and Alexander, 1971). The particulate trehalase in muscle of *Hyalophora cecropia* and *Blaberus discoidalis,* is associated with the 'microsomal' fraction (Gussin and Wyatt, 1965; Gilby *et al.*, 1967). Appreciable activity, of low specific activity, remains with a low-speed fraction that contains myofibrils, nuclei, and mitochondria, but it is reported that isolated mitochondria possess very little activity. In contrast, other investigations indicate that in the cockroach, *Leucophaea maderae*, (Zebe and McShan, 1959), and blowflies, *P. regina,* (Hansen, 1966; Reed and Sacktor, 1971), and *Sarcophaga barbata,* (Clements *et al.*, 1970), trehalase is located in mitochondria. The ultrastructural localization of the enzyme in flight muscle mitochondria of *P. regina* has now been examined by sub-fractionation of mitochondrial membranes and measurements of the distribution of trehalase activity in concert with the distribution of established enzyme markers (Reed and Sacktor, 1971). Localization of trehalase in the mitochondrial matrix or space between the inner and outer membrane has been ruled out. Clean separation of the inner and outer membranes of blowfly flight muscle mitochondria has proved to be difficult, but the hypothesis that trehalase is localized in the inner membrane is still favoured. It is also suggested that trehalase is situated on the outside of the mitochondrial inner membrane.

The marked decrease in the concentration of trehalose in the thorax at the onset of exercise is coincident with the rapid increase in the concentration of glucose (Sacktor and Wormser-Shavit, 1966). These opposite changes, occurring at a time when there has been considerable enhancement of glycolysis, clearly indicates that the cleavage of trehalose to glucose has been greatly facilitated. The elevation of the concentration of glucose is however, transient; within 30 s its concentration has returned to the original level. The changes in the intramuscular concentration of glucose suggest overshoots and periodic fluctuations, indicating that the steady-state levels of metabolites are not reached monotonically but in an oscillatory manner. Such oscillations are indicative of fine adjustments within the regulatory mechanisms of glycolysis in muscle during exercise.

The mechanism for the control of trehalase activity at the rest-to-flight transition has been investigated, but without success. Thoracic muscle trehalase from cockroaches, locusts and moths can be activated several-fold by various physical and chemical treatments, (*i.e.*, freezing and thawing, detergents, phospholipase A) that tend to disrupt lipoprotein structure (Zebe and McShan, 1959; Gussin and Wyatt, 1965; Gilby *et al.*, 1967; Stevenson, 1968a). After activation, the K_m is lowered to approximately one-half. It has been suggested (Gussin and Wyatt, 1965) that this activation may be related to the biological regulation of muscle trehalase, but further studies from the same laboratory cast doubt on this hypothesis (Gilby *et al.*, 1967). They have noted that in the housefly and two species of blowflies, (insects that utilize trehalose for flight), the enzyme is not activated by freezing and thawing. As has been pointed out previously (Sacktor, 1970), the values of trehalase activity in Dipteran muscle preparations (as reported by Gilby *et al.*), are already high. In fact, no further increase should have been expected since they are measuring the decontrolled rate. Vaughan *et al.* (1973) have reported no difference in activities of trehalase from flight muscles of the blowfly, *Sarcophaga barbata,* and the bumblebee, *Bombus hortorum,* when tested at various concentrations of Ca^{2+}. Similar negative results have been obtained for the blowfly, *P. regina,* but these have not been previously quoted (Reed and Sacktor, unpublished observations). We have also noted that high concentrations of ATP are not inhibitory. The demonstration that blowfly flight muscle trehalase is localized in mitochondria, probably on the outer surface of the inner mitochondrial membrane (Reed and Sacktor, 1971), focuses the problem of the control of trehalase activity on the rapid transitions taking place in, or adjacent to the sarcosomes *in situ,* when the insect goes from rest to flight.

Glycolysis

When the blowfly begins to fly, flight muscle glycolytic flux is increased about 100-times and attains rates 30- to 50-fold those of leg and heart muscles of man at maximal activity (Sacktor, 1965; Sacktor and Wormser-Shavit, 1966). The glycolytic potential of insect flight muscles may also be estimated from measurements of maximal activities of hexokinase and phosphofructokinase in muscle extracts (Crabtree and Newsholme, 1972a). These enzymes of carbohydrate metabolism catalyse reactions that are far-displaced from equilibrium in the cell and, thus, may limit the rate of the glycolytic pathway. Reasonable agreement between rates of carbohydrate utilization and these enzymatic activities have been found. For example, locust, cockroach, honey-bee and blowfly flight muscles, and rat heart muscle have metabolic rates equivalent to 14, 15, 32, 59 and 1.2 μmol of hexose/min/g wet wt. of muscle, and hexokinase activities of 12, 18, 29, 35 and 6 and phosphofructokinase activities of 17, 19, 20, 43 and 10 μmol/min/g wet wt. of muscle, respectively. Particularly striking are the findings comparing two species of Lepidoptera and Diptera, one species of each Order having rapid rates of glycolysis, the other not utilizing carbohydrates for flight and presumably having slow rates of glycolytic flux. The hexokinase activities of the flight muscle of the Silver-Y moth, *Plusia gamma,* and the Poplar hawk moth, *Laothoe populi,* are 50 and 3, respectively; the activities of the flight muscles of the blowfly, *Calliphora erythrocephala,* and the Tsetse-fly, *Glossina austeni,* are 35 and 2, respectively. Activity of phosphorylase, also displaced from equilibrium in tissues, is of less predictive value in estimating rates of glycolysis, since some species, such as bees and wasps, rely more on sugars than glycogen as their carbohydrate substrate.

In general, the principal pathway for catabolism of carbohydrate in insect flight muscle is a significant variant of the classical Embden-Meyerhof glycolytic scheme. This sequence of enzymatic reactions comprising glycolysis in insect flight muscle is illustrated diagrammatically in Fig. 1.5. An alternative route, designated as the hexose monophosphate shunt or the pentose phosphate pathway, although prominent in some tissues during the life cycle of the insect and present in muscle (Chefurka, 1965; Sacktor, 1965), probably contributes little to the energetics of flight muscle. Supporting this view on the relative insignificance of the shunt in the bioenergetics of muscle is the finding of Vogell *et al.* (1969) that the activities of glucose-6-P dehydrogenase in locust flight and leg muscles are only 0.1 per cent of those of glycolytic enzymes. Additionally, Agosin *et al.* (1961) have noted that in thoracic extracts of the bug, *Triatoma infestans* 6-phosphogluconate dehydrogenase is even less active than is glucose-6-P dehydrogenase.

The details of previous studies on glycolysis in insects have been described in earlier reviews (Sacktor, 1965; 1970; Chefurka, 1965; Wyatt, 1967; Friedman, 1970) and no attempt will be made here to

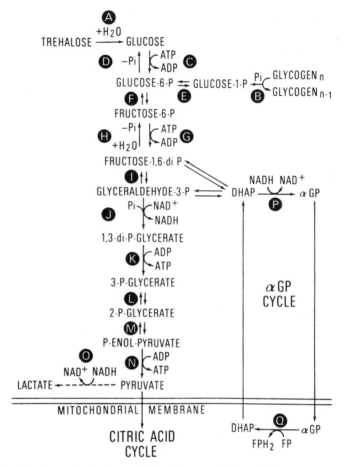

Fig. 1.5. *A schematic diagram of glycolysis and the α-glycerol-P cycle in flight muscle.*

characterize the various steps in the pathway. Rather, a brief description of the overall system in muscle will be considered, and attention will be focused on specific enzymatic reactions in order to discuss the control of glycolysis and to establish a foundation for the discussion, that follows, on how mitochondrial activities and the total metabolic machinery of the muscle are thoroughly integrated and mutually regulated.

The locus of the control of glycolysis in blowfly flight muscle becomes evident from studies in which the concentrations of glycolytic intermediates in the muscle of the insect at rest, and after periods of induced flight are concurrently measured (Sacktor and Wormser-Shavit, 1966). On initiation of flight, the concentration of glucose-6-P decreases and that of fructose-6-P remains essentially constant. In contrast, the concentration of fructose-1, 6-diP increases during the first 5 s of the start of contractions and reaches a peak at 15 s. The concentration of the hexose diphosphate returns to the resting level by the second minute of flight, and from 10 to 60 min of the flight a steady state is attained at a value slightly less than the initial value. A decrease in the concentrations of hexose monophosphates coincident with a rapid accumulation of fructose-1, 6-diP under conditions of maximal glycolytic flux (brought about by initiation of flight), identifies the crossover or control point as the phosphofructokinase reaction (Fig. 1.5, reaction G).

The mechanism whereby the phosphofructokinase reaction is facilitated in the transition from a slowly metabolizing resting muscle to an intensely active working muscle is suggested from earlier studies with the mammalian enzyme *in vitro*. Later this was confirmed with insect preparations, *in vitro,* and from the examination of changes in concentrations of the adenine nucleotides, arginine-P and P_i in *P. regina* flight muscle *in situ* at the onset of flight. Lardy and Parks (1956) discovered that, although ATP is a substrate for phosphofructokinase, an excess of ATP is inhibitory. Passonneau and Lowry (1962) found that this inhibition by ATP may be overcome by either ADP, AMP, P_i, cyclic AMP, fructose-1, 6-diP or, more effectively, by a combination of these activators. Lowry *et al.* (1964) have postulated that whenever formation of ATP does not keep up with its use, then P_i, ADP and, particularly, AMP will increase, and that this combination enhances phosphofructokinase activity autocatalytically. Insect flight muscle phosphofructokinase resembles the mammalian enzyme in that it is inhibited by excess ATP (Grasso and Migliori-Natalizi, 1968; Walker and Bailey, 1969; Vaughan *et al.,* 1973) and that this inhibition is reversed by AMP, cyclic AMP P_i (Walker and Bailey, 1969). However, differences between the mammalian and insect enzymes are evident. Fructose-1, 6-diP, an activator of the mammalian enzyme, inhibits locust flight muscle phosphofructokinase; and citrate, a potent inhibitor of mammalian phosphofructokinases, does not affect the locust enzyme (Walker and Bailey, 1969). Other important metabolites, including P-enolpyruvate, α-glycerol-P, and Ca^{2+}, are also without significant effect. As illustrated in Fig. 1.6, at the onset of flights, the concentrations of ATP and arginine-P in blowfly flight muscle decrease while the levels of P_i, ADP and especially, AMP increase (Sacktor and Hurlbut, 1966). These changes are qualitatively consistent with the theory of Lowry *et al.* (1964) and extend the hypothesis to apply to a working muscle *in vivo.*

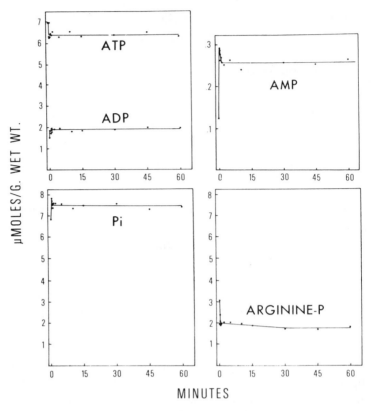

Fig. 1.6. *The sequential changes in the concentrations of ATP, ADP, AMP, P_i and Arginine-P in flight muscle of the blowfly, P. regina, during a one h flight.* (From Sacktor and Hurlbut, 1966.)

It should be pointed out, however, that these experimental findings also suggest that this theory may not account completely for the control of phosphofructokinase. The ATP concentration in flight muscle of the resting blowfly is approximately 6.9 mM and it falls to only 6.2 mM during flight (Fig. 1.6). Both these concentrations are strongly inhibitory to the isolated enzyme. Furthermore, the increase in concentration of AMP during flight is only 2.5-fold, from 0.12 to 0.30 mM. The magnitude of this increase in AMP, relative to the decrease in ATP, is not sufficient to bring about the 100-fold enhancement in the rate of glycolysis that occurs when the insect begins to fly. The existence of a fructose-6-P and fructose-1, 6-diP cycle, which is catalysed by phosphofructokinase and fructose-1, 6-diP phosphatase (see p. 27), could be responsible for an increase in the sensitivity of the rate of fructose-6-P phosphorylation to very small changes in the concentration of AMP (Vaughan *et al.,* 1973).

However, the regulation of phosphofructokinase activity by factors, as yet unknown, seems a more likely possibility, and the mechanism of this regulation remains one of the major questions in the control of flight muscle metabolism.

The pattern of changes in the concentrations of the other glycolytic intermediates in flight muscle of *P. regina* during flight does not reveal any additional primary locus of regulation of glycolysis, although fine adjustments and controls at other enzymatic sites are undoubtedly present (Sacktor and Wormser-Shavit, 1966).

As noted above, hexokinase activity (Fig. 1.5, reaction C) has been measured in flight muscles of a variety of different species. The enzyme from locust muscle is non-specific with respect to sugar and phosphorylates, the D-form of glucose, fructose, mannose and glucosamine (Kerly and Leaback, 1957). The products of the reaction, ADP and glucose-6-P, are inhibitory. In the honey-bee, ADP is a competitive inhibitor, with respect to ATP, and has a K_i value of 9×10^{-4} M (Ruiz-Amil, 1962). Glucose-6-P is a non-competitive inhibitor with respect to glucose. A K_i of 10 mM has been estimated. In rat heart the inhibition by glucose-6-P is relieved by P_i (England and Randle, 1967). This reversal has not been demonstrated in insect flight muscle (Vaughan *et al.,* 1973). Significantly, the concentration of glucose-6-P in blowfly flight muscle at rest greatly exceeds that measured during flight (Sacktor and Wormser-Shavit, 1966). The ratio of glucose-6-P to fructose-6-P approaches the expected value of 3:1 during continuous flight, suggesting that the isomerase reaction (Fig. 1.5, reaction F) may not be at equilibrium in the resting muscle.

Several different types of fructose-1, 6-diP aldolases (Fig. 1.5, reaction I) are present in insects (Brenner-Holzach and Leuthardt, 1967; Bauer and Levenbook, 1969). The enzyme from the flight muscle of *P. regina* has properties resembling the Class 1A isozyme characteristic of rabbit muscle. These include a close similarity in electrophoretic mobility, and a 44-fold greater reaction rate with fructose-1, 6-diP as substrate than with fructose-1-P as substrate. Digestion of the enzyme with carboxypeptidase, a procedure that cleaves the C-terminal tyrosyl residues, decreases the activity towards fructose-1, 6-diP with little effect on the activity towards fructose-1-P (Bauer and Levenbook 1969; Levenbook *et al.,* 1973). The flight muscle enzyme has a molecular wt. of about 160 000 and is probably composed of four sub-units each of 40 000 daltons.

Glyceraldehyde-3-P and dihydroxyacetone-P (DHAP), formed by the action of aldolase, are interconvertible, the reaction being catalyzed by triosephosphate isomerase (Fig. 1.5). The isomerase is the most active glycolytic enzyme of the constant-proportion group of extra-mitochondrial enzymes measured in different muscles (Pette *et al.*, 1962). The molecular weight for triosephosphate isomerase estimated from the housefly, *M. domestica,* and the mosquito, *Aëdes aegypti,* is 60 000 (Chiang, 1972). With glyceraldehyde-3-P as the substrate, housefly isomerase exhibits Michaelis-Menten type kinetics, except for inhibition at high substrate concentrations. A K_m of 1.16 mM is calculated (Chiang, 1972), a value far in excess of the concentration of substrate found in flight muscle (Sacktor and Wormser-Shavit, 1966). With dihydroxyacetone-P as substrate, a sigmoidal curve indicative of allosteric behaviour is observed. The apparent K_m is 1.1 mM and the Hill coefficient is 1.8 (Chiang, 1972). The existence of an allosteric site on the enzyme is suggested additionally by the non-competitive inhibition by P_i at low concentrations of inhibitor, and by competitive inhibition, presumably at the catalytic site, at high concentrations of inhibitor. Other phosphorylated intermediates (*i.e.,* α-glycerol-P, 3-P-glycerate) are also inhibitory, but at K_i values well above their concentrations *in vivo*.

As illustrated in Fig. 1.5, the first oxidative reaction in glycolysis occurs at the triose level; glyceraldehyde-3-P is oxidized to 1, 3-diP-glycerate with the concomitant reduction of NAD^+ to NADH. This reaction has an obligatory requirement for P_i and is catalysed by glyceraldehyde-3-P (triose-P) dehydrogenase (reaction J). The enzyme occurs in muscles of various insects (see, Friedman, 1970) and has been crystallized, and purified to homogeneity by Brosemer and co-workers (Marquardt *et al.*, 1968; Carlson and Brosemer, 1971; Carlson *et al.*, 1971). The dehydrogenase binds NAD^+ firmly, and slightly more than 3 mol of NAD^+ per mole of enzyme is found in the honey-bee. The enzymes from different species vary somewhat in their amino acid composition and other biochemical properties, and these differences have been used to study phylogenetic relationships. It is now generally accepted that -SH reagents, such as iodoacetate, inhibit the enzyme in insect, as well as from mammalian muscles.

Generation of one mol of ATP per 0.5 mol of hexose undergoing glycolysis, occurs in the reaction catalysed by phosphoglycerate kinase (reaction K). A second ATP is formed in the conversion of P-enolpyruvate to pyruvate (reaction N). Regulation of pyruvate kinase in the flight muscle of the locust, *Schistocerca gregaria,* has been studied by Bailey and Walker (1969). Maximal activity is obtained with concentrations of ADP and P-enolpyruvate of 1.0 and

0.1 mM, respectively. Higher concentrations of ADP are slightly inhibitory. Fructose-1, 6-diP, which activates potently by lowering the apparent K_m for P-enolpyruvate in the enzyme from locust fat body, has no effect on the flight muscle enzyme. In this respect, the two enzymes resemble their counterparts from rat liver and muscle. The flight muscle kinase is inhibited by ATP, half-maximal inhibition occurring at about 5 mM. Fructose-1,6-diP does not reverse ATP inhibition, as it does with the fat body enzyme.

Glycolytic-mitochondrial interactions

In early experiments, Sacktor (1961) and Sacktor and Dick (1962) have shown that when NADH is added to suspensions of isolated, intact, insect flight muscle it is not oxidized, even though the mitochondria avidly oxidize added NAD-linked substrates, *e.g.* pyruvate, via intramitochondrial NAD^+. These and other observations have established that mitochondria are impermeable to NADH and NADPH, as well as to their oxidized forms. Thus, a permeability barrier effectively segregates the cytosolic and intramitochondrial pools of pyridine nucleotide. This compartmentation is particularly important because, as noted above, during glycolysis, extra-mitochondrial NADH is formed by glyceraldehyde-3-P dehydro-genase. Unless this NADH is re-oxidized immediately, glycolysis ceases. An explanation of this enigma became a necessity in insects, with the discovery that lactate dehydrogenase (Fig. 1.5, reaction 0) is virtually absent from the flight muscle of flies (Sacktor, 1955).

We now know that during glycolysis in flight muscle, the NADH formed by glyceraldehyde-3-P dehydrogenase is principally re-oxidized concomitantly with the conversion of dihydroxyacetone-P to α-glycerol-P, catalysed by the cytosolic α-glycerol-P dehydro-genase (reaction P) (Zebe and McShan, 1957; Chefurka, 1958; Sacktor and Cochran, 1957a).

$$\text{Dihydroxyacetone-P} + \text{NADH} + H^+ \rightleftarrows \alpha\text{-Glycerol-P} + NAD^+$$

The α-glycerol-P, so formed, is readily accessible to the mito-chondrial α-glycerol-P dehydrogenase and is oxidized, in turn, by the flavoprotein, thereby regenerating dihydroxyacetone-P (reaction Q) (Zebe and McShan, 1957, Sacktor and Cochran, 1957b, 1958; Chance and Sacktor, 1958).

$$\alpha\text{-Glycerol-P} + \text{Flavoprotein} \rightarrow \text{Dihydroxyacetone-P} + \text{Flavoprotein.}H_2$$

This dihydroxyacetone-P is then available for further oxidation of

extramitochondrial NADH. Accordingly, the two reactions constitute the α-glycerol-P cycle (Estabrook and Sacktor, 1958; Bücher
and Klingenberg, 1958). As illustrated diagrammatically in Fig. 1.5,
the cycle is a shuttle system, in which cytosolic NAD-linked
substrates enter and leave the mitochondria in reduced and oxidized
states, respectively. In this way, reducing equivalents from the
extramitochondrial pool of NADH pass the cytosol-mitochondrial
permeability barrier and are oxidized. At the same time, reduced
flavoprotein in the mitochondria donates its reducing equivalents to
the electron transport chain. Further, the cyclic process is self-
generating – only a catalytic quantity of dihydroxyacetone-P being
needed to oxidize the continuously formed NADH (Sacktor and
Dick, 1962). This suggests that most of the dihydroxyacetone-P
produced in the aldolase reaction can be isomerized to glycer-
aldehyde-3-P. Also, all the carbon derived from the carbohydrate
metabolized during prolonged flight can be converted to pyruvate,
and thus is available for further oxidation in the mitochondria, via
the Krebs cycle. This is evident from findings that there is no
accumulation of partially oxidized intermediates (α-glycerol-P,
pyruvate and alanine) during flights of an hour or more by the
blowfly, *P. regina*, (Sacktor and Wormser-Shavit, 1966) and the
locust, *L. migratoria* (Kirsten *et al.*, 1963). At shorter time intervals
such an accumulation does occur and may be of significance (Sacktor
and Wormser-Shavit, 1966; Childress *et al.*, 1967), but these *in vivo*
studies indicate that after an initial lapse, the mitochondria are
capable of oxidizing pyruvate and α-glycerol-P at the rates in which
they are formed. It is important to emphasize that glycolysis in flight
muscle is aerobic and remarkably efficient – the end-product of
glycolysis, (lactate) not accumulating, as it does in exercising
vertebrate muscle.

The glycolytic pathway described above, in which α-glycerol-P and
pyruvate are the end-products, is characteristic of the flight muscle
of insects, especially those of high frequency wingbeat. However,
marked differences are found in this respect between tissues in a
given species and between the same tissue in different orders of
insects. For example, as shown in Table 1.2, flight muscle of the
locust has 167 units of α-glycerol-P dehydrogenase and only 2 units
of lactate dehydrogenase. In contrast, the leg muscle of this insect
has 33 and 117 units of α-glycerol-P and lactate dehydrogenases,
respectively. Extensive studies by Bücher and colleagues have
confirmed this tissue specificity. They point out that the characteristic formation of α-glycerol-P in flight muscle is in accord with its

Table 1.2

Comparison of the α-glycerolphosphate and lactic dehydrogenase activities in different muscles

Muscle	LAD	α-GPD	Reference
	(μmol/g wet wt./min)		
Flight (blowfly)	0	1230	Sacktor and Dick (1962)
Flight (bee)	3	700	Zebe and McShan (1957)
Flight (locust)	2	167	Delbruck *et al.* (1959)
Leg (locust)	117	33	Delbruck *et al.* (1959)
Flight (cockroach)	0.2	48	Chefurka (1958)
Leg (cockroach)	0.1	32	Chefurka (1958)
Flight (praying mantis)	<0.1	11	Kitto and Briggs (1962a,b)
Leg (praying mantis)	1	1	Kitto and Briggs (1962a,b)
Flight (waterbug)	1	51	Crabtree and Newsholme (1972a)
Leg (waterbug)	59	13	Crabtree and Newsholme (1972a)
Flight (cockchafer)	4	103	Crabtree and Newsholme (1972a)
Flight (poplar hawk moth)	3	36	Crabtree and Newsholme (1972a)
Tail (crayfish)	217	5	Zebe and McShan (1957)
Pectoral (pheasant)	542	103	Crabtree and Newsholme (1972a)
Skeletal (rat)	330	50	Bücher and Klingenberg (1958)
Smooth (beef)	25	0.1	Pette *et al.* (1962)

aerobic nature. In contrast, lactate is produced during the re-oxidation of NADH in those muscles that may become temporarily anoxic. Of additional interest in comparing the activities of the two dehydrogenases is the observation of Brosemer (1967), on the flightless grasshopper, *Romalea microptera.* The wing muscle of this insect, which during evolutionary development has lost the ability to fly, no longer has the biochemical characteristics of flight muscle; *i.e.*, high α-glycerol-P and low lactate dehydrogenases. Instead, the activities of the dehydrogenases in the wing muscles are similar to those found in the insect's leg muscles.

In the α-glycerol-P cycle, the difference in redox potential between the two coenzymes, *i.e.*, NAD and FAD, indicates that the shuttle is unidirectional; it transports reducing equivalents into mitochondria. An alternative system, involving the cytosolic and mitochondrial malate dehydrogenases appears to operate in mammalian tissues and may also be found in some insect tissues (Delbruck *et. al.*, 1959; Sacktor, 1960; Borst, 1963). In this scheme, extramitochondrial NADH is oxidized, with reduction of oxaloacetate by the action of the cytosolic malate dehydrogenase, to yield malate. The malate enters the mitochondrion via a carrier (Chappell, 1968), and is oxidized therein to oxaloacetate by the action of the intramitochondrial malate dehydrogenase. The malate shuttle clearly differs from the α-glycerol-P shuttle in that in both compartments the coenzyme is NAD, and the shuttle is reversible functioning in either direction. Krebs *et. al.*, (1967) have shown that electrons from intramitochondrial NADH can pass into the cytosol. In the outward-directed shuttle, intramitochondrial malate dehydrogenase reduces oxaloacetate with concomitant oxidation of intramitochondrial NADH. The malate leaves the mitochondrion and reduces cytosolic NAD^+ by action of the cytosolic malate dehydrogenase. The oxaloacetate formed in the cytosolic compartment can re-enter the mitochondrion after decarboxylation to pyruvate, or transamination with glutamate. However, evidence for the operation of these pathways in insect flight muscle is fragmentary. The presence of the two malate dehydrogenases: in flies and locusts, one, cytosolic and the other, mitochondrial, is well established (Sacktor, 1953a; Delbruck *et al.*, 1959). The dual localizations of glutamate-oxaloacetate and glutamate-pyruvate transaminases has been reported for several different species (Barron and Tahmisian, 1948; Kilby and Neville, 1957; Pette and Luh, 1962), and an active oxaloacetate decarboxylase has been found in houseflies and the Tsetse fly (Lewis and Price, 1956; Bursell, 1965). However, the apparent absence of a malate carrier in Dipteran flight muscle mitochondria argues counter to an operational shuttle in this tissue, although one might predict that the system is functional in tissues wherein active gluconeogenesis takes place, *e.g.*, the fat body of insects.

In a somewhat similar fashion, intramitochondrial reducing equivalents can be transported to form extramitochondrial NADPH, needed for reductive syntheses. In mammalian tissues, intramitochondrial isocitrate may leave the mitochondrion via the citrate carrier and reduce cytosolic $NADP^+$ by the action of the extramitochondrial NADP-linked isocitrate dehydrogenase. The presence of the two forms of isocitrate dehydrogenase in insect muscle is evident from the works of Goebell and Klingenberg (1964), Goebell and Pette (1967) and Ku and Cochran (1971). To what extent this shuttle operates remains to be evaluated. Nevertheless, both the isocitrate and malate shuttles seem to offer effective mechanisms for the regulation and control of the ratio of the intra- and extra-mitochondrial NADH-NAD$^+$ and NADPH-NADP$^+$ couples.

Auxiliary glycolytic reactions

It is highly unlikely that appreciable gluconeogenesis occurs in insect flight muscle, although the synthesis of glycogen from lactate has been demonstrated in vertebrate skeletal muscle (Bendall and Taylor, 1970) and gluconeogenesis is well established in other mammalian tissues. At least four enzymes are crucial for gluconeogenesis: glucose-6-P phosphatase (Fig. 1.5, reaction D), fructose-1,6-diP phosphatase (reaction H), P-enolpyruvate carboxylase and pyruvate carboxylase. Except for the report of P-enolpyruvate carboxylase activity in flight muscle of *Locusta migratoria* (Nolte *et al.*, 1972), there are no significant activities of glucose-6-P phosphatase and P-enolpyruvate carboxylase in insect flight muscle (Newsholme *et al.*, 1972; Crabtree *et al.*, 1972). Pyruvate carboxylase is present in all flight muscles that have been examined (Hansford, 1972a; Crabtree *et al.*, 1972). However, the function of this mitochondrial enzyme in insect flight muscle seems to be to provide the oxaloacetate necessary for the large increase in Krebs cycle activity which occurs at the onset of flight (to be discussed more fully below), rather than to serve in a pathway leading to carbohydrate synthesis. Fructose-1,6-diP phosphatase activity varies greatly, according to the species. For example, it is absent from flight muscles of moths, wasps and honey-bees; moderately active in flight muscles of locusts, blowflies and cockroaches; and extraordinarily active in flight muscles of bumble-bees (Newsholme and Crabtree, 1970; Newsholme *et al.*, 1972; Crabtree *et al.*, 1972). Indeed, the activity of the enzyme on a wet weight basis, in flight muscles of bumble-bees is approximately 30-fold the activity in any other muscle or in rat liver. Furthermore. Newsholme *et al.* (1972) have found that in bumble-bees, the activity of the phosphatase is comparable to the activity of phospho-fructokinase. Moreover, the bumble-bee phosphatase is unaffected by AMP, which is an important specific effector of gluconeogenesis in liver. These authors conclude that the role of fructose-1,6-diP phosphatase in the bumble-bee is not related to the regulation of glycolysis at the phosphofructokinase reaction, as they previously suggested. Instead, they propose that both phosphofructokinase and fructose-1,6-diP phosphatase are simultaneously active and catalyse a substrate cycle between fructose-6-P and fructose-1,6-diP, according to the reactions

$$\text{Fructose-6-P} + \text{ATP} \rightarrow \text{Fructose-1,6-diP} + \text{ADP}$$

$$\text{Fructose-1,6-diP} + H_2O \rightarrow \text{Fructose-6-P} + P_i$$

The sum of these reactions is

$$\text{ATP} + H_2O \longrightarrow \text{ADP} + P_i$$

Such a cycle effects the continuous hydrolysis of ATP, to release energy in the form of heat. Newsholme *et al.* (1972) hypothesize that the generation of heat during short periods of rest between flights in the bumble-bee, helps to maintain the temperature of the thorax at a level suitable for flight. They point out that this hypothesis is consistent with the fact that bumble-bees are able to fly and collect food under cold weather conditions, whereas the honey-bee which has only minimal fructose-1,6-diP phosphatase activity does not fly in inclement weather.

This proposed substrate cycle has recently received substantial support from Lardy and associates (Bloxham *et al.*, 1973; Clark *et al.*, 1973). After establishing an *in vitro* model system, they have estimated the cycling of fructose-6-P through the reactions catalyzed by phosphofructokinase and fructose-1,6-diP phosphatase in bumble-bee, *Bombus affinis*, flight muscle *in vivo*. They have reported that in flight, glucose is metabolized exclusively through glycolysis (20.4 μmol/min/g wet wt.) and there is no evidence for substrate cycling. In the resting bumble-bee exposed to low temperatures (5°C), the pattern of glucose metabolism in the flight muscle is altered so that substrate cycling is high (10.4 μmol/min/g wet wt.), and glycolysis is decreased (5.8 μmol/min/g wet wt.). Moreover, the rate of substrate cycling in the flight muscle of a resting bumble-bee is inversely related to the ambient temperature; at 27°, 21° and 5°C, the rates of substrate cycling are 0, 0.48 and 10.4 μmol/min/g wet wt., repectively. Significantly, Vaughan *et al.* (1973) and Clark *et al.* (1973) have found that Ca^{2+} inhibits fructose-1,6-diP phosphatase at concentrations that are without effect on phosphofructokinase. The latter investigators have shown additionally that this inhibition of fructose-1,6-diP phosphatase is reversed when Ca^{2+} is chelated by an excess of ethanedioxybis-(ethylamine)-tetraacetate (EGTA) and they propose that the rate of fructose-6-P substrate cycling may be regulated by changes in the sarcoplasmic Ca^{2+} concentration associated with the contractile process.

Glycerol, which is formed by the action of lipase on fats, is glycogenic and thus enters the pathway of carbohydrate metabolism. Flight muscles of the housefly, *M. domestica* (Sacktor, 1955) and the silkmoth, *H. cecropia* (Gilbert, 1967), oxidize glycerol at a slow rate. Glycerol is first converted to α-glycerol-P with the utilization of ATP in a reaction catalyzed by glycerol kinase. The low rate of oxidation of glycerol relative to that of α-glycerol-P in flies suggests that the kinase reaction is limiting.

A large number of insect muscles have been surveyed for glycerol kinase activity (Newsholme and Taylor, 1969). On the basis of activity, the different muscles are classified into three groups: muscles that have a low enzyme activity, *i.e.*, <0.3 μmol/min/g, to which belong the flight muscle of the cockroach and tsetse fly, as well as leg mucles of all insects; muscles that have an intermediate activity, *i.e.*, 0.3 to 1.5 μmol/min/g, which include the flight muscles of locusts, moths, cockchafers, and waterbugs; and muscles that have a relatively high enzyme activity, *i.e.*, >1.5 μmol/min/g. Surprisingly, the flight muscles of bees, wasps and some blowflies belong to this group. To explain why the muscles of insects using carbohydrates and not fats have relatively the highest glycerol kinase activity, these authors explored the possibility that the kinase activity may be related to the high rates of glycolysis in these muscles. They proposed that, since the maintenance of glycolysis is dependent on the oxidation of extramitochondrial NADH, via a rapidly functioning α-glycerol-P cycle, then if at any stage of flight (*e.g.*, at initiation), the rate of mitochondrial oxidation of α-glycerol-P is less than the activity of the cytosolic extramitochondrial α-glycerol-P dehydrogenase, α-glycerol-P would accumulate, inhibit the soluble dehydrogenase, and thus inhibit glycolysis. They conjecture that this accumulation of α-glycerol-P can be prevented, by a specific α-glycerol-P phosphatase hydrolysing the intermediate to glycerol. The relatively, active glycerol kinase is needed to rephosphorylate the glycerol after accumulation of α-glycerol-P has ceased. This interesting hypothesis would be strengthened considerably if an active α-glycerol-P phosphatase were found in these flight muscles. To date, there is no evidence for the presence of such a specific enzyme (Sacktor, 1953b).

Oxidation of the end-products of glycolysis and the permeability of mitochondria to metabolic intermediates

As discussed above, in flight muscle of blowflies, and in other species to varying degrees, glycolysis gives rise to an equimolar mixture of α-glycerol-P and pyruvate. Sacktor and Wormser-Shavit (1966) have found that after an hour-long flight of *P. regina*, in which carbohydrates are utilized exclusively, there is no significant accumulation of partially oxidized intermediates. This indicates that flight muscle mitochondria, *in vivo*, oxidizes α-glycerol-P and pyruvate at the rates in which they are formed. (Initially, there is a lag in pyruvate oxidation which lasts several minutes, a phenomenon which will be discussed later.) Since in a prolonged flight, pyruvate is oxidized virtually to completion *via* the citric acid cycle, pyruvate should theoretically elicit a rate of oxygen consumption five times that of α-glycerol-P. Moreover, the rates of oxidation of pyruvate and α-glycerol-P should be additive. To what extent these *in vivo* requisites are met in studies with isolated mitochondria *in vitro*, is discussed below.

When substrates are added to a suspension of mitochondria isolated from flight muscle of flies, rates of oxidation as shown in Table 3 are obtained. It is evident that intact mitochondria oxidize, at appreciable rates, only exogenous α-glycerol-P, pyruvate, acetyl carnitine and, to a lesser extent, proline. In many studies with a variety of insect species, α-glycerol-P is oxidized at a rate greater than that of pyruvate. The relative respiratory rates obtained with these two substrates do differ, however. Other measurements reveal approximately equal rates of oxidation (Gregg, *et al.*, 1960; Van den Bergh and Slater, 1962), and in some cases pyruvate may be oxidized at a rate slightly greater than that for α-glycerol-P (Hansford, 1972a; deKort *et al.*, 1973). In contrast, intermediates of the tricarboxylic acid cycle, such as citrate, isocitrate, α-ketoglutarate, succinate, fumarate and malate, the amino acids glutamate and aspartate, and NADH added to isolated mitochondria are not effective respiratory substrates (Chance and Sacktor, 1958; Van den Bergh and Slater, 1962; Childress and Sacktor, 1966; Childress *et al.*, 1967; Sacktor and Childress, 1967).

An explanation for the low rates of oxidation of intermediates of the Krebs cycle has become evident with the discovery of Van den Bergh and Slater (1962) that flight muscle mitochondria are not readily permeable to these compounds. Subjecting mitochondria to sonic disintegration or freezing-thawing, procedures that disrupt the integrity of the mitochondrial membranes, increases the respiratory rates with these substrates many-fold (Van den Bergh and Slater, 1962; Van den Bergh, 1964; Sacktor and Childress, 1967). In contrast, the oxidations of α-glycerol-P and pyruvate are not stimulated, and that of proline may be markedly decreased by these and other disruptive procedures (Sacktor and Childress, 1967; Norden and Venturas, 1972).

Other techniques have also been employed to examine the permeability of flight muscle mitochondria to respiratory substrates. Chappell and Crofts (1966) have reported that when mitochondria are suspended in iso-osmotic solutions of ammonium salts they swell, measurable by a decrease in absorbance, only if the anion is capable of penetrating the inner mitochondrial membrane. By this criterion, too, it is concluded that intermediates of the citric acid cycle do not penetrate the inner membrane of flight muscle mitochondria: *i.e.*, ammonium salts of succinate, malate, citrate and isocitrate do not cause swelling in housefly or locust mitochondria (Van den Bergh, 1967); and ammonium salts of succinate and malate are inactive both in the presence and absence of 5 mM phosphate in this test with mitochondria from the blowfly *Sarcophaga barbata* (Donnellan *et al.*, 1970). Thus, these experiments provide no evidence for the presence of the specific dicarboxylate and tricarboxylate anion permeases for the exchange-diffusion of Krebs cycle substrates, as described in mammalian mitochondria (Chappell and Haarhoff, 1967). In apparent contradiction, Tulp and Van Dam (1969) have claimed that succinate can exchange for malate in housefly mitochondria. However, the carrier is saturated by low concentrations of phosphate, so that in the presence of phosphate little succinate can enter. In the absence of phosphate, or in the presence of phosphate and mersalyl, an inhibitor of phosphate transport in insect mitochondria (Carafoli *et al.*, 1971), succinate readily enters and is rapidly oxidized. A satisfactory explanation for the difference between the findings of Tulp and Van Dam (1969) with houseflies, and those of Donnellan *et al.* (1970) with blowflies, as well as those of Hansford (1971) with the 17-year cicada, is not evident at this time.

Additional observations by Hansford (1971) indicate that mitochondria from the flight muscle of the periodical cicada, *Magicicada septendecim*, differ from preparations of other insects in that they are permeable to glutamate and α-ketoglutarate, but not readily permeable to other members of the Krebs cycle. He proposes that specific membrane carriers are genetically determined, and functionally related to the physiology of the tissue. As an extension to this general concept, one may postulate that the absence of dicarboxylate and tricarboxylate anion carriers in flight muscle of most insects, is an adaptation to prevent the efflux of these intermediates from the mitochondrion, rather than simply to inhibit their entry. In flight muscle, Krebs cycle intermediates are needed

almost exclusively for respiratory function, as precursors of oxaloacetate for catabolism of pyruvate, and not for their participation in
various biosynthetic processes outside the mitochondrion, as found
in tissues such as the mammalian liver.

Application of the ammonium salt-swelling test to α-glycerol-P and
pyruvate is somewhat more complex. Both of these anions, when
added exogenously, are rapidly oxidized. However, ammonium
α-glycerol-P does not cause the mitochondria to swell (Hansford,
1968; Donnellan et al., 1970). This is readily rationalized; α-
glycerol-P does not have to enter the osmotically active space to be
oxidized. The substrate has access to its dehydrogenase located on
the outside of the inner membrane and electrons are transferred to
the respiratory chain, which are also located in the inner membrane.
The situation with ammonium pyruvate is unclear. Donnellan et al.
(1970) have reported that swelling of blowfly, S. barbata, mitochondria in ammonium pyruvate is rapid and extensive. In contrast,
housefly, M. domestica, blowfly, M. vomitoria, and locust mitochondria do not swell (Van den Bergh, 1967; Hansford, 1968). Aside
from the conflicting experimental findings, at least two possible
explanations have been proposed for the failure (if verified) of
pyruvate to penetrate the inner mitochondrial membrane but,
nevertheless, to be oxidized (Hansford and Sacktor, 1971). One is
that the hypothetical carrier is energy-linked, and therefore no
transport occurs in the non-metabolizing mitochondrion, as is the
case in the ammonium swelling test. The other, is that pyruvate has
access to pyruvate dehydrogenase, whose orientation in the inner
membrane is not known.

As noted above, the ability of blowfly flight muscle mitochondria
to simultaneously oxidize pyruvate and α-glycerol-P, the end-
products of glycolysis in this tissue, is a prime requisite for these
mitochondria in situ. Yet, when isolated mitochondria are incubated
in the presence of both pyruvate and α-glycerol-P, the resultant
respiratory rate usually shows only partial summation; that is, the
rate of oxidation is much less than the sum of the two rates obtained
with each substrate separately. This strongly suggests that in this in
vitro situation the electron transport system is limiting and that
pyruvate and α-glycerol-P dehydrogenases interact competitively
with a component(s) of the respiratory chain.

The highest value for the oxidation of pyruvate in relation to that
of α-glycerol-P has been reported to be two to one for isolated
Dipteran flight muscle mitochondria (Hansford and Sacktor, 1971).
However, even this ratio falls far short of the 5-fold greater oxygen
consumption with pyruvate relative to α-glycerol-P that must be

attained *in vivo*, during a prolonged flight. This suggests that the conditions which will yield maximal respiration when isolated mitochondria are oxidizing exogenous pyruvate, have yet to be discovered. However, an alternative hypothesis for the failure to obtain the 5:1 ratio cannot be ignored (Sacktor, 1973). This proposal suggests that the maximal rate of oxidation obtained using α-glycerol-P as substrate in *in vitro* systems is higher than the rate at which α-glycerol-P is oxidized *in vivo*, even during prodigious muscular work. This seeming paradox may be due to the fact that in *in vitro* experiments with isolated mitochondria, the concentration of α-glycerol-P normally used is one that yields maximal rates of oxygen uptake. The effective concentration of α-glycerol-P available to the mitochondrial dehydrogenase *in situ* may be less. It is known that the concentration of α-glycerol-P in blowfly flight muscle is much below that needed to saturate the enzyme (Sacktor and Wormser-Shavit, 1966). Moreover, it is possible that the dehydrogenase is never more than partially saturated, even in the presence of Ca^{2+} (this action of Ca^{2+} will be discussed in a subsequent part of the review).

Additional reasoning also tends to support the view that the measured rate of pyruvate oxidation obtained with mitochondrial suspensions is adequate to meet the metabolic requirements of the muscle *in situ*. Chance and Sacktor (1958) determined that the turnover numbers of flavoprotein and cytochromes in isolated mitochondria oxidizing α-glycerol-P, approach the values estimated

Table 1.3
Respiratory activities of mitochondria from blowfly flight muscle

Substrate	Rate
α-Glycerol-P	1.35
Pyruvate	0.98
Acetyl carnitine	0.49
Proline	0.19
Citrate	0.01
α-Ketoglutarate	0.07
Succinate	0.09
Fumarate	0.05
Malate	0.04
Glutamate	0.04
Aspartate	0.01

Respiratory rate = μg atoms oxygen/min/mg mitochondrial protein. Data compiled from Sacktor and Childress (1967), Childress *et al.* (1967), Childress and Sacktor (1966), Chance and Sacktor (1958), and Bulos *et al.* (1972).

to exist *in vivo* during flight. From this it can be deduced that the *in vitro* rate of oxidation of α-glycerol-P, by itself is of the right order of magnitude to account for the oxygen consumption of the flying insect (Chance and Sacktor, 1958). It can be seen from the data in Table 1.3, as well as those reported elsewhere, *i.e.*, Hansford (1972a), that the rate of pyruvate oxidation is comparable to that obtained maximally with α-glycerol-P. Therefore, it follows that the rate of oxidation of pyruvate, as measured presently, also satisfies the respiratory rate of the fly in flight. Moreover, a rate of oxidation of α-glycerol-P only one-fifth this value would suffice to satisfy the physiological demands of the tissue in the flying insect.

Utilization of amino acids

Free amino acids are found in insect tissues, including muscle, in extraordinary high concentrations (Florkin, 1958; Kermack and Stein, 1959; Kirsten *et al.*, 1963). In addition, flight muscle mitochondria are capable of oxidizing several amino acids. However, with a few notable exceptions that are discussed below, these observations do not imply that amino acids serve as a major reserve of metabolic energy (Sacktor and Childress, 1967; Van den Bergh, 1964). For example, in early experiments Wigglesworth (1949) has shown that glycine and alanine, when fed to exhausted fruit flies, are unable to support sustained flight, and Clements (1959) has reported that flight muscle of locusts produces only a negligible amount of $^{14}CO_2$ when muscle was incubated with labeled glycine and leucine.

On the other hand, flight muscle homogenates of the housefly do oxidize selected amino acids, including proline, glutamate and cysteine (Sacktor, 1955). Moreover, these early studies have demonstrated that the oxidation of glutamate is coupled to synthesis of ATP (Rees, 1954; Sacktor and Cochran, 1956).

As shown in Table 1.3 and as discovered earlier (Sacktor, 1955), mitochondria isolated from flight muscle of flies oxidize proline at a notably high rate. The oxidation of proline by flight muscle preparations has now been reported for all the major orders of insects (Crabtree and Newsholme, 1970). Proline is oxidized by mitochondria to glutamate via Δ'-pyrroline-5-carboxylate as an intermediate (Sacktor *et al.*, 1965; Brosemer and Veerabhadrappa, 1965, Sacktor and Childress, 1967). Significantly, flight muscle rapidly metabolizes proline *in vivo*, as shown in the tsetse fly (Bursell, 1963), locust (Kirsten *et al.*, 1963), blowfly (Sacktor and Wormser-Shavit, 1966), and honey-bee (Barker and Lehner, 1972).

The physiological significance of the oxidation of proline in the flight muscle of the blowfly, particularly at the initiation of flight, is apparently the provision of tricarboxylic acid cycle intermediates needed for the maximal rate of oxidation of pyruvate (Sacktor and Wormser-Shavit, 1966; Childress and Sacktor, 1966; Sacktor and Childress, 1967). This suggestion is supported by experiments showing that isolated mitochondria rapidly lose their ability to oxidize pyruvate, and that this loss is reversed by proline but not by Krebs cycle intermediates nor glutamate (Sacktor and Childress, 1967). This finding is consistent with the ability of blowfly mitochondria to oxidize exogenous proline but not the other metabolites, when added exogenously (Table 1.3). Proline is converted to glutamate by the actions of proline and Δ'-pyrroline-5-carboxylate dehydrogenases. Transamination of glutamate with pyruvate gives rise to alanine and α-ketoglutarate. The intramitochondrial α-ketoglutarate is further metabolized via the Krebs cycle to form oxaloacetate (Sacktor and Childress, 1967; Bursell, 1967). This oxaloacetate can then condense with acetyl CoA (derived from pyruvate), forming citrate, thus, effecting the complete oxidation of pyruvate by means of the tricarboxylic acid cycle. This concept receives additional substantiation from studies of metabolite levels during rest and flight of the blowfly *P. regina* (Sacktor and Wormser-Shavit, 1966). During the first few seconds of flight the concentration of proline in flight muscle decreases abruptly. At the same time there is a stoichiometric accumulation of alanine, suggesting that it is only when proline oxidation has yielded sufficient oxaloacetate does pyruvate begin to be oxidized as rapidly as it is being produced by glycolysis. In the interim, pyruvate transaminates with glutamate to form alanine (Sacktor and Childress, 1967), or forms acetyl CoA, which gives rise to acetyl carnitine (Childress *et al.*, 1967) or is deacylated to yield acetate (Tulp and Van Dam, 1970).

A complete explanation for the unique role of proline in providing precursors of oxaloacetate has yet to be formulated. Two possibilities seem attractive at this time. One is that proline, because it is electrically neutral can pass through the mitochondrial membrane permeability barrier, whereas glutamate, aspartate and citric acid cycle intermediates, which carry net charges, are unable to do in the absence of specific permeases. The second possibility is that proline dehydrogenase is located on the outer surface of the inner mitochondrial membrane. Thus, like α-glycerol-P, there would be no barrier to proline. This hypothesis receives some support from the observation that ADP activates proline dehydrogenase in the presence of atractyloside (Hansford and Sacktor, 1970b), suggesting that the action of ADP on the dehydrogenase is external to the ADP translocase, an inner mitochondrial enzyme.

Although the suggested physiological importance of the oxidation of proline in flight muscle of the blowfly is the provision of oxaloacetate on the initiation of flight, the oxidation of proline may play a larger role in flight muscle metabolism of the tsetse fly, *Glossina morsitans*. In an elegant series of papers Bursell (1963, 1965, 1966, 1967) has shown that the concentration of proline decreases markedly during the flight of the tsetse fly and that the content of alanine rises concomitantly and nearly stoichiometrically. He proposed a metabolic cycle

Proline $\longrightarrow$ Δ'-Pyrroline-5-carboxylate $\longrightarrow$ Glutamate

Glutamate + Pyruvate $\longrightarrow$ α-Ketoglutarate + Alanine

α-Ketoglutarate $\longrightarrow$ Succinyl CoA $\longrightarrow$ Succinate $\longrightarrow$
   Fumarate $\longrightarrow$ Malate $\longrightarrow$ Oxaloacetate

Oxaloacetate $\longrightarrow$ Pyruvate + CO_2.

The sum of these four equations is

Proline $\longrightarrow$ Alanine + 2 CO_2 + 3 NADH + 2 Flavoprotein.H_2.

It is noted that the number of moles of ATP that can be produced from the conversion of one mole of proline to alanine by this pathway is 14, a value comparing favourably with the 15 moles of ATP produced by the complete oxidation of one mole of pyruvate via the citric acid cycle. Further, the scheme envisages the use of only a segment of the Krebs cycle, and it also requires the presence of an active oxaloacetate decarboxylase. This has been confirmed (Bursell, 1965). The report that this enzyme is found in the soluble fraction of the muscle (Bursell, 1965), rather than in the mitochondria as the metabolic scheme demands, may, reflect mitochondrial lysis in the distilled water used in the extraction procedure (see Hansford and Sacktor, 1971). Oxaloacetate decarboxylase is localized in mitochondria of the blowfly, *P. regina* (Sacktor and Childress, 1967). A study with labelled amino acids provides evidence consistent with the proposed scheme (Bursell, 1966). Thus, ^{14}C from proline appears in glutamate in high specific activity on initiation of flight, and subsequently in alanine. Label from glutamate is diluted rapidly when flight begins, which is consistent with formation of unlabelled glutamate from the large proline pool. In a later study (Bursell, 1967), the conversion of glutamate into alanine is demonstrated directly, using specifically labelled glutamate, and the label distribution found in alanine is essentially that required by the proposed pathway.

The singular importance of proline oxidation and its associated metabolic pathways in the tsetse fly gains added credence from the findings that the glycogen content of the tsetse fly is barely detectable and that sugars, normally present in insects, *e.g.*, trehalose and glucose, are found only in very low concentrations (Norden and Patterson, 1969; Geigy *et al.*, 1959). Moreover, the enzymes concerned with carbohydrate catabolism, including phosphorylase, trehalase, phosphoglucomutase, hexosephosphate isomerase, phosphofructokinase and aldolase, have low activities in tsetse fly flight muscle as compared to the activities present in the blowfly (Norden and Patterson, 1969). In contrast, succinic dehydrogenase, which according to the proposed scheme functions in the tsetse fly as well as in the blowfly, has similar activities in the two species.

The recent discoveries of remarkably high activities of proline dehydrogenase in the flight muscle of the Colorado potato beetle, *Leptinotarsa decemlineata* (deKort *et al.*, 1973), and the dehydrogenase and alanine-α-ketoglutarate transaminase in the cockchafer, *Melontha melontha*, flight muscle (Crabtree and Newsholme, 1970) may suggest that the oxidation of proline has a greater and more widespread relevance than previously supposed. In the Colorado potato beetle, proline is oxidized at a rate several-fold those of pyruvate and α-glycerol-P. In the cockchafer, the activity of proline dehydrogenase is similar in magnitude to the activities of two important enzymes of the Krebs cycle, *i.e.*, NAD-linked isocitrate dehydrogenase and succinic dehydrogenase.

In addition to proline, other amino acids are metabolized by a variety of insect flight muscle preparations, and the early studies have been summarized (Chefurka, 1965; Sacktor, 1965). In general, amino acids are catabolized by two major processes, oxidative deamination and transamination. In the case of glutamate, Crabtree and Newsholme (1970) have surveyed flight muscles of a relatively large number of insect species for glutamate dehydrogenase, alanine-α-ketoglutarate aminotransferase and aspartate-α-ketoglutarate aminotransferase. The enzymes for both systems are found in all preparations, although the relative activities of the enzymes vary in different species.

Oxidative deamination of glutamate is catalyzed by glutamic dehydrogenase

Glutamate $+ NAD^+ + H_2O \rightleftharpoons$
$$\alpha\text{-Ketoglutarate} + NH_4^+ + NADH$$

This reaction is known to be reversible, formation of glutamate from α-ketoglutarate and NH_4^+ having been demonstrated in the mitochondria of the flight muscle of houseflies (Van den Bergh, 1964). Mills and Cochran (1963) have partially purified the dehydrogenase from mitochondria of cockroach muscle and have found the enzyme to be specific with respect to glutamate and NAD^+. In contrast, the mitochondrial enzyme from locusts, as well as from many mammalian tissues, is able to reduce $NADP^+$ at a rate 60 percent that of NAD^+ (Klingenberg and Pette, 1962).

Glutamate is also initially transaminated and its product, α-ketoglutarate, is subsequently oxidized

Glutamate + Oxaloacetate $\rightleftharpoons$ α-Ketoglutarate + Aspartate
α-Ketoglutarate $+ 1\frac{1}{2}O_2 \longrightarrow$ Oxaloacetate $+ CO_2 + H_2O$

The pathway between α-ketoglutarate and oxaloacetate is part of the Krebs cycle. Aspartate is a principal product of the oxidation of glutamate by this mechanism. Indeed, Mills and Cochran (1963) have found that, although both glutamic dehydrogenase and aspartate-α-ketoglutarate aminotransferase are present in cockroach muscle, the mitochondrial catabolism of glutamate leads to formation of aspartate, in agreement with the findings of Krebs and Bellamy (1960) with most mammalian preparations. In contrast, Van den Bergh (1964) has reported that the product of the oxidation of glutamate by *M. domestica* flight muscle mitochondria is α-ketoglutarate; aspartate is never observed. Subsequently, however, aspartate has been found as a product of proline oxidation in flight muscle mitochondria of *P. regina* (Sacktor and Childress, 1967).

As noted above, cicada flight muscle mitochondria are permeable to glutamate and α-ketoglutarate. Hansford (1971) has observed that the highest rates of α-ketoglutarate oxidation are found with insects late in their life. At this time there appears to be a diminution in the mass of flight muscle and a significant decrease in flight activity. He suggests that flight muscle autolysis may be providing substrates to support egg-laying, and that the permeabilities and oxidations of glutamate and α-ketoglutarate may be related to this. A similar suggestion, the utilization of amino acids to support flight during egg-laying periods in the mosquito, *Culex pipiens*, has been made previously (Clements, 1955).

Oxidation of fat

As in the discussion of carbohydrate and amino acid metabolism, the present examination of the oxidation of fat will be concerned primarily with the utilization of fatty acid in flight muscle. The important questions of the biosynthesis, mobilization and transport of fat to the muscle will not be considered in this Chapter. However, the significance of these processes in relation to flight metabolism should be emphasized. For example, it has been estimated that a flying locust consumes fat at a rate of 4.1 mg/hour. The fat content of the muscle is about 3 mg, whereas the fat body contains more than 18 mg (Beenakkers, 1965). Since the locust can fly continuously for 7 to 8 hours and its reserves of carbohydrate can last for only 1 to 2 hours, it is obvious that the fat in the fat body provides most of the fuel for the flight, and that its mobilization and transport by way of the haemolymph to the muscle is crucial. For additional discussion of the physiological and biochemical mechanisms of the release of fat from its storage depots and its transfer to the muscle, the reader is referred to previous comprehensive reviews by Gilbert (1967) and Sacktor (1970), and the review by Bailey in this volume.

The early studies of the physiology of insect flight have demonstrated that in some species fat can serve as the metabolic fuel for intensely active muscles. In these cases, the importance of lipid should not be underestimated. Fat is the most concentrated source of biological energy, yielding per gram over twice as many calories as do carbohydrates and amino acids. On the other hand, it should be re-asserted that locusts and cockroaches, species that utilize fats in a sustained flight, will first consume much of their carbohydrates (Bücher and Klingenberg, 1958; Hofmanova *et al.*, 1966). Additionally, the mitochondria isolated from the flight muscle of Southern armyworm moth, *Prodenia eridania*, metabolize pyruvate + malate at a rate comparable to that measured with mitochondria from flies (Stevenson, 1968a). Conversely, the ability of the mitochondria, from flight muscle of flies to oxidize fatty acids, is extremely limited (Sacktor, 1955; Childress *et al.*, 1967); yet, during development of the muscle in the pharate imago, fatty acids are oxidized at low, but significant rates (D'Costa and Birt, 1969).

Little is known of the processes by which lipids in the blood are transported into the muscle or of the mechanisms by which exogenous and endogenous glycerides are hydrolyzed to yield fatty acids by muscle lipases. George and Bhakthan (1960a,b, 1961) and

Crabtree and Newsholme (1972b) have surveyed a variety of species, representing the major orders, for lipase activities. Considerably greater activities are found in flight muscles of moths, locusts, dragonflies, cockchafer and waterbug than in those of flies, wasps and bees. This difference between species appears to be correlated with the use of fat as a major metabolic fuel during sustained flights. In general, the maximum rates of hydrolysis of glycerides increases as the number of acyl groups decreases, from tri-, di-, to mono-glycerides. This is particularly marked in moths and the cockchafer (Gilbert *et al.*, 1965; Stevenson, 1972; Crabtree and Newsholme, 1972b). However, in the locust and waterbug, which utilize fat for flight, and in the fly and bee, which do not, the rates of hydrolysis of mono- and diglycerides are about equal, but rates of hydrolysis of both these glycerides are significantly higher than that for tri-glycerides (Crabtree and Newsholme, 1972b). It is claimed by Crabtree and Newsholme (1972b) that the activities of diglyceride lipases and the calculated rates of fatty acid oxidation are in reasonable agreement for the locust, moth and butterfly and that this comparison is consistent with the suggestion that diglycerides are the principal substrate for flight in these insects (Gilbert, 1967; Tietz, 1967). On the other hand, Stevenson (1969) has shown that for at least some moths, diglyceride lipase activity in flight muscle is far too low to provide free fatty acid at the rate required to support flight. Indeed, Crabtree and Newsholme (1972a) have also reported results showing that in several other moths the rate of hydrolysis of monoglycerides by flight muscles is at least 10-fold that of diglycerides.

The activity of locust flight muscle triglyceride lipase is unaffected by high concentrations of salt, suggesting that the activity is not due to a lipoprotein lipase (Crabtree and Newsholme, 1972b). The diglyceride lipase from flight muscle of the moth, *Hyalophora cecropia*, is not activated by Ca^{2+} (Gilbert *et al.*, 1965). Triglyceride lipase from flight muscles of the locust, waterbug and blowfly are not activated by cyclic AMP (in the presence of ATP, Mg^{2+} and caffeine), suggesting either that the activities of the lipase are not stimulated by the nucleotide or that the enzymes are present in the extracts in the activated form (Crabtree and Newsholme, 1972b).

The discoveries of Friedman and Fraenkel (1955) and Fritz (1955) on the role of carnitine in the transport of fatty acids have provided the basis for the demonstration of significant rates of fatty acid oxidation by insect muscle preparations. These earlier investigations have described the reversible acyl transfer between CoA and carnitine and the stimulation, by carnitine, of the rate of oxidation of fatty acids in mammalian tissues. From these and other studies, Fritz and Marquis (1965) have proposed that fatty acyl CoA thioesters do not readily penetrate mitochondrial membranes, whereas fatty acyl carnitine esters do. Also the formation of carnitine esters by acyl transferases effects the translocation of fatty acyl groups to the site of fatty acid oxidation. In accord with this hypothesis, Beenakkers (1963a) and Bode and Klingenberg (1964) have shown that added carnitine markedly stimulates oxidation of fatty acids in locust flight muscle and that fatty acids supplied as acyl carnitine esters are metabolized at even greater rates. The requirement for carnitine for oxidation of fatty acids is strongly correlated with the presence in the muscle of active carnitine-acetyl and -palmitoyl transferases (Beenakkers and Klingenberg, 1964, Beenakkers *et al.*, 1967). Particularly striking are the differences in transacetylase activity between flight muscles of the locust, which oxidize fatty acids, and those of the bee, which oxidize only carbohydrates in flight. This enzyme is virtually absent from flight muscle of the bee, whereas it is very active in locust flight muscle. A similar species distribution for carnitine-palmitoyl transferase is found for insects that do, and do not, oxidize fatty acids (Crabtree and Newsholme, 1972b). The activities of carnitine-acyl transferases and that of 3-hydroxyacyl-CoA dehydrogenase – a component of the β-oxidation pathway for fatty acid oxidation, form a constant proportion group in different tissues (Beenakkers *et al.*, 1967).

Carnitine-acetyl transferase in locust flight muscle is localized entirely in the mitochondria (Beenakkers and Henderson, 1967). These investigators found that isolated mitochondria oxidize acetyl CoA in the presence of carnitine, but not in its absence. This indicates that acetyl CoA cannot penetrate mitochondria and that there must be carnitine-acetyl transferase activity external to this diffusion barrier, which is presumably the inner mitochondrial membrane. Logically, the activity must be duplicated on the inside of the barrier to convert acetyl carnitine back to acetyl CoA. Based on experiments with mammalian tissues (Norum and Bremer, 1966), the same localization is assumed in insects for the long-chain carnitine-acyl transferases.

Conflicting with this view for an essential role of carnitine in fatty acid oxidation in insects, however, are the findings of Stevenson (1966, 1968b) with flight muscle of two species of moths. For example, he has reported that mitochondria from *Prodenia eridania* flight muscle oxidize palmitate, in the absence of carnitine, at a very high rate, 715 μl O_2 mg protein/hour. The addition of carnitine does not enhance this rate, nor is there any evidence for the presence of carnitine-palmitoyl transferase. There is sufficient activity of an ATP-dependent fatty acyl-CoA synthetase to explain the high rates of oxygen uptake. Thus, in this moth the free acid is apparently able to enter the mitochondrion. A grossly damaged membrane in the isolated mitochondria seems to be an unlikely explanation because of the high respiratory control and P:O ratios of these preparations. The argument for the absolute requirement for carnitine in fatty acid oxidation in Lepidoptera, is further weakened by the calculations of Crabtree and Newsholme (1972b). These (to the reviewer), indicate that the activities of carnitine-palmitoyl transferase in flight muscles of the Peacock butterfly, *Vanessa io*, and the Silver-Y moth, *Plusia gamma*, although high, fail to meet the rate of lipid utilization during flight.

Another apparent anomaly with respect to carnitine has been found in the blowfly, *P. regina*. Flight muscle of the blowfly, which like the bee is deficient in fatty acid oxidase and has only a negligible capacity to oxidize palmitoyl carnitine, has an exceedingly high content of carnitine and a very active carnitine-acetyl transferase (Childress *et al.*, 1967). It has been found that in the blowfly, but not in the bee, carnitine is involved in carbohydrate utilization. The carnitine-acetyl transferase in mitochondria from *P. regina* flight muscle catalyzes the synthesis of acetyl carnitine from carnitine, and acetyl CoA derived from pyruvate. Formation of acetyl carnitine occurs *in vitro* and *in vivo*; on initiation of flight its concentration in flight muscle increases four-fold, paralleling the increase in concentration of pyruvate (Childress *et al.*, 1967).

Approximately 90 percent of the carnitine-acetyl transferase in *P. regina* flight muscle is found in isolated mitochondria (Childress *et al.*, 1967). Exogenous acetyl CoA, in the presence of carnitine, is not oxidized by these mitochondria, although acetyl carnitine is oxidized with a rate of about 0.5 μg atoms oxygen/min/mg protein (Table 1.3). This indicates that the blowfly mitochondrial inner membrane is not permeable to the thioester and neither does the mitochondrial carnitine-acetyl transferase transfer acetyl groups from extramitochondrial acetyl CoA to carnitine and, therefore, not into the mitochondrial matrix. Instead, the evidence suggests that the mitochondrial enzyme mediates the transfer of acetyl groups to the outside. However, the apparent presence of about 10 percent of the carnitine-acetyl transferase activity in the post-mitochondrial supernatant of muscle homogenates, leaves open the possibility that *in situ* carnitine may be acetylated extramitochondrially with subsequent transport of the acetyl carnitine across the mitochondrial barrier.

Based on known values of oxygen uptake in locusts during flight, Beenakkers (1965) has calculated that the flight muscle will consume fatty acids at a rate of 23 mg/g wet wt. of muscle/hour. *In vitro* studies by Meyer *et al.* (1960), Domroese and Gilbert (1964) and Bode and Klingenberg (1964) show the utilization of fatty acids by flight muscle at rates only a fraction of this calculated *in vivo* rate. On the other hand, Stevenson (1966, 1968b), using flight muscle mitochondria of the adult Southern armyworm, has reported respiratory rates greater than 700 μl O_2 mg mitochondrial protein/ hour, a value approaching the *in vivo* rate for this species.

The requirement for ATP, Mg^{2+}, CoA and an intermediate of the Krebs cycle (for priming) for maximal rates of oxidation of fatty acids by flight muscle preparations (Meyer *et al.*, 1960; Domroese and Gilbert, 1964) suggests that in insects activation of fatty acid to its acyl CoA derivative and its subsequent catabolism via the β-oxidation pathway is the same as that established in mammalian tissues. The findings of β-ketoacylthiolase and β-hydroxyacyl dehydrogenase in locust flight muscle (Zebe, 1960; Beenakkers, 1963a,b) and fatty acyl CoA synthetase in moths (Domroese and Gilbert, 1964; Stevenson, 1968b) strongly support the presence of the entire sequence of reactions in insect flight muscle. The enzymes are located largely in the mitochondria of the muscle (Beenakkers, 1963b; Stevenson, 1968b). Successive repetition of the β-oxidation cycle results in the complete degradation of even-numbered fatty acids to acetyl CoA. The acetyl CoA enters the tricarboxylic acid cycle and is degraded in one turn of the cycle to carbon dioxide and water.

THE CONTROL OF FLIGHT MUSCLE MITOCHONDRIAL METABOLISM

As pointed out in the previous discussion the large increase in the rate of oxygen uptake upon initiation of flight, over 100-times in some cases, indicates that there is an exceptionally high degree of respiratory control in flight muscle mitochondria, *in situ*. In the sections that follow, studies suggesting possible mechanisms contributing to this regulation will be reviewed.

Morphological and biochemical organization of flight muscle mitochondria

It has become evident that the complexities of the regulation of mitochondrial function are best understood in terms of the ultra-structure and molecular organization of the mitochondrion as well as its role in the overall metabolism of the cell. The general organization of insect flight muscle has been ably described by Smith (1961a,b, 1966). The fine structure of flight muscle of the blowfly, *P. regina*, representing an insect having an asynchronous type of excitation-contraction coupling, is illustrated in Fig. 1.7. The cylindrical fibrils are very large, approximately 2 μm in diameter and 3 μm in length. The mitochondria are ovoid and irregular in shape, up to 4 μm in length and are not precisely aligned with respect to the myofibrillar striations (Smith and Sacktor, 1970). Mitochondria comprise about 40 percent of the total muscle mass (Levenbook and Williams, 1956), and half the total protein (Sacktor, 1953a). It is estimated that each milligram of wet weight of *P. regina* flight muscle contains 1.1×10^8 mitochondria (Levenbook and Williams, 1956). The sarcoplasmic reticulum (SR) of the asynchronous *P. regina* flight muscle is markedly reduced and occupies a volume far less than that of the corresponding system in synchronous muscles. On the other hand, the T-system has developed profusely in the asynchronous muscle. Each tracheole, the vehicle for the supply of respiratory oxygen to the flight muscle, carries with it into the fibre a concentric sheath of muscle plasma membrane. This sheath, together with the tubular extensions from it, are associated with the SR in the dyad (Smith and Sacktor, 1970). Ferritin particles injected into the haemolymph of the blowfly *in vivo* are rapidly distributed throughout the extracellular space surrounding the tracheoles, and also in the T-system tubules throughout the fibres and wherever these tubules are involved with the SR in dyads (Smith and Sacktor, 1970). Thus, the flight muscle is deeply impregnated with a relatively immense volume of extracellular milieu abutting, or in close proximity to the fibrillar and nonfibrillar membrane components of the cell.

Mitochondria have two membranes, an outer limiting membrane that is smooth and an inner membrane that has many inward folds, called cristae (Fig. 1.8). In blowfly, flight muscle mitochondria the cristae are exceedingly numerous, packed very tight, and nearly fill the entire lumen of the mitochondrion. Lehninger (1970) has calculated that the inner membrane of blowfly flight muscle mitochondria has a surface in excess of 400 square metres/gram of

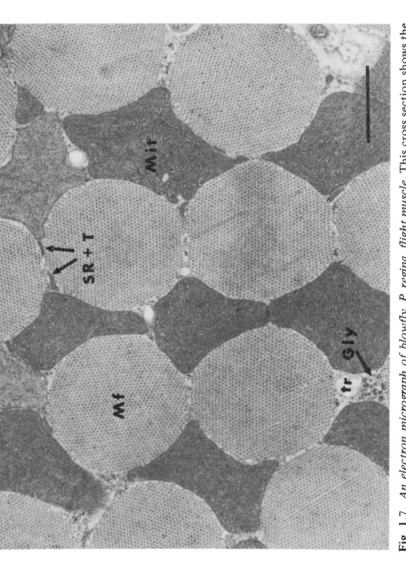

Fig. 1.7. *An electron micrograph of blowfly, P. regina, flight muscle.* This cross section shows the arrangement of mitochondria (Mit) and myofibrils (Mf). Other membranous structures evident are elements of the sarcoplasmic reticulum (SR) and the T-system (T). The T-system is comprised, in part, of elements of the plasma membrane which surround the invaginating tracheoles (tr). Glycogen rosettes (Gly) fill the sarcoplasmic spaces. The bar represents 1 μm. (From Reed and Sacktor, 1971.)

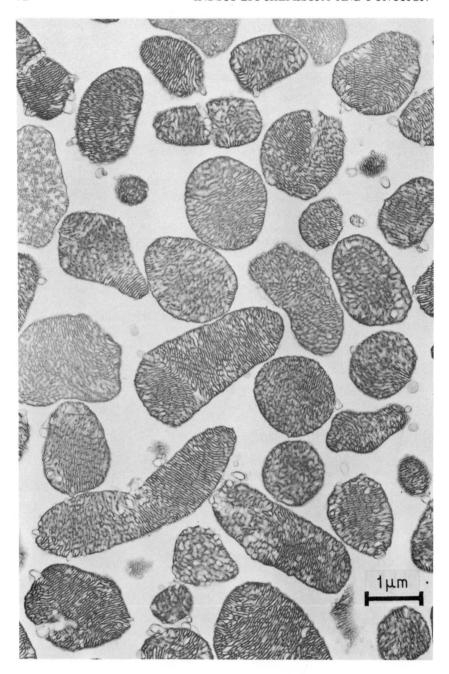

Fig. 1.8. *An electron micrograph of a typical preparation of flight muscle mitochondria.* Mitochondria were isolated from the blowfly, *P. regina,* by the Nagarse procedure. Note the thin outer membrane and the inner membrane infolded into numerous cristae.

mitochondrial protein. The cristae are arrayed as parallel plates, 30–35 cristae per micron, running transversely to the longitudinal axis of the mitochondrion (Smith, 1963; Gregory *et al.*, 1968). The membranes of the cristae are about 70 Å in thickness and the intervening spaces, the matrix of the mitochondrion and the intracristal spaces are each approximately 70–100 Å wide. The cristae are fenestrated and the perforations are aligned between one cristae and the next to form cylindrical channels within the mitochondrial matrix. This intramitochondrial arrangement, as found in blowfly flight muscle mitochondria, differs significantly from that found in other kinds of mitochondria, i.e. those from mammalian liver. In rat liver, the cristae are sparse and irregular, the inner membrane has a surface of only 40 square metres per gram and the inner compartment (matrix) is large. The differences between the ultrastructure of mitochondria from blowfly flight muscle and those from rat liver, reflect differences in the known function of the two organelles. Flight muscle mitochondria have a requirement for a large number of respiratory assemblies corresponding to the intense respiratory capability of the mitochondria which is expressed during flight. In contrast, mitochondria from liver cells need fewer respiratory assemblies corresponding to their lower rates of oxygen uptake. On the other hand, neither the biosynthetic activities nor the urea cycle enzymes of rat liver mitochondria are present in blowfly flight muscle mitochondria, thus, there is less need for matrix space.

The outer and inner mitochondrial membranes have different permeability properties. The outer membrane is generally permeable to sucrose, trehalose, respiratory substrates and nucleotides, so that the outer surface of the inner membrane is the surface over which the exchange between the cytosol and the mitochondrion takes place (Van den Bergh and Slater, 1962; Childress and Sacktor, 1966; Sacktor and Childress, 1967; Klingenberg and Pfaff, 1966; Hansford and Chappell, 1968; Reed and Sacktor, 1971; Hansford and Sacktor, 1971).

It is generally accepted that cytochrome oxidase and the other respiratory chain enzymes as well as ATP synthetase are located in the inner mitochondrial membrane (Hansford and Sacktor, 1971). There are other enzymes which remain particulate after subfractionation and cosediment with respiratory chain components. These are assigned to the inner membrane and include: α-glycerol-P dehydrogenase (Zebe and McShan, 1957, Reed and Sacktor, 1971), proline dehydrogenase (Brosemer and Veerabhadrappa, 1965; Sacktor and Childress, 1967), and trehalase (Reed and Sacktor, 1971).

The inner membrane, itself, has an outer and an inner surface. The presence of knob-like structures, (each 80–95 Å in diameter, connected to the cristae by a cylindrical stalk of 30–40 Å in diameter by 40–50 Å in length) on the inner surface implies that the ATP synthetase is oriented towards the matrix in the intact mitochondrion (Smith, 1963; Greville *et al.*, 1965; Walker and Birt, 1969). Earlier experiments showing low rates of oxidation of succinate and NADH with intact mitochondria, but high rates of respiration with these same substrates with disrupted mitochondria (Chance and Sacktor, 1958; Van den Bergh and Slater, 1962; Sacktor and Childress, 1967). From these experiments one can deduce that succinic and NADH dehydrogenases are localized on the inside of the inner mitochondrial membrane. Other evidence suggests the α-glycerol-P dehydrogenase (Hansford and Chappell, 1967; Donnellan *et al.*, 1970; Klingenberg, 1970), proline dehydrogenase (Hansford and Sacktor, 1970a), and trehalase (Reed and Sacktor, 1971) are positioned on the outer side of the inner membrane – accessible to metabolites in the cytosol (the outer mitochondrial membrane being readily permeable).

Those enzymes that are released from flight muscle mitochondria after disruption fit the operational definition of soluble matrix proteins. Among them are: citrate synthetase (Beenakkers *et al.*, 1967), NAD-linked isocitrate dehydrogenase (Goebell and Klingenberg, 1963; Hansford, 1972a), NADP-linked isocitrate dehydrogenase (Goebell and Klingenberg, 1963), malate dehydrogenase (Delbruck *et al.*, 1959; Reed and Sacktor, 1971), alanine and aspartate amino transferases (Brosemer *et al.*, 1963), 3-hydroxyacyl CoA dehydrogenase (Beenakkers *et al.*, 1967), and carnitine acetyl and palmitoyl transferases (Beenakkers and Henderson, 1967; Beenakkers *et al.*, 1967; Childress *et al.*, 1967).

A suitable enzyme marker for the outer membrane of insect flight muscle mitochondria has yet to be described. For the blowfly, *P. regina*, Reed and Sacktor (1971) have used the finding that rat brain mitochondrial hexokinase, when solubilized, will rebind preferentially to the outer membranes of various mitochondria (Kropp and Wilson, 1970). Adenylate kinase serves as a marker enzyme for the space between the outer and inner membranes in blowfly mitochondria (Reed and Sacktor, 1971). The precise ultrastructural localization of pyruvate and α-ketoglutarate dehydrogenase complexes is uncertain at this time. The complexes are functionally intimately associated with the respiratory chain in the inner membrane and with citrate synthetase and succinyl thiokinase in the matrix. Moreover, they are released from blowfly mitochondria by procedures similar to those used to solubilize the matrix enzymes (Chiang and Sacktor, 1973). The pyruvate dehydrogenase complex remains particulate, however, and after removal of inner membrane components by low-speed centrifugation the complex can be largely sedimented at moderate gravitational forces (100 000 x g, for 30 min). It is tentatively suggested that the α-keto acid dehydrogenase complexes are localized on the inner surface of the inner membrane, having sites reactive with both the membrane and the matrix, but easily dissociated from the membrane. The localization of enzymes within the mitochondrion, as indicated from studies with insect flight muscle mitochondria, supplemented with information derived from work with mammalian mitochondria (Lehninger, 1970), are listed in Table 1.4.

Table 1.4
Ultrastructural localization of mitochondrial enzymes

Outer Membrane:
    Monoamine oxidase
    Rotenone-insensitive cytochrome *c* reductase
    Hexokinase binding

Space between the Membranes:
    Adenylate kinase
    Nucleoside diphosphokinase

Matrix:
    Citrate synthetase
    Aconitase
    Isocitrate dehydrogenases
    Fumarase
    Malate dehydrogenase
    Alanine and Aspartate aminotransferases
    Carnitine, Acetyl and Palmityl transferases
    Fatty acid oxidation enzymes

Inner Membrane:
    Respiratory chain enzymes
    ATP-synthesizing enzymes
    α-Glycerolphosphate dehydrogenase
    Succinate dehydrogenase
    Proline dehydrogenase
    α-Keto acid dehydrogenases
    D-β-Hydroxybutyrate dehydrogenase
    Trehalase

(From Sacktor 1974)

Oxidative phosphorylation

A general discussion of the current views on the mechanisms of oxidative phosphorylation is beyond the scope of this review and can be found elsewhere; *e.g.*, Lehninger (1970). For the present, it is noted that there are three major hypotheses for the mechanism of energy conservation during electron transport. These postulates propose that the coupling enzymes conserve respiratory energy in such a way that three different kinds of work can be carried out: (1) the chemical work of oxidative phosphorylation, *i.e.*, the chemical coupling hypothesis; (2) the osmotic work required for accumulation of ions, *i.e.*, the chemi-osmotic coupling hypothesis; and (3) the mechanical work of conformational change in the mitochondrial membranes, *i.e.*, the mechano-chemical coupling hypothesis. These three hypotheses differ in the primary form in which the oxido-reduction energy is conserved.

Oxidative phosphorylation in insects was first demonstrated by Sacktor (1954) using mitochondria from flight muscles of the housefly, *M. domestica*. This was soon followed by the independent efforts of Lewis and Slater (1954). Since then, as summarized by Sacktor (1965), the process has been confirmed by others and extended to mitochondria from the major orders of insects and coupled variously to the oxidation of pyruvate and intermediates of the citric acid cycle, α-glycerol-P, amino acids and fatty acids.

Numerous studies, with mammalian and insect mitochondrial preparations (Hansford and Sacktor, 1971), have shown that when a pair of electrons pass from substrate to oxygen via NAD, three molecules of ATP are generated. However, when the pyridine nucleotide is not involved in the oxidation, as in the case of α-glycerol-P and succinate, only two molecues of ATP are formed. Therefore, the number of molecules of ATP that may be formed for each molecule of pyruvate oxidized to completion via the tri-carboxylic acid cycle can be estimated. In the conversion of pyruvate to carbon dioxide and water there are four NAD-linked oxidative steps, namely, the oxidation of pyruvate to acetyl CoA, isocitrate to α-ketoglutarate, α-ketoglutarate to succinyl CoA, and malate to oxaloacetate. Each of these reactions is accompanied by the formation of three molecules of ATP. In addition, the oxidation of succinate to fumarate, when reducing equivalents enter the respiratory chain subsequent to the first phosphorylation step, yields two molecules of ATP. Another molecule of ATP is formed by the substrate-level phosphorylation at the α-ketoglutarate step. Thus, for each molecule of pyruvate that is oxidized to completion 15 molecules of ATP are formed. The overall equation is depicted as

$$\text{Pyruvate} + 2.5O_2 + 15P_i + 15ADP \longrightarrow$$
$$3CO_2 + 15ATP + 17H_2O$$

In terms of energy balance, this yield of ATP represents the conservation of approximately 39 percent of the free energy of pyruvate oxidation.

For the complete oxidation of glucose to carbon dioxide and water via glycolysis and the Krebs cycle, the following equations apply

$$\text{Glucose} + 2P_i + 2ADP + 2NAD^+ \longrightarrow$$
$$2\text{ Pyruvate} + 2NADH + 2H^+ + 2ATP + 2H_2O$$

$$2\text{ Pyruvate} + 5O_2 + 30ADP + 30P_i \longrightarrow$$
$$6CO_2 + 30ATP + 34H_2O$$

to which is added the overall equation for the oxidation via the α-glycerol-P cycle of the two molecules of extramitochondrial NADH formed in the aerobic conversion of glucose to pyruvate, a process which generates two molecules of ATP per pair of electrons

$$2NADH + 2H^+ + O_2 + 4P_i + 4ADP \longrightarrow$$
$$2NAD^+ + 4ATP + 6H_2O$$

The sum of these three equations is

$$\text{Glucose} + 6O_2 + 36P_i + 36ADP \longrightarrow$$
$$6CO_2 + 36ATP + 42H_2O$$

The overall efficiency of energy recovery is approximately 39 percent.

With these computations in mind, it is of some interest to re-examine respiratory rates in terms of rates of formation of ATP. One of the highest rates of oxygen uptake for isolated mitochondria is that of the blowfly oxidizing the two products of glucose metabolism pyruvate and α-glycerol-P simultaneously. A rate of 2.67 μg atoms of oxygen/min/mg/mitochondrial protein at 25°C is found (Hansford and Sacktor, 1971). It is estimated that the rate of synthesis of ATP concomitant with this rate of oxidation is about 6.6 μmol/min/mg, a truly remarkable value.

When substrate is not limiting, the mitochondrial respiratory rate, and thus the rate of ATP production, is determined primarily by the relative concentrations of ADP, P_i and ATP. In general, the maximal rate of oxygen uptake is found when the concentrations of ADP and

P_i in the medium are high and the concentration of ATP is low. Conversely, when the concentration of ATP is high and the levels of ADP and P_i are low, the rate of mitochondrial respiration is minimal. A typical experiment illustrating this classical mechanism of respiratory control by ADP is shown in Fig. 1.9. In the presence of substrate, pyruvate + proline, and P_i the rate of oxidation by *P. regina* flight muscle mitochondria is negligible until ADP is added. Addition of ADP, 0.465 µmol in this experiment, evokes essentially a maximal rate of respiration, which continues until nearly all the ADP is phosphorylated to ATP. At this time the respiratory rate abruptly returns to the low pre-stimulated level. A second addition of ADP initiates another burst of oxygen uptake. Again, when the ADP is all

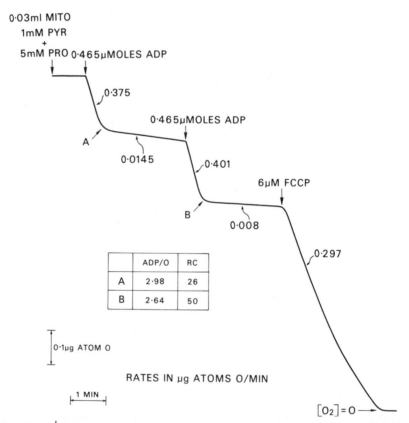

Fig. 1.9. *A typical experiment illustrating the control of the oxidation of pyruvate + proline by ADP.* Mitochondria were isolated from the flight muscle of the blowfly, *P. regina*. Oxygen uptake is measured polarographically with a Clark electrode at 30°C. The reaction medium contains 0.15M KCl, 0.01M tris chloride, 0.001M EDTA, 0.02M K P_i, and 0.5 per cent defatted bovine serum albumin, adjusted to pH 7.4, in a final volume of 2 ml. (From Sacktor, 1974.)

converted to ATP, respiration is sharply diminished. Finally, with the addition of the uncoupling agent, FCCP, respiration proceeds until the system goes anaerobic. The high rates of respiration exhibited by mitochondria in the presence of substrate, oxygen, P_i and ADP is termed the State 3 rate; the slow rate of oxygen uptake found in the absence of ADP is termed the State 4 rate (Chance and Williams, 1956). The respiratory control ratio, defined as the ratio of the rate of oxygen uptake in the presence of added ADP (State 3) to the rate of respiration after the added ADP has been completely phosphory-lated (State 4), is remarkably high in this experiment, over 25. Experiments, such as illustrated in Fig. 9, are also used to determine the ADP:O ratio. As calculated in this example, ADP:O values for the oxidation of pyruvate approach the theoretical limit of 3.

The dependence of the respiratory rate of mitochondria on the concentration of ADP may have profound physiological relevance. For example, a muscle at rest and using little ATP may have a relatively high concentration of ATP and low concentration of ADP, and so would have a very low rate of respiration. If the muscle is stimulated to contract, its ATP is rapidly broken down to ADP and P_i. This sudden increase in ADP and P_i may provide a signal to initiate an increase in the rate of respiration, which is accompanied by phosphorylation of ADP to ATP. Respiration continues at a high rate as long as ADP and P_i are generated by the ATP-requiring actomyosin contractile system. When contractions stop, dephos-phorylation of ATP ceases, and the rate of oxidation is rapidly lowered to the resting level. To what extent this control system applies to flight muscle of insects is discussed below.

The respiratory control ratios obtained with flight muscle mitochondrial preparations from different insects during the oxid-ation of pyruvate or α-glycerol-P have been compiled by Hansford and Sacktor (1971). For the oxidation of pyruvate, the reported ratios mostly range from 5 to 30. In contrast, for the oxidation of α-glycerol-P by these same preparations, the ratios are significantly lower, ranging only from 1.2 to 3.3. Another important distinction between the oxidations of pyruvate and α-glycerol-P has been found (Hansford and Chappell, 1968). This concerns the effects of the concentrations of ADP and P_i on the respiratory rates with the two substrates. In the blowfly, *P. regina*, the apparent K_m values for ADP and P_i in the oxidation of α-glycerol-P are 0.034 mM and 0.46 mM, respectively (Bulos *et al.*, 1972). In the oxidation of pyruvate + proline, the apparent K_m values for ADP and P_i are 0.21 mM and 1.22 mM, respectively. Maximal rates of oxidation of α-glycerol-P are obtained with about 0.1 mM ADP and 5 mM P_i whereas V_{max} for the oxidation of pyruvate + proline requires considerably higher concentrations.

The apparent discrepancies between the respiratory control ratios and effects of the concentrations of ADP and P_i with α-glycerol-P and pyruvate are somewhat enigmatic. As noted, there are three phosphorylation steps involved in the oxidation of NADH (from pyruvate) and only two in the oxidation of α-glycerol-P, so perhaps some differences may be expected, but the ratios should be very much closer than in fact they are. One possible explanation, which does not involve any differential uncoupling between phosphorylation sites, has been suggested by Hansford and Sacktor (1971). If a mitochondrial preparation contains intact mitochondria plus some partially damaged mitochondria, both the intact and broken mitochondria will oxidize α-glycerol-P rapidly. However, oxidation by the damaged mitochondria is without control and is not responsive to ADP and P_i. In contrast, oxidation of pyruvate is much more labile. Partially disrupted mitochondria having a lesion in the Krebs cycle, such as loss of NAD, CoA, oxaloacetate, or one of the soluble enzyme components of the cycle, will not oxidize pyruvate. Therefore, the presence of these damaged mitochondria will not lower the respiratory control ratio, except that they may contribute ATPase activity. Accordingly, Hansford and Sacktor (1971) have calculated that if a mitochondrial suspension contains mitochondria damaged to the extent of 20 per cent, then an intrinsic respiratory control of 10 would appear as only 3.5 with α-glycerol-P, whereas the ratio would be essentially unaltered with pyruvate. It is to be noted, however, that this stimulation of respiration by ADP, although of considerable significance, is much too small in itself, to account for the physiological control of respiration at the onset of insect flight.

An alternative explanation for the greater respiratory control ratio and the need for higher concentrations of ADP and P_i with pyruvate or pyruvate + proline, as substrate, relative to that with α-glycerol-P is that in the case of pyruvate (+ proline) there is control by ADP and P_i at a site(s) in addition to that reversing inhibition of electron transport through the respiratory chain in the classical manner (Chance and Williams, 1956). That is, sites of control are also to be found at the dehydrogenase level. In the discussions that follow, we shall examine how the respiratory rate of insect flight muscle mitochondria responds to effectors acting at the dehydrogenase level.

Control of the mitochondrial oxidation of α-glycerol-P

As described above, α-glycerol-P is one of the end-products of glycolysis in insect flight muscle. Its concentration in the muscle of the blowfly, *P. regina*, at rest, is relatively high and remains constant during flight (Sacktor and Wormser-Shavit, 1966). This indicates that the additional α-glycerol-P formed during the increased glycolytic flux associated with flight is immediately oxidized. These and other findings have suggested that the mitochondrial oxidation of α-glycerol-P is activated during the transition of the muscle from the resting to the active state and this activation represents one of the biochemical control points in the metabolism of the muscle (Sacktor and Wormser-Shavit, 1966).

In earlier studies, Chance and Sacktor (1958) have hypothesized that the mode of control of α-glycerol-P oxidation in flight muscle may be novel, in that the dehydrogenase rather than the respiratory chain is seemingly limiting. Subsequently, Estabrook and Sacktor (1958) have found that EDTA inhibits the oxidation of α-glycerol-P and that the inhibition is reversed by addition of Ca^{2+} or Mg^{2+}, and by excess substrate. The locus of inhibition has been identified as the dehydrogenase itself. These findings have prompted the proposal that regulation of α-glycerol-P oxidation involves the reversal of an inhibited resting state by the release of a divalent cation by a metal sequestering system in the muscle, coincident with the initiation of flight. The accumulation and release of Ca^{2+} by the sarcoplasmic reticulum, as found in locust flight muscle (Tsukamoto *et al.*, 1966) may be one such mechanism, despite the marked reduction of the SR in the asynchronous flight muscle (Smith, 1963).

This proposal receives strong support with the investigations of Hansford and Chappell (1967). They have shown that Ca^{2+} stimulates the α-glycerol-P dehydrogenase. Moreover, the concentration of free Ca^{2+} that is needed, is within the physiological range. Half-maximal rates of oxidation are obtained using concentrations less than 10^{-7} M (Fig. 1.10). The activation of α-glycerol-P dehydrogenase by Ca^{2+} in the μM range has since been confirmed with flight muscle mitochondria and submitochondrial particles from a variety of insect species (Donnellan and Beechey, 1969; Carafoli and Sacktor, 1972). Significantly, the concentration of Ca^{2+} necessary for activation of α-glycerol-P dehydrogenase is very similar to that required to activate the actomyosin ATPase (Chaplain, 1967) and the phosphorylase *b* kinase (Hansford and Sacktor, 1970a) of flight muscle.

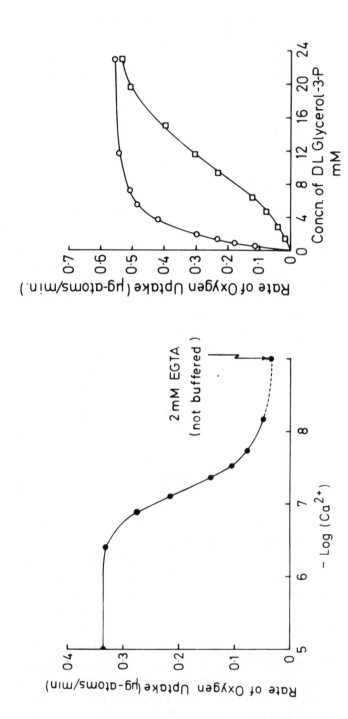

Fig. 1.10. *The control of α-glycerol-P oxidation by Ca²⁺ with mitochondria isolated from flight muscle of the blowfly, Calliphora vomitora. Left:* The effect of the concentration of free Ca²⁺, established with EGTA-Ca²⁺ buffers, on the rate of oxidation. *Right:* The effect of the concentration of α-glycerol-P on the rate of oxidation, in the presence and absence of Ca²⁺. (From Hansford and Chappell, 1967.)

As also shown in Fig. 1.10, the mechanism of action of Ca^{2+} in activating α-glycerol-P dehydrogenase is to lower the apparent K_m for α-glycerol-P (Hansford and Chappell, 1967). A plot of enzyme activity versus concentration of substrate is sigmoidal in the absence of Ca^{2+}. At 2 mM α-glycerol-P, which is the concentration found in flight muscle (Sacktor and Wormser-Shavit, 1966), Ca^{2+} effects a 10-fold increase in rate of oxidation. This is greater than the 3-fold enhancement of the respiratory rate, obtained by the addition of ADP to mitochondrial reactions oxidizing α-glycerol-P in controlled (State 4) systems.

Control of the mitochondrial oxidation of proline

The concentration of proline in flight muscle of the blowfly, *P. regina*, is extraordinarily high, nearly 7 mM. This level drops abruptly on initiation of flight. It has been suggested that the mitochondrial oxidation of proline is facilitated by the rest-flight transition. Therefore its oxidation appears to be of physiological importance in providing intermediates of the tricarboxylic acid cycle, necessary for the rapid and complete oxidation of pyruvate (Sacktor and Wormser-Shavit, 1966).

As shown in Fig. 1.11, oxidation of proline by isolated flight muscle mitochondria of the blowfly, *P. regina*, is activated by ADP in the presence of uncoupling agents as well as in the presence of oligomycin or atractyloside (Hansford and Sacktor, 1970b). Stimulation by ADP is also seen when sonically prepared submitochondrial particles are used, or when phenazine methosulfate is used as the electron acceptor to bypass the respiratory chain. These findings rule out the possibilities that the activation by ADP is related to the penetration of proline into the mitochondrion and that ADP is acting at the level of the respiratory chain. Rather, these results indicate that the site of action of ADP is proline dehydrogenase itself. Also illustrated in Fig. 11 are experiments showing that ADP is an allosteric effector of the dehydrogenase and its mode of action is to lower the apparent K_m for proline (Hansford and Sacktor, 1970b). Significantly, the apparent K_m is lowered from 33 mM to 6 mM, the latter approximating the concentration of proline found in the resting muscle of the blowfly (Sacktor and Wormser-Shavit, 1966). This activation by ADP, of proline dehydrogenase has been confirmed using mitochondria from flight muscle of the tsetse fly, *G. morsitans*; the blowfly, *Sarcophaga nodosa*; and the housefly, *M.*

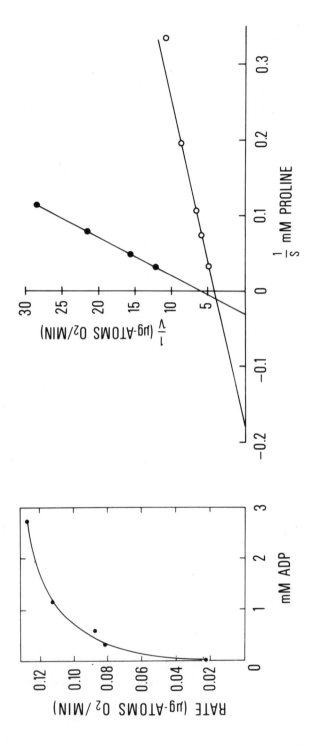

Fig. 1.11. *The control of proline oxidation by ADP with mitochondria isolated from flight muscle of the blowfly, P. regina. Left:* The effect of the concentration of ADP on the rate of oxidation. The uncoupling agent, FCCP, is in the reaction mixture. *Right:* The kinetics of the oxidation of proline. ●— in the absence of ADP; ○— 2.3 mM ADP added. (From Hansford and Sacktor, 1970b.)

domestica (Norden and Venturas, 1972). The nucleotide apparently has no effect on the enzyme of the locust, *Schistocerca gregaria.*

In the presence of ADP, the rate of oxidation of proline is additionally enhanced by high concentrations of P_i (Hansford and Sacktor, 1970b). In the absence of ADP, the level of P_i is of little significance. In the blowfly, the affinity of proline dehydrogenase for its substrate is also increased by pyruvate (Hansford and Sacktor, 1970b). The several-fold increase in the concentration of pyruvate in flight muscle of the blowfly at the initiation of flight (Sacktor and Wormser-Shavit, 1966), suggests an alternative way of stimulating proline oxidation at the dehydrogenase level.

The rate of proline oxidation is dependent on the relative proportions of the adenine nucleotides, as well as the absolute level of ADP (Hansford and Sacktor, 1970b). The rate is particularly sensitive to small increases in ADP in the presence of a high percentage of ATP. This suggests that the relatively small changes in concentrations of ATP and ADP in blowfly flight muscle at the beginning of flight, (Sacktor and Hurlbut, 1966, see Fig. 6) may lead to appreciable increases in the rate of proline oxidation.

Proline dehydrogenase is also strongly inhibited by glutamate (Norden and Venturas, 1972). The inhibition is non-competitive and the K_i is 3.6 mM. In the tsetse fly, but not in the blowfly, housefly or locust, this inhibition is relieved by ADP. The lack of a substantial, negative feedback in tsetse fly mitochondria, in the presence of ADP may reflect the fact that proline serves as a primary substrate for flight, rather than as a precursor of the intermediates of the Krebs cycle required for pyruvate oxidation (Norden and Venturas, 1972). In the tsetse fly, a significant share of the energy derived from the oxidation of proline, is liberated in the conversion of proline to glutamate. The lack of feedback inhibition ensures the continued metabolism of proline despite a temporary accumulation of glutamate.

Control of the mitochondrial oxidation of pyruvate and the citric acid cycle

As noted above, with the start of flight, the respiratory rate of the flying insect may be increased to 100-times that of the same insect at rest. Since the oxidation of pyruvate via the tricarboxylic acid cycle accounts for five-sixths of this increment (Van den Bergh and Slater, 1962; Sacktor and Wormser-Shavit, 1966), it is evident that the metabolism of pyruvate is subject to exceptionally tight regulation.

Another indication that there is control on the oxidation of pyruvate comes from findings that at the beginning of flight of the blowfly, *P. regina*, the concentration of pyruvate increases and alanine and acetyl carnitine accumulate in the flight muscle (Sacktor and Wormser-Shavit, 1966; Childress *et al.*, 1967). This demonstrates that initially pyruvate is not oxidized (via the Krebs cycle) as fast as it is formed (by glycolysis), and that the inhibition is relieved shortly after the onset of flight. One plausible explanation for this control is that mitochondria both isolated and *in vivo*, are deficient in the tricarboxylic acid cycle intermediates necessary for the generation of oxaloacetate, which, in turn is necessary, for condensation with acetyl CoA. This limitation is relieved when the precursors of oxaloacetate (derived from the oxidation of proline), become available (Sacktor and Childress, 1967). An alternative pathway for the formation of oxaloacetate, is via the pyruvate carboxylase reaction (Hansford, 1972a; Crabtree *et al.*, 1972).

Another mechanism, previously described for mammalian tissues (Linn *et al.*, 1969), has the potential for modulating the oxidation of pyruvate. It acts at the level of the pyruvate dehydrogenase complex and has recently been demonstrated in the flight muscle mitochondria of the blowfly, *P. regina* (Chiang and Sacktor, 1973). In a manner analogous to glycogen phosphorylase and glycogen synthetase, pyruvate dehydrogenase can exist in either the phosphorylated or the dephosphorylated form. Only the dephosphorylated form is catalytically active. In the presence of ATP and Mg^{2+}, pyruvate dehydrogenase kinase converts the active enzyme to its inactive, phosphorylated form. Of interest, the kinase is inhibited by pyruvate. Thus, in situations in which the concentration of pyruvate rises, such as at the initiation of flight, its oxidation is facilitated. A specific phosphatase cleaves the protein-phosphate and converts the enzyme into an active form. Significantly, the phosphatase is Ca^{2+}-activated. The physiological relevance of these enzymes and how their actions are correlated with the other mechanisms (discussed below) for controlling the rate of pyruvate oxidation in flight muscle remains to be established.

Van den Bergh (1964) has shown that maximal oxidation of pyruvate by housefly, *M. domestica*, flight muscle mitochondria requires high concentrations of P_i and ADP. In subsequent studies with the blowfly, *Calliphora vomitora*, Hansford and Chappell (1968) have confirmed that the rate of pyruvate oxidation is markedly influenced by P_i; in fact, they have reported that more than 25 μM P_i is needed for a maximal State 3 rate. In the presence of ADP, the rate of oxygen uptake with pyruvate increases by a factor of about 10 as the concentration of P_i is raised from 1.3 to 25 mM. This requirement becomes evident particularly at higher pH values (Hansford, 1972a), but it is not as apparent in the uncoupled oxidation of pyruvate by *P. regina* mitochondria (Bulos *et al.*, 1973). It is unlikely, however, that the requirement for high concentrations of P_i has anything to do with respiratory chain phosphorylation, for a considerably lower concentration is needed for the ADP-stimulated oxidation of α-glycerol-P; *e.g.* 5 mM (Bulos *et al.*, 1972).

These findings imply that the enzyme catalyzing the rate limiting step of the Krebs cycle requires a high level of P_i. Van den Bergh (1964) has suggested that this is succinyl CoA synthetase. This was found not to be the case, however, for Hansford (1968) has shown that the oxidation of α-ketoglutarate has no need for this exceptionally high concentration of P_i. On the other hand, Hansford and Chappell (1968) found that the NAD-dependent isocitrate dehydrogenase of blowfly mitochondria is almost dependent on P_i. Moreover, the requirement for P_i is very high; at 5 mM only 5 percent of the maximal activity is expressed and at 30 mM activity of the dehydrogenase is 80 percent of maximum.

Further support for the view that isocitrate dehydrogenase is subject to tight regulation and may limit pyruvate oxidation in insect mitochondria has come from the finding that the enzyme is activated by ADP (Goebell and Klingenberg, 1964). In kinetic studies, these workers have demonstrated that the NAD-linked isocitrate dehydrogenase from locust flight muscle mitochondria is markedly activated by ADP and by isocitrate, and is inhibited by ATP. Later work with various insect species has extended these findings, showing additionally that isocitrate dehydrogenase is activated by citrate and H^+ as well as by ADP, P_i and isocitrate, and is inhibited by NADH and Ca^{2+} in addition to ATP (Lennie and Birt, 1967; Hansford and Chappell, 1968; Vaughan and Newsholme, 1969; Ku and Cochran, 1971; Zahavi and Tahori, 1972; Hansford, 1972a).

The mode of action of ADP is to lower the K_m for isocitrate (Chen and Plaut, 1963; Goebell and Klingenberg, 1964; Hansford and Chappell, 1968; Vaughan and Newsholme, 1969). In the presence of a concentration of isocitrate approximating that found in the mitochondrion, there is a 20-fold increase in isocitrate dehydrogenase activity on adding ADP (Hansford and Chappell, 1968). Since the stimulation by ADP and inhibition by ATP are dependent on the concentrations of these effectors, the activity of the dehydrogenase is largely determined by the relative proportions of the two nucleotides in a fixed total concentration of adenine nucleotide. The concentration of adenine nucleotide in fly mitochondria, based on the content determined by Price and Lewis (1959), divided by the matrix space estimated by Hansford and Lehninger (1972) is 5 to 6 mM. If, as illustrated in Fig. 1.6, one assumes that in mitochondria from resting flight muscle most of the adenine nucleotide is ATP (Sacktor and Hurlbut, 1966), then the control of isocitrate dehydrogenase and, in turn, pyruvate oxidation must be quite rigorous. This has been substantiated for both the dehydrogenase and the oxidation of pyruvate for a pH range of 6.8 to 6.9 (Hansford, 1972a).

An important factor in the regulation of isocitrate dehydrogenase is the finding of Vaughan and Newsholme (1969). They observed that the effect of ADP on the enzyme in crude extracts of flight muscles from locusts, water bugs and blowflies is dependent on the concentration of Ca^{2+}. At a minimal Ca^{2+} concentration, approximately 10^{-9} M, the dehydrogenase is maximally active (at a given isocitrate concentration) in the absence of added ADP. At 10^{-5} M Ca^{2+}, however, and in the absence of added ADP, activity is extremely low but can be increased by the addition of ADP. Both Ca^{2+} and ADP affect the K_m of the dehydrogenase for isocitrate. However, raising the concentration of ADP decreases the K_m whereas raising the concentration of Ca^{2+} increases the K_m. The effects of Ca^{2+} and ADP on the enzyme are independent. This action of Ca^{2+} on isocitrate dehydrogenase takes on added significance with the recent findings of Bulos et al. (1973) that Ca^{2+} at low, physiological concentrations inhibits the oxidation of pyruvate in the mitochondria of the blowfly, P. regina, (Fig. 1.12). Inhibition of pyruvate oxidation by higher (mM) concentrations of Ca^{2+} has also been reported (Carafoli et al., 1971; Hansford, 1972a).

A series of critical investigations, indicating isocitrate dehydrogenase as the primary limit to the oxidation of pyruvate via the Krebs cycle, has been carried out recently (Hansford 1973 and Johnson and Hansford 1973). They have measured the concentrations of many of the intermediates of the citric acid cycle both in isolated blowfly, P. regina, flight muscle mitochondria during the State 4 to 3 transition and in flight muscle during the rest → flight transition. They found that the concentration of metabolites preceeding the isocitrate dehydrogenase reaction decrease on activation of respiration, whereas the concentration of intermediates subsequent to the dehydrogenase increase in concentration during the transition. This cross-over at the isocitrate dehydrogenase

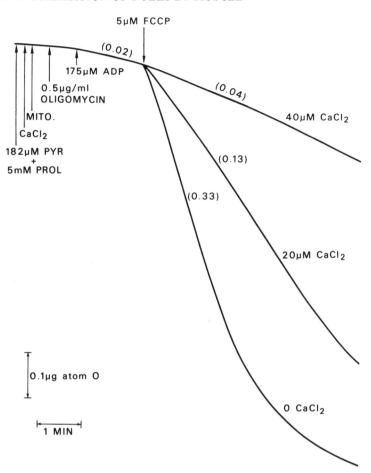

Fig. 1.12. *The inhibition by physiological concentrations of Ca²⁺ of the oxidation of pyruvate by blowfly, P. regina, flight muscle mitochondria.* Additions to the reaction mixture are indicated by arrows. Numbers along each line show rates of respiration in μg atoms of oxygen per minute. (From Bulos *et al.*, 1973.)

reaction, concomitant with an increased flux through the system, clearly identifies the point of control.

The phosphorylation state of the intramitochondrial adenine nucleotide pool regulates the activity of α-ketoglutarate dehydrogenase as well as that of isocitrate dehydrogenase. Blowfly flight muscle mitochondria oxidize α-ketoglutarate at a maximal rate when the pool is largely ADP (Hansford, 1972b). ATP inhibits the dehydrogenase; its mechanism of action is to decrease the affinity of the enzyme for α-ketoglutarate. The inhibition by ATP is reversed by

ADP or AMP. Neither of the nucleotides activates the enzyme in the absence of ATP. These findings are of significance in that control of this segment of the Krebs cycle may also be effected by the 'energy charge' of the system (Atkinson, 1968).

It is evident from the discussion of regulatory mechanisms for controlling the oxidations of proline, pyruvate and α-ketoglutarate that the phosphorylation state of the adenine nucleotides is of fundamental importance. As discussed above, the relative propor-tions of adenine nucleotides are also crucial in the control of glycogenolysis, at the phosphorylase site, and glycolysis, at the phosphofructokinase reaction. In general, the internal mitochondrial adenine nucleotide composition is manipulated through changes in the composition of the extramitochondrial mileau, the total level of internal nucleotide remaining constant, being fixed by that present initially within the mitochondrion (Klingenberg and Pfaff, 1966). As illustrated in Fig. 1.6, the concentrations of ATP, ADP and AMP in flight muscle of the blowfly, *P. regina,* at rest, are 6.9, 1.5 and 0.13 μmol/g wet weight, respectively (Sacktor and Hurlbut, 1966). With the initiation of muscular contraction and respiratory stimu-lation, the concentration of ATP decreases whereas those of ADP, AMP and P_i increase. The changes in concentration are extremely rapid and are evident as early as 5 s, the first measurement made after the start of flight. These patterns of change are in accord with the postulated mechanisms of mitochondrial regulation as well as of glycogenolysis and glycolysis.

Mitochondrial ion movements and the special role of Ca^{2+} in the regulation of flight muscle metabolism

The ability of mitochondria to establish a proton gradient during respiration is basic to the chemi-osmotic coupling hypothesis for oxidative phosphorylation, and the reader is referred to the works of Mitchell (1966) and others for comprehensive reviews on this topic.

As described above, maximal oxidation of pyruvate by flight muscle mitochondria requires a high concentration of P_i, and this has been attributed to the dependence of NAD-isocitrate dehydrogenase on P_i (Hansford and Chappell, 1968). The activity of proline dehydrogenase is also enhanced relatively high levels of P_i (Hansford and Sacktor, 1970a). It is noteworthy, therefore, that isolated blowfly mitochondria are able to accumulate P_i in a reversible energy-dependent fashion (Hansford and Chappell, 1968). The concentration of P_i in the intramitochondrial water may reach four-times that in the suspending medium, and intramitochondrial concentrations as high as 60 mM are attained experimentally. The uptake of P_i is supported by pyruvate and α-glycerol-P oxidation and accumulation is inhibited as well as reversed by the uncoupler, FCCP. The uptake of these large concentrations of P_i must involve the uptake of a counter cation, or exchange for an anion already contained within the mitochondria. Since no appreciable swelling occurs under the conditions that P_i is accumulated, these investigators have suggested that anion exchange occurs and that the bicarbonate anion may be involved.

In addition to P_i, blowfly, *P. regina,* flight muscle mitochondria respiring in State 4 take up K^+, Na^+, Cl^- and choline ions (Hansford and Lehninger, 1972). Uptake of cations is accompanied by an increase in the volume of the mitochondrial matrix. The rapid and large entry of K^+ requires the presence of a permeant anion, and is probably an electrophoretic process rather than an exchange for H^+. No increase in permeability to K^+ or to other cations occurs in State 3 respiration.

The significant role of Ca^{2+} in regulating the metabolism of insect flight muscle has been emphasized above. Therefore, the capacity of mitochondria to translocate Ca^{2+} may be of utmost importance. This is even more crucial in the asynchronous flight muscle, wherein the sarcoplasmic reticulum is very sparce (Smith, 1966), and there is a major question as to the identity of the organelles participating in Ca^{2+} segregation and release.

Carafoli *et al.* (1971) have found that Ca^{2+} is accumulated by respiring blowfly, *P. regina,* mitochondria. As illustrated in Fig. 1.13, the uptake of Ca^{2+} is supported by the oxidations of α-glycerol-P and pyruvate + proline, requires the presence of a permeant anion, and is prevented by respiratory chain inhibitors, uncouplers, ruthenium red (Carafoli and Sacktor, 1972), but not by oligomycin. At high levels of Ca^{2+}, P_i supports Ca^{2+} accumulation better than acetate, but at low Ca^{2+} levels, P_i and acetate are equally effective. Since acetate penetrates blowfly mitochondria, it is probable that respiration-coupled uptake of Ca^{2+} is accompanied by passive entry of matching anion.

The accumulation of Ca^{2+} by *P. regina* mitochondria does not exhibit saturation kinetics, so differing from the kinetics shown by mammalian mitochondria. The interaction of Ca^{2+} with blowfly mitochondria contrasts additionally with that with mammalian mitochondria in that the State 4 respiration of the former is not stimulated by Ca^{2+}, the uptake of Ca^{2+} is not associated with the

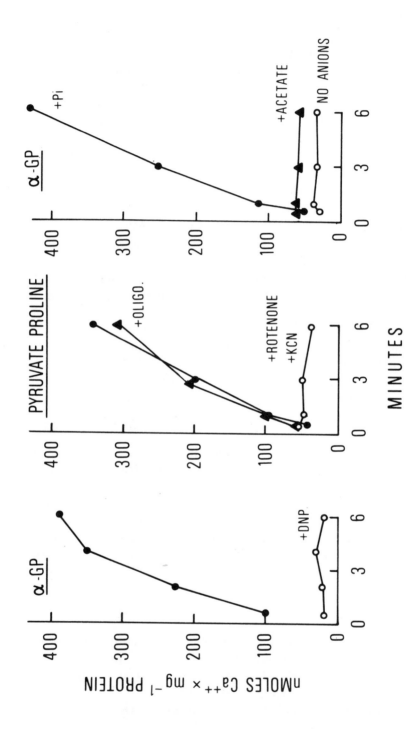

Fig. 1.13. *The energy-dependent uptake of Ca²⁺ by blowfly, P. regina, flight muscle mitochondria.* In the experiment indicated as 'no anions,' the non-penetrant, Cl⁻, was used. (From Carofoli *et al.*, 1971.)

ejection of H^+, nor does Ca^{2+} induce a reversible shift in the oxidation-reduction state of the cytochromes (Carafoli *et al.,* 1971). Moreover, no high affinity Ca^{2+} binding sites characteristic of a specific Ca^{2+} carrier are detected. These findings suggest that the respiration-dependent accumulation of Ca^{2+} in blowfly mitochondria is not mediated by a specific carrier, but occurs through slow, concentration-dependent diffusion through the membrane, in response to an electrochemical gradient generated by electron transport.

The uptake of Ca^{2+} by locust flight muscle mitochondria (Dawson *et al.,* 1971) shows many similarities with the uptake by blowfly preparations. They differ, however, in that in the locust there is apparently a limited accumulation of Ca^{2+} in the absence of a permeant anion, and Ca^{2+} may stimulate State 4 respiration. Ca^{2+} uptake also elicits an 'oxygen jump' in mitochondria of the cicada (Hansford, 1971). It is suggested that the difference in response to Ca^{2+} by the mitochondria of cicadas and locusts, on one hand, and those of blowflys, on the other, reflects the presence or absence of a Ca^{2+} carrier. This in turn is related to the different physiology of synchronous and asynchronous flight muscles. In other studies (Balbono, 1972), it is found that the H^+-linked inhibition of α-glycerol-P oxidation in the flight muscle mitochondria of the honey-bee, is released by Ca^{2+}, in a reaction accompanied by an uptake of Ca^{2+} and an ejection of H^+.

In view of the rate of Ca^{2+} uptake into isolated blowfly mitochondria at physiological concentrations of Ca^{2+}, Carafoli *et al.* (1971) reasoned that a mechanism must exist *in situ* which either prevents the uptake from continuing indefinitely or induces the discharge of the accumulated Ca^{2+}, or both of these. It is possible that the rate of uptake may be slowed by a decrease in the level of extramitochondrial Ca^{2+} to below 1 μM, by the binding of Ca^{2+} to other organelles; *e.g.,* either the remnants of the sarcoplasmic reticulum or the dyads (Fig. 1.7), or to Ca^{2+}-binding enzymes. As for Ca^{2+} release, Carafoli *et al.* (1971) have speculated on the possible role of depolarization of the mitochondrial membrane brought about by a nerve impulse. Indeed, assuming that the uptake of Ca^{2+} is driven by an electrochemical gradient across the mitochondrial membrane (Mitchell, 1966), it is conceivable that depolarization of the membrane induces its release. This is not unreasonable, as the plasma membrane of blowfly flight muscle invaginates with an extensive and profuse T-system and tracheoles, and lies in close apposition to each mitochondrion (Smith and Sacktor, 1970).

Another possible mechanism for the release of intramitochondrial Ca^{2+}, is that ADP produced by myofibrillar ATPase during contraction, discharges the electrochemical gradient across the mitochondrial membrane by its interaction with mitochondrial ATP synthetase.

Whatever the mechanism that induces Ca^{2+} release, it is apparent that the movements of Ca^{2+} into and out of blowfly flight muscle mitochondria, may be too slow to play a primary role in the rapid single contraction-relaxation cycles of this muscle during flight. Moreover, since the blowfly muscle has a much reduced sarcoplasmic reticulum, it seems that segregation of Ca^{2+} within these membrane vesicles is not prerequisite for the single contraction-relaxation cycle of this asynchronous muscle. On the other hand, it may be possible for Ca^{2+} to accumulate slowly during a train of successive contraction-relaxation cycles and then to be released when the next nerve impulse arrives (Carafoli et al., 1971).

The apparent absence of a Ca^{2+} binding carrier in blowfly flight muscle mitochondria as well as the rather singular permeability properties of these mitochondria towards di- and tricarboxylic acid intermediates of the Krebs cycle, prompted Guarnieri and Sacktor (1973) to analyze P. regina flight muscle mitochondria for their lipid and phospholipid content. These mitochondria were found to have a pattern of phospholipid remarkably different from mammalian mitochondria, confirming the reports on the phospholipid composition of housefly mitochondria (Crone, 1964; Chan, 1970). In the blowfly, the ratio of phosphatidyl ethanolamine, phosphatidyl choline, phosphatidyl diglycerol, and phosphatidyl inositol is approximately 60:15:15:10, in contrast to a ratio of about 36:40:15:3 for most mammalian mitochondria (Fleischer et al., 1967). Interestingly, the phospholipid composition of the flight muscle mitochondria of the silkmoth, Hyalophora cecropia, resembles that of mammalian rather than of dipteran muscle (Thomas and Gilbert, 1967). This biochemical difference may reflect the physiological differences between the asynchronous and synchronous types of flight muscle.

As previously discussed, at least five of the enzymes participating in the regulation of the rest → flight transition of blowfly muscle (Sacktor and Wormser-Shavit, 1966), are now known to be sensitive to Ca^{2+} at the level of the divalent cation; i.e., 800 nmol/g wet wt. (Carafoli et al., 1971). As illustrated in Fig. 1.14, two of these, actomyosin ATPase (Sacktor, 1953b; Jewell and Ruegg, 1966) and phosphorylase b kinase (Hansford and Sacktor, 1970a; Sacktor et al.,

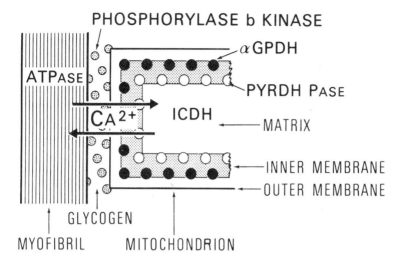

Fig. 1.14. *Diagramatic illustration of the molecular orientation of the five enzymes that regulate flight muscle metabolism whose activities are sensitive to physiological concentrations of Ca²⁺.* Glycolytic flux and respiration would be maximal when the extramitochondrial concentration of Ca^{2+} is high and the intramitochondrial concentration is low. Conversely they would be inhibited when the extramitochondrial concentration of Ca^{2+} is low and the intramitochondrial concentration is high.

1974) are extramitochondrial and activated by Ca^{2+}. The other three Ca^{2+}-dependent enzymes are mitochondrial; namely α-glycerol-P dehydrogenase (Estabrook and Sacktor, 1958, Hansford and Chappell, 1967), pyruvate dehydrogenase phosphatase (Chiang and Sacktor, 1973) and NAD-linked isocitrate dehydrogenase (Vaughan and Newsholme, 1969). As shown in Fig. 1.14, the mitochondrial α-glycerol-P dehydrogenase is located on the outer surface of the inner membrane (Klingenberg, 1970; Donnellan *et al.*, 1970) and the NAD-linked isocitrate dehydrogenase is present in the mitochondrial matrix (Goebell and Klingenberg, 1963). The location of the pyruvate dehydrogenase phosphatase is unknown at present, but it is tentatively placed at the inner surface of the cristal membrane. It is important to note, however, that for each of the two dehydro-genases, of known location, the control by Ca^{2+} is different. Ca^{2+} is an allosteric activator of α-glycerol-P dehydrogenase, whereas the divalent cation is an allosteric inhibitor of isocitrate dehydrogenase. The opposing effects of Ca^{2+} on these two mitochondrial dehydro-genases, which must be activated coincidentally with the initiation of flight, may depend on the topological position of their allosteric sites. α-Glycerol-P dehydrogenase is activated by Ca^{2+} external to the inner membranes, presumably sharing a common pool with phosphorylase b kinase and actomyosin ATPase; whereas isocitrate dehydrogenase, which is intramitochondrial, may be controlled by Ca^{2+} in the matrix. Glycolytic flux and respiration would therefore be maximal when the extramitochondrial concentration of Ca^{2+} is high and the intramitochondrial concentration is low. Conversely they would be inhibited when the extramitochondrial concentration of Ca^{2+} is low and the intramitochondrial concentration is high.

The oxidation-reduction state of the respiratory components in mitochondria and the control of metabolism

The tight regulation of the oxidations of α-glycerol-P, proline and pyruvate by the α-glycerol-P, proline and isocitrate dehydrogenase, respectively, raises the fundamental question as to the control of flight muscle respiration. It could be exerted at the substrate (dehydrogenase) level or it could be effected by the release of inhibition at a coupling site between the respiratory components and ATP synthetase, as proposed for mammalian tissues, (Chance and Williams, 1956) or control could be achieved by the concerted action of both processes. Since the State 3 rates of oxidation, of pyruvate + proline and α-glycerol-P, show at least partial summation when the substrates are added together, it is clear that the respiratory chain does not limit the maximal rate of oxidation of either substrate alone.

Spectroscopic observations have contributed effectually to the problem as to whether the dehydrogenases or the respiratory chain is activated on initiation of active respiration. The respiratory chain in insect flight muscle mitochondria is comprised of a system of pyridine nucleotide-linked dehydrogenases, flavoproteins, non-haem iron proteins, quinones and cytochromes all of which catalyze electron transport. The spectral identification, chemical characteristics and concentrations of the components in flight muscle mitochondria, as well as the sequence of these carriers in electron transport have been described elsewhere (Sacktor, 1974). In the pioneering experiments on the cytochromes in living insects, Keilin (1925), using an ocular spectroscope, has observed that when the insect is at rest the absorption bands of the cytochromes are not visible. During muscular activity the cytochrome bands become detectable, but they are never so strongly absorbing as in specimens exposed to pure nitrogen or cyanide. He has concluded that under resting conditions the cytochromes are in an oxidized form and that they become only partially reduced during exercise, however strenuous. Thus, the spectroscopic condition of the cytochromes in the organism denotes the resultant of their rates of oxidation and reduction at that particular time.

Using sensitive spectrophotometers, Chance and Sacktor (1958) have examined the respiratory components in mitochondria isolated from flight muscle of the housefly, *M. domestica.* Typical absorption spectra representing the difference between aerobic and anaerobic mitochondria, and between aerobic mitochondria and those oxidizing α-glycerol-P in the steady state are shown in Fig. 1.15. Spectral peaks are observed at 605, 563 and 550 nm, representing the α bands of reduced cytochromes *a*, *b* and $c + c_1$, respectively. The absorption bands at 520–530 nm represent a mixture of β bands of the reduced cytochromes. The trough at 470 nm represents flavoprotein. (Flavoproteins when reduced undergo bleaching with loss of the 470 nm absorption.) The peak of cytochrome a_3 is at 445 nm. The Soret bands of cytochromes *b* and *c* at 430 and 419 nm, respectively, are distinguished as shoulders on the spectrophotometric trace. With fly mitochondria, a defined peak, at 340 nm representing reduced pyridine nucleotide, is not evident because of interference by cytochromes at this wavelength. With locust flight muscle mitochondria, the peak of pyridine nucleotide at 340 nm is seen (Klingenberg and Bücher, 1959). Measurement of intramitochondrial pyridine nucleotide oxidation-reduction by fluorescence, *e.g.,* Hansford (1972a) in the blowfly, is probably the method of choice for this component of the respiratory chain.

Because the oxidation-reduction steady state of a respiratory component is a function of two opposing tendencies, reduction by substrates and oxidation by oxygen, the change in the redox state on initiation of flight, or on the addition of effector to an *in vitro* reaction containing isolated mitochondria, should indicate which of these processes has been facilitated the more. In early experiments with fly and locust mitochondria, using α-glycerol-P as substrate (Chance and Sacktor, 1958; Sacktor and Packer, 1961; Klingenberg and Bücher, 1961), addition of substrate to phosphorylating media causes an abrupt reduction of intramitochondrial cytochrome. The subsequent addition of ADP, or an uncoupling agent, induces the oxidation of the respiratory component and effects an increase in respiratory rate. Observations such as these suggest that the phosphate acceptor reverses an inhibition of electron transport through the

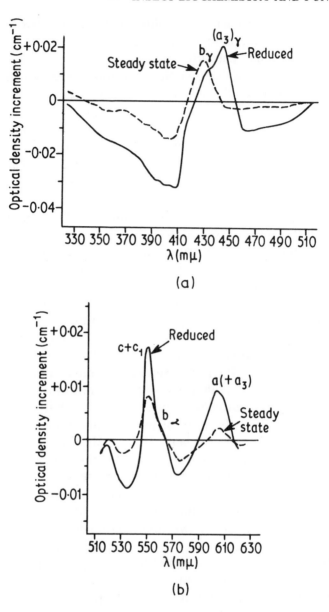

Fig. 1.15. *Absorption spectra showing reduction of cytochromes in isolated mitochondria from Musca domestica.* Spectra representing the difference between aerobic mitochondria and steady state reduced mitochondria (dashed curve) and anaerobic mitochondria (solid curve). The split beam recording spectrophotometer was used to record the difference between two samples of aerobic mitochondria, one of which contained 5 mM α-glycerol-P. The solid trace was taken when the oxygen had been exhausted in the cuvette containing substrate (from Chance and Sacktor, 1958.)

carriers. This appears to be consistent with the hypothesis proposed by Chance and Williams (1956).

However as pointed out by Hansford and Sacktor (1971), these studies do not allow an unequivocal assignment of the control mechanism, for in all these investigations, the ADP (uncoupler) has been added to a reaction in which the dehydrogenase has been maximally activated by Ca^{2+} and α-glycerol-P. When the experiments are repeated, such that the effects of both Ca^{2+} and ADP (uncoupler) are investigated (Hansford and Sacktor, 1971), it is apparent that Ca^{2+} causes an enormous reduction of cytochrome c, and ADP, a small re-oxidation. Thus, cytochrome c is 8.4 per cent reduced in the presence of 2.5 mM α-glycerol-P, 5 mM P_i and 1 mM EGTA; 39 per cent reduced upon addition of Ca^{2+}, resulting in 10^{-5} M concentration of free Ca^{2+} and 24.5 per cent reduced, on the subsequent addition of ADP. It appears, therefore, that for the oxidation of α-glycerol-P, the dehydrogenase is probably limiting. Measurements of oxygen consumption are consistent with the spectroscopic observations and lead to the same conclusion. At physiological concentrations of α-glycerol-P, the absence of Ca^{2+} lowers the rate of oxygen uptake below that obtained by the lack of ADP. Thus, the high rate of oxidation of α-glycerol-P that is initiated by the transition of the blowfly from rest $\rightarrow$ flight (Sacktor and Wormser-Shavit, 1966) may be attributed to Ca^{2+}, presumably as the divalent cation becomes available as a consequence of depolarization. It should be emphasized that this hypothesis does not preclude a supplementary increase in the rate of α-glycerol-P oxidation by ADP, interacting with the electron transport system at the coupling site.

Analogous oxidation-reduction steady state experiments have also been carried out using pyruvate as the substrate (Hansford and Chappel, 1968; Hansford, 1972a). Despite the tremendous increase in respiratory rate that is observed when ADP is added to reaction mixtures of mitochondria oxidizing pyruvate (Fig. 1.9), the changes in redox states of NAD and cytochrome c are very small (Hansford and Chappell, 1968) or, more typically, there is an increased reduction for both NAD and cytochrome c (Hansford, 1972a). As shown in Fig. 1.16, the addition of ADP causes a 20–25 per cent reduction of both NAD and cytochrome c, which persists until the ADP is phosphorylated. This response is most pronounced in the presence of ATP (Hansford, 1972a). In other experiments, the addition of P_i, also effects a reduction (Hansford, 1968). These findings strongly support the concept that the oxidation of pyruvate

by blowfly flight muscle mitochondria is limited by the dehydrogenase level and not by the respiratory chain. As previously
discussed, the marked dependency of both pyruvate oxidation and
isocitrate dehydrogenase on ADP and P_i, the inhibition of both
pyruvate oxidation and isocitrate dehydrogenase by low levels of
Ca^{2+}, and the cross-over evident from assays of Krebs cycle
intermediates during controlled and active respiration are all consistent with the hypothesis that the rate-limiting enzyme for
pyruvate oxidation is the NAD-linked isocitrate dehydrogenase. It is
the activation of this enzyme by ADP that brings about the redox
changes illustrated in Fig. 1.16. The demonstration of pyruvate
dehydrogenase kinase in blowfly flight muscle and the control of
α-ketoglutarate dehydrogenase by adenine nucleotides, suggest other
possible, supplementary mechanisms for the control of pyruvate
oxidation at the dehydrogenase level. As in the oxidation of

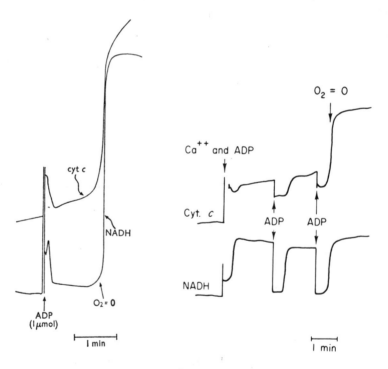

Fig. 1.16. *Effect of ADP and Ca^{2+} on respiratory chain components of
mitochondria from blowfly flight muscle.* NAD was measured fluorimetrically,
and cytochrome c was measured spectrophotometrically. Left: ADP was added
to a medium containing pyruvate, P_i, $NaHCO_3$ and ATP (from Hansford 1972a).
Right: ADP and Ca^{2+} was added to a medium containing pyruvate, α-glycerol-P,
P_i, bicarbonate, ATP, and EGTA (from Hansford and Sacktor, 1971.)

α-glycerol-P, activation by ADP and P_i at the dehydrogenase level does not preclude their concurrent activation of the respiratory chain.

An *in vitro* experiment attempting to simulate the *in situ* conditions at the initiation of flight, (Hansford and Sacktor, 1971), is shown in Fig. 1.16. The oxidation-reduction steady states of cytochrome *c* and NAD are monitored as the blowfly mitochondria oxidize both α-glycerol-P and pyruvate. When both substrates are present, as is probably the situation *in vivo* (Sacktor and Wormser-Shavit, 1966), the simultaneous addition of ADP and Ca^{2+}, simulating the increased availability of both effectors concomitant with excitation and contraction, causes a net reduction of the respiratory carriers. This is the resultant of three opposing tendencies: a reduction caused by the action of Ca^{2+} on α-glycerol-P dehydrogenase and perhaps pyruvate dehydrogenase phosphatase; a reduction caused by the action of ADP on isocitrate dehydrogenase (and on proline dehydrogenase, when proline is added initially to provide intramitochondrial oxaloacetate); and an oxidation caused by the action of ADP on the respiratory chain. The fact that the respiratory chain components show a net reduction indicates that controls at the dehydrogenase level are dominant. It may be recalled that Keilin (1925) has observed that the reduced cytochrome bands appear in the flight muscle, *in situ,* when the waxmoth starts to flap its wings. Thus, the *in vitro* experiment, illustrated in Fig. 16, is in accord with the pioneering observation.

When flight muscle is at rest its rate of respiration is slow. A dynamic steady state is operative, so that energy-requiring functions, *e.g.,* maintenance of membrane polarity and biosyntheses of glycogen, are precisely matched with the generation of low levels of proton motive force, reducing equivalents and ATP. Initiation of contraction sets off a chain of biochemical events so that a new dynamic steady state is established. This now permits glycogenolysis, glycolysis, respiration, and synthesis of ATP to proceed at the tremendous rates needed to support flight. The change from one steady state to the other is achieved by the concerted actions of effectors at several independent loci of control, perhaps by the mechanisms described in this review.

REFERENCES

Agosin, M., Scaramell, N. and Neghme, A. (1961) *Comp. Biochem. Physiol.,* 2, 143–159.
Atkinson D. E. (1968) *Biochemistry,* 7, 4030–4034.

Bailey, E. and Walker, P. R. (1969) *Biochem. J.,* 111, 359–364.

Balboni, E. R. (1972) *J. Insect Physiol.,* 18, 355–358.

Barker, R. J. and Lehner, Y. (1972) *Comp. Biochem. Physiol.,* 43B 163–169.

Barron, E. S. G. and Tahmisian, T. N. (1948) *J. Cell. Comp. Physiol.,* 32, 57–76.

Bauer, A. C. and Levenbook, L. (1969) *Comp. Biochem. Physiol.,* 28, 619–632.

Beenakkers, A. M. T. (1963a) *Acta Physiol. Pharmacol. Neerl.,* 12, 332–335.

Beenakkers, A. M. T. (1963b) *Biochem. Z.,* 337, 436–439.

Beenakkers, A. M. T. (1965) *J. Insect Physiol.,* 11, 879–888.

Beenakkers, A. M. T. (1969) *J. Insect Physiol.,* 15, 353–361.

Beenakkers, A. M. T. and Henderson, P. T. (1967) *Eur. J. Biochem.,* 1, 187–192.

Beenakkers, A. M. T. and Klingenberg, M. (1964) *Biochim. Biophys. Acta,* 84, 205–207.

Beenakkers, A. M. T., Dewaide, J. H., Henderson, P. T. and Lutgerhorst, A., (1967) *Comp. Biochem. Physiol.,* 22, 675–682.

Bendall, J. R. and Taylor, A. A. (1970) *Biochem. J.,* 118, 887–893.

Bloxham, D. P., Clark, M. G., Holland, P. C. and Lardy H. A. (1973) *Biochem. J.,* 134, 581–587.

Bode, C. and Klingenberg, M. (1964) *Biochim. Biophys. Acta,* 84, 93–95.

Borst, P. (1963) In *'Functionelle und Morphologische Organization der Zelle',* (P. Karlson, ed.), p. 137, Springer, Berlin.

Bowers, W. S. and Friedman, S. (1963) *Nature,* London, 198, p. 685.

Brenner-Holzach, O. and Leuthardt, F. (1967) *Helv. Chim. Acta,* 50, 1366–1372.

Brosemer, R. W. (1967) *J. Insect Physiol.,* 13, 685–690.

Brosemer, R. W. and Veerabhadrappa, P. S. (1965) *Biochim. Biophys. Acta,* 110, 102–112.

Brosemer, R. W., Vogell, W. and Bücher, T. (1963) *Biochem. Z.,* 338, 854–910.

Bücher, T. and Klingenberg, M. (1958) *Angew. Chem.,* 70, 552–570.

Bulos, B., Shukla, S. and Sacktor, B. (1972) *Arch. Biochem. Biophys.* 149, 461–469.

Bulos, B. A., Thomas, B. J. and Sacktor, B. (1973) unpublished observations.

Bursell, E. (1963) *J. Insect Physiol.,* 9, 439–452.

Bursell, E. (1965) *Comp. Biochem. Physiol.,* 16, 259–266.

Bursell, E. (1966) *Comp. Biochem. Physiol.,* 19, 809–818.

Bursell, E. (1967) *Comp. Biochem. Physiol.,* 23, 825–829.

Candy, D. J. (1970) *J. Insect Physiol.,* 16, 531–543.

Carafoli, E., Hansford, R. G., Sacktor, B. and Lehninger, A. L. (1971) *J. Biol. Chem.,* 246, 964–972.

Carafoli, E. and Sacktor, B. (1972) *Biochem. Biophys. Res. Commun.,* 49, 1498–1503.

Carlson, C. W. and Brosemer, R. W. (1971) *Biochemistry,* 11, 2113–2119.

Carlson, C. W., Fink, S. C. and Brosemer, R. W. (1971) *Arch. Biochem. Biophys.,* 144, 107–114.

Chadwick, L. E. (1953) In *'Insect Physiology',* (K. D. Roeder, ed.), pp. 615–636, John Wiley and Sons, New York.

Chan, S. K. (1970) *J. Insect Physiol.,* 16, 1575–1577.

Chance, B. and Sacktor, B. (1958) *Arch. Biochem. Biophys.,* 76, 509–531.

Chance, B. and Williams, G. R. (1956) *Advan. Enzymol.,* 17, 65–134.

Chaplain, R. A. (1967) *Biochim. Biophys. Acta,* 131, 385–392.
Chappell, J. B. (1968) *Brit. Med. Bull.,* 24, 150–157.
Chappell, J. B. and Crofts, A. R. (1966) in *'Regulation of Metabolic Processes in Mitochondria',* (J. M. Tager *et al.*, eds.), p. 293, Elsevier, Amsterdam.
Chappell, J. B. and Haarhoff, K. N. (1967) In *'Biochemistry of Mitochondria',* (E. C. Slater *et al.,* eds.), p. 75, Academic Press, London.
Chefurka, W. (1958) *Biochim. Biophys. Acta,* 28, 660–661.
Chefurka, W. (1965) In *'The Physiology of Insecta',* (M. Rockstein, ed.), Vol. II, pp. 581–768, Academic Press, New York.
Chen, R. F. and Plaut, G. W. E. (1963) *Biochemistry,* 2, 1023–1032.
Chiang, P. K. (1972) *Insect Biochem.,* 2, 257–278.
Chiang, P. and Sacktor, B. (1973) unpublished observations.
Childress, C. C. and Sacktor, B. (1966) *Science,* 154, 268–270.
Childress, C. C. and Sacktor, B. (1970) *J. Biol. Chem.,* 245, 2927–2936.
Childress, C. C., Sacktor, B. and Traynor, D. R. (1967) *J. Biol. Chem.,* 242 754–760.
Childress, C. C., Sacktor, B., Grossman, I. W. and Bueding, E. (1970) *J. Cell Biol.,* 45, 83–90.
Clark, M. G., Bloxham, D. P., Holland, P. C. and Lardy, H. A. (1973) *Biochem. J.,* 134, 589–597.
Clegg, J. S. and Evans, D. R. (1961), *J. Exp. Biol.,* 38, 771–792.
Clements, A. N. (1955), *J. Exp. Biol.,* 32, 547–554.
Clements, A. N. (1959), *J. Exp. Biol.,* 36, 665–675.
Clements, A. N., Page, J., Borck, K. and van Ooyen, A. J. J. (1970), *J. Insect Physiol.,* 16, 1389–1404.
Crabtree, B. and Newsholme, E. A. (1970), *Biochem. J.,* 117, 1019–1021.
Crabtree, B. and Newsholme, E. A. (1972a) *Biochem. J.,* 126, 49–58.
Crabtree, B. and Newsholme, E. A. (1972b) *Biochem. J.,* 130, 697–705.
Crabtree, B., Higgins, S. J. and Newsholme, E. A. (1972) *Biochem. J.,* 130, 391–396.
Crone, H. D. (1964) *J. Insect Physiol.,* 10, 499–507.
Davis, R. A. and Fraenkel, G. (1940) *J. Exp. Biol.,* 17, 402–407.
Dawson, A. P., Dunnett, S. J. and Selwyn, M. J. (1971) *Eur. J. Biochem.,* 21, 42–47.
D'Costa, M. A. and Brit, L. M. (1969) *J. Insect Physiol.,* 15, 1629–1645.
DeKort, C. A. D., Bartelink A. K. M. and Schuurmans, R. R. (1973) *Insect Biochem.,* 3, 11–17.
Delbruck, A., Zebe, E. and Bücher, T. (1959) *Biochem. Z.,* 331, 273–296.
Domroese, K. A. and Gilbert, L. I. (1964) *J. Exp. Biol.,* 41, 573–590.
Donnellan, J. F. and Beechey, R. B. (1969) *J. Insect Physiol.,* 15, 367–372.
Donnellan, J. F., Barker, M. D., Wood, J. and Beechey, R. B. (1970) *Biochem. J.,* 120, 467–478.
England, P. J. and Randle, P. J. (1967) *Biochem. J.,* 105, 907–920.
Estabrook, R. W. and Sacktor, B. (1958) *J. Biol. Chem.,* 233, 1014–1019.
Fleischer, S., Rouser, G., Fleischer, B., Casu, A. and Kritchevsky, G. (1967) *J. Lipid Res.,* 8, 170–180.
Florkin, M. (1958) *Proc. Intern. Congr. Biochem.,* Vienna, 4th, 12, 63–77.
Friedman, S. (1970) In *'Chemical Zoology',* (M. Florkin and B. T. Scheer, eds.), Vol. V, pp. 167–197, Academic Press, New York.

Friedman, S. and Alexander, S. (1971) *Biochem. Biophys. Res. Commun.*, 42, 818—823.
Friedman, S. and Fraenkel, G. (1955) *Arch. Biochem. Biophys.*, 59, 491—501.
Fritz, I. B. (1955) *Acta Physiol. Scand.*, 34, 367—385.
Fritz, I. B. and Marquis, N. R. (1965) *Proc. Nat. Acad. Sci. USA*, 54, 1226—1233.
Geigy, R., Huber, M. Weinman, D. and Wyatt, G. R. (1959) *Acta Trop.*, 16, 255—262.
George, J. C. and Bhakthan, N. M. G. (1960a), *J. Exp. Biol.*, 37, 308—315.
George, J. C. and Bhakthan, N. M. G. (1960b) *J. Anim. Morph. Physiol.*, 7, 141—149.
George, J. C. and Bhakthan, N. M. G. (1961) *Nature,* London, 192, p. 356.
Gilbert, L. I. (1967) *Adv. Insect Physiol.*, 4, 69—211.
Gilbert, L. I., Chino, H. and Domroese, K. (1965) *J. Insect Physiol.*, 11, 1057—1070.
Gilby, A. R., Wyatt, S. S. and Wyatt, G. R. (1967) *Acta Biochim. Polon.*, 14, 83—100.
Goebell, H. and Klingenberg, M. (1963) *Biochem. Biophys. Res. Commun.*, 13, 209—212.
Goebell, H. and Klingenberg, M. (1964) *Biochem. Z.*, 340, 441—464.
Goebell, H. and Pette, D. (1967) *Enzymol. Biol. Clin.*, 8, 161—175.
Goldsworthy, G. J. (1970) *Gen. Comp. Endocrinol.*, 14, 78—85.
Grasso, A. and Migliori-Natalizi, G. (1968), *Comp. Biochem. Physiol.*, 26, 979—984.
Gregg, C. T., Heisler, C. R. and Remmert, L. F. (1960) *Biochim. Biophys. Acta,* 45, 561—570.
Gregory, D. W., Lennie, R. W. and Birt, L. M. (1968) *J. Roy. Microsc. Soc.*, 88, 151—175.
Greville, G. D., Munn, E. A. and Smith, D. S. (1965) *Proc. Roy. Soc. B*, 161, 403—420.
Guarnieri, M. and Sacktor, B. (1973) Unpublished observations.
Gussin, A. E. S. and Wyatt, G. R. (1965) *Arch, Biochem. Biophys.*, 112, 626—634.
Hansen, K. (1966) *Biochem. Z.*, 344, 15—25.
Hansford, R. G. (1968) University of Bristol, Ph. D. Thesis.
Hansford, R. G. (1971) *Biochem. J.*, 121, 771—780.
Hansford, R. G. (1972a) *Biochem. J.*, 127, 271—283.
Hansford, R. G. (1972b) *FEBS Lett.*, 21, 129—141.
Hansford, R. G. (1973) Unpublished observations.
Hansford, R. G. and Chappel, J. B. (1967) *Biochem. Biophys. Res. Commun.*, 27, 686—692.
Hansford, R. G. and Chappell, J. B. (1968) *Biochem. Biophys. Res. Commun.*, 30, 643—648.
Hansford, R. G. and Lehninger, A. L. (1972) *Biochem. J.*, 126, 689—700.
Hansford, R. G. and Sacktor, B. (1970a) *FEBS Lett.*, 7, 183—187.
Hansford, R. G. and Sacktor, B. (1970b) *J. Biol. Chem.*, 245, 991—994.
Hansford, R. G. and Sacktor, B. (1971) In 'Chemical Zoology' (M. Florkin and B. T. Scheer, eds.) Vol. VI, pp. 213—247, Academic Press, New York.

Heilmeyer, L. M. G., Meyer, F., Haschke, R. H. and Fischer, E. H. (1970), *J. Biol. Chem.*, **245**, 6649–6656.

Hess, R. and Pearse, A. G. E. (1961), *Enzymol. Biol. Clin.*, **1**, 15–33.

Hofmanova, O., Cerkasovova, A., Foustka, M. and Kubista, V. (1966) *Acta Univ. Carol, Biol.*, 183–189.

Huston, R. B. and Krebs, E. G. (1968) *Biochemistry*, **7**, 2116–2122.

Jewell, B. R. and Ruegg, J. C. (1966) *Proc. Roy. Soc. B.*, **164**, 428–459.

Johnson, R. N. and Hansford, R. G. (1973) Unpublished observations.

Keilin, D. (1925) *Proc. Roy. Soc. B*, **98**, 312–339.

Kerly, M. and Leaback, D. H. (1957) *Biochem. J.*, **67**, 245–250.

Kermack, W. O. and Stein, J. M. (1959) *Biochem. J.*, **71**, 648–653.

Kilby, B. A. and Neville, E. (1957) *J. Exp. Biol.*, **34**, 276–289.

Kirsten, E., Kirsten, R. and Arese, P. (1963) *Biochem. Z.*, **337**, 167–178.

Kitto, G. B. and Briggs, M. H. (1962a) *Nature*, **193**, 1003–1004.

Kitto, G. B. and Briggs, M. H. (1962b) *Science*, **135**, 918.

Klingenberg, M. (1970) *Eur. J. Biochem.*, **13**, 247–252.

Klingenberg, M. and Bücher, T. (1959) *Biochem. J.*, **331**, 312–333.

Klingenberg, M. and Bücher, T. (1961) *Biochem. J.*, **334**, 1–17.

Klingenberg, M. and Pette, D. (1962) *Biochem. Biophys. Res. Commun.*, **7**, 430–432.

Klingenberg, M. and Pfaff, E. (1966) In '*Regulation of Metabolic Processes in Mitochondria*', (J. M. Tager *et al.*, eds.), p. 180, Elsevier, Amsterdam.

Krebs, E. G., Love, D. S., Brotvold, G. E., Trayser, K., Meyer, W. L. and Fischer, E. H. (1964) *Biochemistry*, **3**, 1022–1033.

Krebs, H. A. and Bellamy, D. (1960) *Biochem. J.*, **75**, 523–529.

Krebs, H. A., Gascoyne, T. and Notron, B. M. (1967) *Biochem. J.*, **102**, 275–282.

Krogh, A. and Weis-Fogh, T. (1951) *J. Exp. Biol.*, **28**, 344–357.

Kropp, E. S. and Wilson, J. E. (1970) *Biochem. Biophys. Res. Commun.*, **38**, 74–79.

Ku, T. Y. and Cochran, D. G. (1971) *Insect Biochem.*, **1**, 81–96.

Lardy, H. A., and Parks, R. E., Jr. (1956) In *Enzymes: Units of Biological Structure and Function* (O. H. Gaebler, ed.), pp. 584–587, Academic Press, New York.

Larner, J. and Villar-Palasi, C. (1971) In *Current Topics in Cellular Regulation* (B. L. Horecker and E. R. Stadtman, eds.), pp. 195–236, Academic Press, New York.

Lehninger, A. L. (1970) *Biochemistry. The Molecular Basis of Cell Structure and Function*. Worth Publishers, New York.

Lennie, R. W. and Birt, L. M. (1967) *Biochem. J.*, **102**, 338–350.

Levenbook, L. and Williams, C. M. (1956) *J. Gen. Physiol.* **39**, 497–512.

Levenbook, L., Bauer, A. C. and Shigematsu, H. (1973) *Arch. Biochem. Biophys.*, **157**, 625–631.

Lewis, S. E. and Price, G. M. (1956) *Nature*, **177**, 842–843.

Lewis, S. E. and Slater, E. C. (1954) *Biochem. J.*, **58**, 207–217.

Linn, T. C., Pettit, F. H., Hucho, F. and Reed, L. G. (1969) *Proc. Nat. Acad. Sci. USA*, **64**, 227–234.

Liu, T. P. and Davies, D. M. (1971) *Can J. Zool.*, **49**, 219–221.

Lowry, O. H., Passonneau, J. V., Hasselberger, F. X. and Schultz, D. W. (1964) *J. Biol. Chem.*, **239**, 18–30.

Marquardt, R. R., Carlson, C. W. and Brosemer, R. W. (1968) *J. Insect Physiol.*, **14**, 317–333.

Maruyama, K., Pringle, J. W. S. and Tregear, R. T. (1968) *Proc. Roy. Soc. B*, **169**, 229–240.

Merlevede, W. and Riley, G. A. (1966) *J. Biol. Chem.*, **241**, 3517–3524.

Meyer, H., Preiss, B. and Bauer, S. (1960) *Biochem J..*, **76**, 27–35.

Mills, R. R. and Cochran, D. G. (1963) *Biochim. Biophys. Acta*, **73**, 213–221.

Mitchell, P. (1966) *Biol. Rev.*, **41**, 445–502.

Murphy, T. A. and Wyatt, G. R. (1965) *J. Biol. Chem.*, **240**, 1500–1508.

Newsholme, E. A. and Crabtree, B. (1970) *FEBS Lett.*, **7**, 195–198.

Newsholme, E. A. and Taylor, K. (1969) *Biochem J.*, **112**, 465–474.

Newsholme, E. A., Crabtree, B., Higgins, S. J., Thornton, S. D. and Start, C. (1972) *Biochem. J.*, **128**, 89–97.

Nolte, J., Brdiczka, D. and Pette, D. (1972) *Biochim. Biophys. Acta*, **284**, 497–507.

Norden, D. A. and Patterson, D. J. (1969) *Comp. Biochem. Physiol.*, **31**, 819–827.

Norden, D. A. and Venturas, D. J. (1972) *Insect Biochem.*, **2**, 226–234.

Norum, K. and Bremer, J. (1966) *Abst. 3rd Meeting Federation European Biochem. Soc.*, Warsaw, p. 119.

Ozawa, E., Hosoi, K. and Ebashi, S. (1967) *J. Biochem.*, **61**, 531–533.

Passonneau, J. V. and Lowry, O. H. (1962) *Biochem. Biophys. Res. Commun.*, **7**, 10–15.

Pearson, O. P. (1950) *Condor*, **52**, 145–152.

Pette, D. and Luh, W. (1962) *Biochem. Biophys. Res. Commun..*, **8**, 283–287.

Pette, D., Luh, W. and Bücher T. (1962) *Biochem. Biophys. Res. Commun.*, **7**, 419–424.

Piras, R., Rothman, L. B. and Cabib, E. (1968) *Biochemistry*, **7**, 56–66.

Polacek, I. and Kubista, V. (1960) *Physiol. Bohemoslov.*, **9**, 228–234.

Price, G. M. and Lewis, S. E. (1959) *Biochem. J.*, **71**, 176–185.

Pringle, J. W. S. and Tregear, R. T. (1969) *Proc. Roy. Soc. B*, **174**, 33–50.

Reed, W. D. and Sacktor, B. (1971) *Arch. Biochem. Biophys.*, **145**, 392–401.

Rees, K. R. (1954) *Biochem. J.*, **58**, 196–202.

Rosell-Perez, M. and Larner, J. (1964) *Biochemistry*, **3**, 75–81.

Ruiz-Amil, M. (1962) *J. Insect Physiol.*, **8**, 259–265.

Sacktor, B. (1953a) *Arch. Biochem. Biophys.*, **45**, 349–365.

Sacktor, B. (1953b) *J. Gen. Physiol.*, **36**, 371–387.

Sacktor, B. (1954) *J. Gen. Physiol.*, **37**, 343–359.

Sacktor, B. (1955) *J. Biophys. Biochem. Cytol.*, **1**, 29–46.

Sacktor, B. (1960) *Proc. XI Int. Congr. Entomol. Symp. Wien*, **3**, 180–183.

Sacktor, B. (1961) *Ann. Rev. Entomol.*, **6**, 103–130.

Sacktor, B. (1965) In *The Physiology of Insecta*, (M. Rockstein, ed.), Vol. 2, pp. 438–580, Academic Press, New York.

Sacktor, B. (1970) *Advan. Insect Physiol.*, **7**, 267–348.

Sacktor, B. (1974) In *Physiology of Insecta* (M. Rockstein, ed.), Vol. 4, pp. 271–353, Academic Press, New York.

Sacktor, B. and Childress, C. C. (1967) *Arch. Biochem. Biophys.*, **120**, 583–588.

Sacktor, B, and Cochran, D. G. (1956) *J. Amer. Chem. Soc.*, **78**, 3227.

Sacktor, B. and Cochran, D. G. (1957a) *Biochim. Biophys. Acta*, **25**, 649.

Sacktor, B, and Cochran, D. G. (1957b) *Biochim. Biophys. Acta*, **26**, 200–201.

Sacktor, B. and Cochran, D. G. (1958) *Arch. Biochem. Biophys.*, **74**, 266–276.

Sacktor, B. and Dick, A. (1962) *J. Biol. Chem.*, **237**, 3259–3263.

Sacktor, B. and Hurlbut, E. C. (1966) *J. Biol. Chem.*, **241**, 632–634.

Sacktor, B. and Packer, L. (1961) *Biochim. Biophys. Acta*, **49**, 402–404.

Sacktor, B. and Shimada, Y. (1972) *J. Cell Biol.*, **52**, 465–477.

Sacktor, B. and Wormser-Shavit, E. (1966) *J. Biol. Chem.*, **241**, 624–631.

Sacktor, B., Wormser-Shavit, E. and White, J. I. (1965) *J. Biol. Chem.*, **240**, 2678–2681.

Sacktor, B., Wu, N-C., Lescure, O. and Reed, W. D. (1974) *Biochem. J.*, **137**, 535–542.

Seiss, M. and Pette, D. (1960) *Biochem. Z.*, **232**, 495–502.

Smith, D.S. (1961a) *J. Biochem. Biophys. Cytol.*, **10**, 123–158.

Smith, D. S. (1961b) *J. Biochem. Biophys. Cytol* **11**, 119–145.

Smith, D. S. (1963) *J. Cell Biol.*, **19**, 115–138.

Smith, D. S. (1966) In *Progress in Biophysics and Molecular Biology* (J. A. Butler and H. E. Huxley, eds.) Vol. **16**, pp. 107–142, Pergamon Press, New York.

Smith, D.S. and Sacktor, B. (1970) *Tissue and Cell*, **2**, 355–374.

Srivastava, P. N. and Rockstein, M. (1969) *J. Insect Physiol.*, **15**, 1181–1186.

Stevenson, E. (1966) *Biochim. Biophys. Acta*, **128**, 29–33.

Stevenson, E. (1968a) *J. Insect Physiol.*, **14**, 179–198.

Stevenson, E. (1968b) *Biochem. J.*, **110**, 105–110.

Stevenson, E. (1969) *J. Insect Physiol.*, **15**, 1537–1550.

Stevenson, E. (1972) *J. Insect Physiol.*, **18**, 1751–1756.

Thomas, K. K. and Gilbert, L. I. (1967) *Comp. Biochem. Physiol.*, **21**, 279–290.

Tietz, A. (1967) *Eur. J. Biochem.*, **2**, 236–242.

Trivelloni, J. C. (1960) *Arch. Biochem. Biophys.*, **89**, 149–150.

Tsukamoto, M., Nagai, Y., Maruyama, K. and Akita, Y. (1966) *Comp. Biochem. Physiol.*, **17**, 569–581.

Tulp, A. and Van Dam, K. (1969) *Biochim. Biophys. Acta*, **189**, 337–341.

Tulp, A. and Van Dam, K. (1970) *FEBS Lett.*, **10**, 292–294.

Van den Bergh, S. G. (1964) *Biochem. J.*, **93**, 128–136.

Van den Bergh, S. G. (1967) In *Mitochondrial Structure and Compartmentation* (E. Quagliariello, S. Papa, E. C. Slater and J. M. Tager, eds.), pp. 203–206, Bari.

Van den Bergh, S. G. and Slater, E. C. (1962) *Biochem. J.*, **82**, 362–371.

Van Handel, E. and Nayar, J. K. (1972) *Insect Biochem.*, **2**, 203–208.

Vardanis, A. (1967) *J. Biol. Chem.*, **242**, 2306–2311.

Vaughan, H. and Newsholme, E. A. (1969) *FEBS Lett.*, **5**, 124–126.

Vaughan, H., Thornton, S. D. and Newsholme, E. A. (1973) *Biochem. J.*, **132**, 527–535.

Vogell, W., Bischai, F. R., Bücher, T. and Klingenberg, M. (1959) *Biochem. Z.*, **332**, 81–117.

Walker, A. C. and Birt, L. M. (1969) *J. Insect Physiol.*, **15**, 519–527.

Walker, P. R. and Bailey, E. (1969) *Biochem. J.*, **111**, 365–369.

Weis-Fogh, T. (1952) *Phil. Trans. Roy. Soc. B*, **237**, 1–36.

Weis-Fogh, T. (1964) *J. Exp. Biol.,* **41**, 229–256.
Weis-Fogh, T. (1967) *J. Exp. Biol.,* **47**, 561–587.
Wiens, A. W. and Gilbert, L. I. (1967) *Comp. Biochem. Physiol.,* **21**, 145–159.
Wigglesworth, V. B. (1949) *J. Exp. Biol.,* **26**, 150–163.
Wyatt, G. R. (1967) *Adv. Insect Physiol.,* **4**, 237–360.
Zahavi, M. and Tahori, A. S. (1972) *J. Insect Physiol.,* **18**, 609–614.
Zebe, E. (1954) *Z. Vergl. Physiol.,* **36**, 290–317.
Zebe, E. (1960) *Biochem. Z.,* **332**, 328–332.
Zebe, E. C. and McShan, W. H. (1957), *J. Gen. Physiol.,* **40**, 779–790.
Zebe, E. C. and McShan, W. H. (1959), *J. Cell Comp. Physiol.* **53**, 21–30.

2 Biochemistry of Insect Flight
Part 2 Fuel Supply

E. BAILEY

INTRODUCTION

The flight of an insect involves very rapid oxidation of respiratory fuels by the flight muscle (see reviews by Sacktor 1965, 1970 and Chapter 1). High, overall levels of metabolism of working flight muscle have been calculated (Weis-Fogh, 1952) and flight is known to be accompanied by a substantial increase in oxygen uptake (Davis and Fraenkel, 1940; Krogh and Weis-Fogh, 1951). Clearly in order to maintain very high rates of respiration, the flight muscles must be well supplied with both oxygen and the appropriate fuels. The rapid increase in respiration that occurs at the onset of flight, and the decrease in respiration rate that occurs at the end of flight, necessitates the existence of systems to control the rate of flight muscle metabolism and the supply of oxygen and substrates for respiration. The previous article has discussed the oxidation of substrates by flight muscles and the control of such processes. This review is concerned with the supply of respiratory fuels to the flight muscles and the control of such supply.

Several difficulties are encountered when assessing the literature concerned with the supply of fuels for flight. Because of the small size of most insects it is difficult to obtain individual insect tissues in quantities sufficient for experimental study. Consequently, many biochemists still resort to using homogenates of whole insects. Although in the past such experiments have yielded much useful general information, their true value is of doubtful significance. There is no reason to suppose that the metabolism of insects is any simpler than that of higher animals. Work with the latter has shown that a proper understanding of metabolic processes requires not only an understanding of the biochemistry of particular tissues but also an understanding of the biochemistry of the different types of cells within tissues and also the regions within cells. The small size of insects should not present a bar to progress since present day methods allow the use of very small amounts of material; for example, Kato and Lowry (1973a, b) have recently studied the distribution of the activities of nine enzymes between the nucleus and cytoplasm of single dorsal root ganglion cells of the rabbit.

Difficulties also arise due to the wide variety of insect species that have been used, since it is likely that differences in metabolism are species specific and work with mammals indicates that not only species differences but also strain differences are reflected in metabolism (Bartley *et al.*, 1967). Clearly generalizations drawn from results of experiments using different insect species may be misleading. Finally, difficulties may arise from different workers using the same species of insect but at different stages of development or of a different sex. Occasionally neither the sex nor the age of the insect is indicated. Considerable metabolic differences are to be expected at different developmental stages and indeed make a fascinating study, but changes also occur with age at the same developmental stage; for instance, the lipogenic capacity of the fat body of the male desert locust, *Schistocerca gregaria*, varies considerably during adult development and determination of lipogenic enzyme activities at various ages can yield very different results (Walker and Bailey 1969a, 1970a, b). The work of Van Handel and Lum (1961) has shown clearly that in the mosquito, *Aëdes aegypti*, the sex of the insect is a very important factor in determining the rate of lipogenesis. Major metabolic changes must occur in the female adult insect in relation to reproduction and unless this important aspect of insect metabolism is being studied, the male would be better experimental material. It is hoped that this review will recognize the deficiencies in current knowledge and help to eliminate them by stimulating further work.

THE NATURE OF RESPIRATORY FUELS

Lipids, carbohydrates and, in some insects, certain amino acids are used as respiratory fuels to supply the energy for flight (Sacktor, 1965, 1970). Measurements of respiratory quotients of flying insects indicate that some insects in orders such as Diptera and Hymenoptera, oxidize carbohydrates as the major source of energy whereas others, orders Lepidoptera and Orthoptera, utilize lipids. However several Orthopteran species studied use both lipid and carbohydrate as fuels for flight. Polacek and Kubista (1960) showed that carbohydrate is the major energy source for flight of the cockroach, *Periplaneta americana*, the observed low respiratory quotient of 0.64 being due to incomplete carbohydrate oxidation rather than to lipid oxidation. In the desert locust, *Schistocerca gregaria*, lipid seems to be the main energy source, but carbohydrate is also used, particularly in the early stages of flight (Weis-Fogh, 1952). The amino acid proline may also be used occasionally (Kirsten *et al.*, 1963; Mayer and Candy, 1969a). The use of carbohydrate during early flight, and lipid for prolonged flight has also been indicated for the male Douglas-Fir beetle, *Dendroctonus pseudotsugae* (Coleoptera), by the work of Thompson and Bennett (1971). Zebe (1959) and Beenakkers (1969a) have categorized three physiological types of flight muscle;

the carbohydrate utilizers as in dipterous insects (Chadwick, 1947); the lipid utilizers, as in lepidopterous insects (Zebe, 1954); and the combination utilizers, as in orthopterous insects (Krogh and Weis-Fogh, 1951). Although the energy for flight in the mosquito, *Culex pipiens*, (Diptera), is usually provided by glycogen reserves in the insect, such energy can also be provided by amino acids produced by the digestion of a high protein content blood meal (Clements, 1955). Further, the work of Van Handel and Nayar (1972a) indicates that Lepidopterans may use carbohydrate as well as lipid as energy sources for flight. These workers found that in the moth, *Spodoptera frugiperda*, glucose, trehalose and glycogen could all be used for flight energy without prior conversion to lipid.

Although lipids and carbohydrates are the major sources of energy for flight in most insects, in some, various amino acids can also act in the same capacity. Bursell (1963, 1966) using the tsetse fly, *Glossina morsitans* (Diptera) and De Kort *et al.* (1973) using the colorado beetle, *Leptinotarsa decemlineata* (Coleoptera) have shown that proline in particular is a very important energy source. The use of amino acids as an energy source in some species of Diptera and Orthoptera has previously been mentioned.

STORES OF RESPIRATORY FUELS

Insect flight muscle contains its own limited stores of respiratory fuels which can be quickly mobilized to meet the energy requirements of the muscle. The flight muscles of the blowfly, *Phormia regina*, contain quite large quantities of glycogen, 5 to 16 mg/g wet wt. of muscle, and a little trehalose, 0.3 mg/g wet wt. of muscle (Sacktor, 1961, 1965; Clegg and Evans, 1961). Both glucose (0.14 to 10 mg/g wet wt. of muscle) and trehalose (1.4 to 7.2 mg/g wet wt. of muscle) have been found in the flight muscles of *Locusta migratoria* (Bücher and Klingenberg, 1958). High concentrations of amino acids are also found in flight muscles of a variety of insects (Price, 1961; Bursell, 1963; Kirsten *et al.*, 1963; Sacktor and Wormser-Shavit, 1966; De Kort *et al.*, 1973). Cockbain (1961) showed that fat occurs inbetween fibrils of flight muscle of aphids and is utilized during flight. Walker *et al.* (1970) found that flight muscle of the male desert locust, *Schistocerca gregaria*, contained 4 to 4.5 mg lipid per 100 mg wet wt. of tissue at all stages of adult development. Flight muscles contain enough fuels to initiate flight but the endogenous reserves are too small to sustain prolonged flight for which respiratory fuels must be supplied to the muscles from exogenous sources.

The flight muscles derive their exogenous supply of respiratory fuels from the haemolymph, which itself contains an important reserve of organic materials. However, the major energy storage site in the insect is the fat body which is often considered the equivalent of the mammalian liver and adipose tissue combined (Kilby, 1965), and which may contain large quantities of glycogen, triglycerides and proteins. Particularly after feeding, the gut of the insect can supply the flight muscle with respiratory fuels and also provide the fat body with substrates to convert into energy reserves. Beenakkers (1965) has calculated that the haemolymph, fat body and flight muscles of the locust, *Locusta migratoria*, contain enough energy reserves for five hours of flight and if substrates from other parts of the body, such as the gut, are utilized, the locust can continue to fly for a further two hours. The interrelationship between the flight muscles, the haemolymph, the gut and the fat body are illustrated in Fig. 2.1. Clearly, as with all living systems, the metabolism of proteins, lipids and carbohydrates is intimately intertwined. However, for the sake of convenience and clarity, the supplies of lipid, carbohydrate and amino acid energy sources to the flight muscles and the control of such supplies are dealt with in separate sections. The close interrelationship between these areas of metabolism is exemplified by the fact that the author has chosen to include a brief discussion of fat body glycolysis in the section dealing with lipids for although the glycolytic pathway is an important route for carbohydrate degradation, it can also be considered as the initial part of the pathway of conversion of carbohydrate to lipid. Similarly, although the pentose phosphate pathway is a pathway of carbohydrate metabolism, it is also involved in the provision of NADPH for fatty acid biosynthesis and is discussed in this review in that connection. The tricarboxylic acid cycle, the respiratory chain and oxidative phosphorylation are common to lipid, carbohydrate and protein metabolism, and for convenience are discussed in the section dealing with lipids.

LIPIDS AND THEIR METABOLISM

Some aspects of the topics dealt with in this section have been discussed in previous reviews by Kilby (1963), Tietz-Devir (1963), Tietz (1965), Fast (1964, 1970), Chefurka (1965a, b), Gilby (1965), Gilmour (1965), Gilbert (1967a) and Sacktor (1965, 1970). Lipids are very suitable energy sources in insects since an isocaloric quantity of triglyceride occupies much less storage space than the

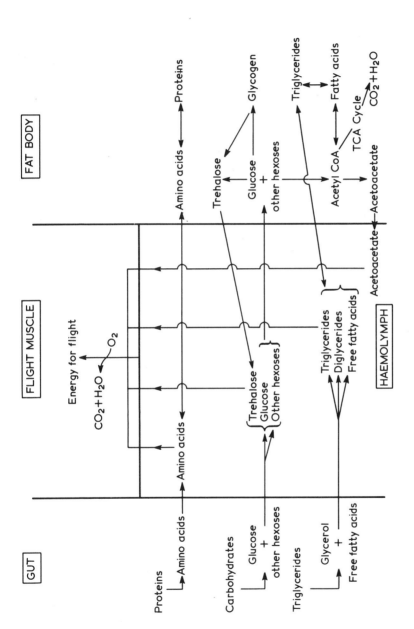

Fig. 2.1. Metabolic interrelationships between various insect tissues.

equivalent amount of glycogen. Clearly this is extremely important
in insects which have to fly for long periods of time; for example the
migratory monarch butterfly, *Danaus plexipus*, utilizes lipid during
flight and lays down a store of fat (as do migrating birds) prior to
migration (Beall, 1948). Lipid reserves are also used as energy sources
for processes other than flight. Many insects characteristically
accumulate lipid in high concentrations at physiological stages of
development preceding periods of non-feeding, such as pupation or
diapause and also in maturing females for deposition in eggs. Insects
which feed during adult life can acquire lipid reserves after adult
emergence, as occurs in the desert locust, *Schistocerca gregaria*
(Walker *et al.*, 1970), whereas insects which do not feed as adults
must acquire their lipid reserves at an earlier stage, as in the
Angoumois grain moth, *Sitotraga cerealella* (Chippendale, 1973).
Insect lipids are derived from the diet and/or synthesized from
non-lipid precursors such as carbohydrates and proteins which are
present in the diet or stored in the tissues.

Tissue content of lipids

Whole insects
Much of the earlier work was concerned with the extraction of total
lipid from whole insects. The total lipid content of a large number of
insects at various stages of development has been described by Fast
(1964) and discussed by Gilbert (1967a). There is a wide variation in
lipid content of insects of different orders and often within a single
family. Lipids can comprise as much as 50 percent of the wet weight
of the beetle *Packymerus* (Niemierko, 1959). As discussed by
Scoggin and Tauber (1950), many factors influence the lipid content
of insects, including stage of development, nutrition, environmental
temperature, sex, starvation, diapause, cold hardiness, whether
migratory or not and finally the systematic position of the organisms
under study. Clearly only a discussion of lipid content in relation to
flight is pertinent to this review. In this connection it is of interest
that although, in general, female insects contain more lipid than
males (probably related to the use of lipid in egg development)
nevertheless in many families of Lepidoptera (see Niemierko *et al.*,
1956; Demyanovsky and Zubova, 1957; Gilbert and Schneiderman,
1961) newly emerged males contain considerably more lipid than
newly emerged females. Gilbert (1967a) has suggested that the high
lipid content of male *Hyalophora cecropia* is related to mating
behaviour, since the male moth flies considerable distances in search

of the virgin female and hence requires a suitable stored energy source since it does not feed. In contrast the female flies little after emergence and therefore needs only limited amounts of fuel reserves for flight. No correlation is apparent between total lipid content of adult insects and the species preference for major fuel for flight. This is not surprising since lipid is used for a variety of purposes in adult insects other than as a fuel for flight.

A decrease in total body lipid during flight was first reported by Fulton and Romney (1940) during the migratory flight of the leafhopper, *Euttetix tenellus*. A similar observation was made by Beall (1948) for the migratory flight of the monarch butterfly, *Danaus plexipus*. Atkins (1966) demonstrated a positive correlation between the fat content and the inclination to fly in the Douglas-fir beetle, *Dendroctonus pseudotsugae*. Moreover, flown beetles had significantly less lipid than non-flown control beetles (Atkins, 1969). Recently Kerukize (1973) has investigated the changes in total lipid and glycogen content during flight in two species of the tettizoniid genus *Homorocaryphus* (Orthoptera), *H. nitidulus vicinus* and *H. subvittatus*. The mean fat content of individuals of *H. nitidulus vicinus* from flying swarms averaged 467 mg/g dry wt. and this was depleted during flight. Only traces of glycogen could be detected. The fat content of *H. subvittatus* was low (146 mg/g dry wt.) and the insects were disinclined to fly. A species difference is clearly apparent from these results.

Triglycerides probably form the major part of the lipid content of insects at all developmental stages. This was shown in earlier work on extracts of whole insects summarized by Fast (1964). Some very early reports (e.g. Giral, 1946) suggested the occurrence of large quantities of free fatty acids, but faulty handling of the material leading to hydrolysis of the glycerides was almost certainly the cause. The fatty acid compositions of whole insects has been reviewed by Fast (1964) and Gilbert (1967a) and will not be dealt with here except to mention that detailed work on the fatty acid composition of the flies *Ceratitis capitata* and *Dacus oleae* and their individual lipids has recently been carried out by Municio and colleagues (Barroso *et al.*, 1969; Madariaga *et al.*, 1970, a, b; Fernandez-Souza *et al.*, 1971). The major fatty acid component of insects are usually, as in other forms of life, the C16 and C18 saturated and unsaturated acids. Work with whole insects gives some indication that during flight selective oxidation of fatty acids may take place since Thompson and Bennett (1971) have shown that in the Douglas-fir beetle, *Dendroctonus pseudotsugae*, the mono-unsaturated acids —

palmitoleic (C16:1) and oleic (C18:1) – were utilized at the greatest rate, followed by the saturated ones – palmitic (C16), stearic (C18) and myristic (C14). However, the oxidation of a particular fatty acid is probably related to its relative abundance.

Fat body

The majority of insect lipids are usually found in the fat body, and the lipid composition of whole insects probably reflects the lipid composition of the fat body (Kilby, 1963). Occasionally the fat body only contains a small proportion of the body lipids, e.g. in the termite queen, *Macrotermes natalensis*, only 22 percent of the total body lipids is contributed by the fat body while 72 percent comes from the reproductive organs. In *Macrotermes goliath* the amount of lipid from the fat body is even smaller (11 percent) while 83 percent is contained in the reproductive organs (Cmelik, 1969). However, the large quantities of lipids in the reproductive organs are located in the eggs which comprise the bulk of the weight of the organs. Usually the fat body stores large amounts of lipids and in some cases the fat content of the tissue may exceed 50 percent of the total dry weight of the animal (Bursell, 1970).

The proportion of the fat body which is lipid varies but can be particularly high in reproducing female insects. In adult, female *Pyrrhocoris apterus* lipid accounts for 75–84 percent of the fat body dry weight (Martin 1969a). Orr (1964) found that during the reproductive cycle in the female blowfly, *Phormia regina* lipid accounted for 52–59 percent of the dry weight of the fat body, Hill *et al.* (1968) found that the fat body of the desert locust *Schistocerca gregaria*, was 56–71 percent lipid during the first reproduction cycle and Gilbert (1967b) found that during oogenesis in the female cockroach, *Leucophaea maderae*, the lipid content of the fat body varied from 29–46 percent of the dry weight. The lipid content of the adult male insect fat body may also be high, e.g. during adult development of the male desert locust, *Schistocerca gregaria* the lipid content of the fat body varies from 10–30 percent of the wet weight (Walker *et al.*, 1970). Beenakkers (1965) has shown that the fat body of the locust, *Locusta migratoria*, contains 43.5 percent of its dry weight as esterified fatty acids of which 42 percent is palmitate (C16). There are also many examples of high fat body lipid contents at other stages of development, e.g. lipid comprises 75 percent of the dry weight of the fat body of fully grown female larvae of the Angoumois grain moth, *Sittotroga cerealella*, (Chippendale, 1973).

Weis-Fogh (1952) found that in the desert locust, *Schistocerca gregaria*, 80–85 percent of the total energy expended in the first five hours of flight was derived from lipid and the fat body was responsible for providing 85–90 percent of the energy expended by the flight muscles. Lipid accumulation in the fat body of the adult desert locust after emergence is probably related to flight. In the laboratory, the desert locust after emergence undergoes a period of intensive feeding activity and lipid accumulation in the fat body before sexual maturation occurs (Hill *et al.*, 1968; Walker *et al.*, 1970) – see Fig. 2.2. It is likely that in the field, large reserves of lipid need to be built up as a potential energy source for migration (see also pp. 135–136).

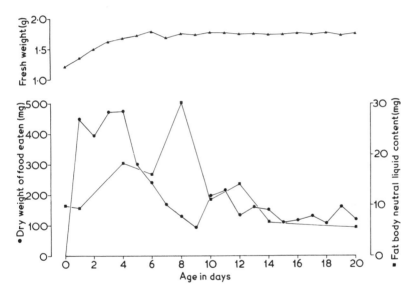

Fig. 2.2. Food intake growth and total lipid of fat body of the adult male desert locust. (Data modified from Walker *et al.*, 1970.)

The major lipid component of all fat body lipid extracts so far examined is triglyceride as it is with total insect lipids. Other neutral lipids and phospholipids (presumably mainly structural components of cell membranes) occur in much smaller amounts. Fat body lipids were first quantitatively analysed by Chino and Gilbert (1965a) and since their work relatively few analyses have been made despite the rapid separations of lipids that can be achieved using thin layer and gas chromatographic techniques. A summary of the available data is given in Table 2.1. Clearly triglycerides are the major lipid store in

Table 2.1

Lipid composition of Fat body

					Lipid (%
Insect	*Sex*	*Developmental stage*	*Phospholipid*	*Monoglyceride*	*Diglyceride*
Hyalophora cecropia	male male	pupa adult		0.7 0.2	1.3 2.3
Locusta migratoria	female	adult			2
Locusta migratoria	male	adult		0.1	1–2
Periplaneta americana	male and female	adult		0.1–1.8	0.4–1.8
Periplaneta americana	female	adult		5	2
Hyalophora cecropia	male male	pupa adult	4 0.8	0.15 0.08	0.20 0.18
Pyrrhocoris apterus	female	adult			
Schistocerca gregaria	male	adult	12	3	2
Diatraea grandiosella	male and female	pupa larva			5 5
Manduca sexta	? ?	larva pupa		2.9 4.2	10 12.5
Prodenia eridania	male	adult		8	8
Sitotroga cerealella	female	larva			2.1

100

total lipid)

Triglyceride	Free fatty acids	Sterols	Sterol esters	Hydrocarbons	References
95		1.1			Chino and Gilbert
96.4		0.6			(1965a)
97					Tietz (1967)
98	0.4				Beenakkers and Scheres (1971)
97+	0.3–0.7				Nelson *et al.* (1967)
93					Cook and Eddington (1967)
95	0.24				Beenakkers and
98.7	0.25				Gilbert (1968)
95+					Martin (1969a)
78	5				Walker *et al.* (1970)
90			5		Chippendale (1971)
90			5		
84.2	2.9				Chang and
79.1	4.2				Friedman (1971)
83	1				Stevenson (1972)
85.4		7.6	3.1	1.8	Chippendale (1973)

fat body, always accounting for 78 percent or more of the total lipids. The amounts of the other neutral lipids detected vary considerably from insect to insect but they are always minor components. Walker *et al.* (1970) studied changes in fat body lipid composition during development of the adult male desert locust, *Schistocerca gregaria*, and some of the results obtained are summarized in Table 2.2. Apart from a decrease in the triglyceride content and a concomitant increase in the content of the other lipid components in the day following emergence, only minor changes in lipid composition occur. Triglycerides account for more than 78 percent of the total fat body lipids at all stages.

Table 2.2
Changes in lipid composition of the fat body during development of the adult male desert locust
(data modified from Walker *et al.*, 1970)

Age (days)	Lipid (% Total lipids)				
	Phospholipid	Monoglyceride	Diglyceride	Free Fatty Acid	Triglyceride
0	2.5	1.0	1.0	0.5	95
1	6.5	2.3	2.0	3.2	86
4	5.0	3.3	1.7	7.5	87
6	7.5	4.0	3.0	6.5	79
8	4.0	4.0	2.5	4.0	85.5
10	9.0	3.5	2.0	5.5	80
12	9.0	3.0	2.0	4.0	82
14	9.0	3.5	3.5	6.0	78
20	12.0	3.0	2.0	5.0	78

Some analyses have been carried out on the fatty acid composition of total fat body lipids. Martin (1969a) reported data for adult female *Pyrrhocoris apterus* and Cmelik (1969) studied queens of the termites *Macrotermes natalensis* and *M. goliath*. Stephen and Gilbert (1970) investigated the fatty acid composition of total phospholipid and total neutral lipid from the fat bodies of pupae, pharate adults and adult *Hyalophora cecropia*. Some workers have studied the fatty acid composition of the individual fat body lipids. Nelson *et al.* (1967) carried out analyses of two, four and six month old adult cockroaches, *Periplaneta americana*, and Beenakkers and Gilbert (1968) reported data for pupae and pharate adults of the silkmoth, *Hyalophora cecropia.* More recently, Beenakkers and Scheres (1971) studied the effect of the fatty acid composition of dietary lipid on the fatty acid composition of neutral lipids in the fat body of adult male *Locusta migratoria.* Considerable variations in fatty acid compositions of fat body lipids were found as had previously been found for total insect lipid analyses (see Gilbert 1967a). Variations in fatty acid composition occur from insect to insect and also in the same insect at different developmental stages. Clearly from the work of Beenakkers and Scheres (1971) much of the variation may be due to differences in dietary lipid fatty acid composition.

Haemolymph

Although the lipid content of the haemolymph is usually small compared with that of the fat body nevertheless it constitutes an important energy reserve. Larval haemolymph of *Phormia regina* contains about 10 mg per ml lipid (Hopf, 1940) and 1.37 mg per ml of fatty acids and sterols are present in the haemolymph of *Gastrophilus intestinalis* larvae (Levenbook, 1950). In *Papillia japonica* larvae, the total lipid is 42 mg per ml in the haemolymph (Ludwig and Wugmeister, 1953) and in adult *Hydrous piceus*, the total fatty acids amount to 55 mg per ml with an unsaponifiable fraction of 22 mg per ml (Wyatt, 1961). Tietz-Devir (1963) demonstrated the presence of 28 micro-equivalents per ml of glycerides (aproximately 9 mg per ml) and 0.2 μmol of free fatty acids per ml of haemolymph from *Locusta migratoria.* More detailed analysis (Tietz, 1967a) revealed 12 micro-equivalents of phospholipids, 21 micro-equivalents of diglycerides and 0.6 μmol of free fatty acids per ml of locust haemolymph. The work of Chino and Gilbert (1965a) demonstrated the presence of 3.8 mg per ml of lipid in the haemolymph of *Hyalophora cecropia* pupae, diglyceride being the major component but triglyceride, monoglyceride and sterols also being present. In contrast, Sridhara and Bhat (1965b) found only minute amounts of diglyceride in the haemolymph of *Bombyx mori.* Wlodawer *et al.* (1966) detected 23 mg total lipid per ml in the haemolymph larvae of the wax moth *Galleria mellonella,* the major components being sterol esters, hydrocarbons, triglycerides, diglycerides and phospholipids. Monoglycerides, free fatty acids and free sterols were also detected.

The haemolymph of 1 to 3 week old female cockroaches (*Periplaneta americana*), contains 0.22 mg triglycerides, 1.94 mg diglyceride and 0.23 mg monoglyceride per ml (Cook and Eddington, 1967). However in the adult of this insect, the composition of the haemolymph, including the lipid content varies with age and sex

(Nelson *et al.*, 1967). Haemolymph triglyceride accounted for 54 percent of the haemolymph lipids at 2 months, increasing to 83 and 86 percent at 4 months in males and females respectively, and decreasing to 36 and 26 percent at 6 months. In contrast the haemolymph diglyceride decreased from 38 percent at 2 months to 10 and 12 percent at 4 months for males and females respectively. Monoglycerides and free fatty acids were present in small amounts at all ages and in both sexes. In 2 month old male *Periplaneta americana* diglyceride was found to be the major haemolymph lipid component (approximately 65 percent of the total), with considerable amounts of triglyceride (27 percent of total) and trace amounts of mono-glyceride and free fatty acids (Downer and Steele, 1969, 1972). Goldsworthy *et al.* (1972) obtained figures of 6–13.1 mg total lipid per ml haemolymph of adult male *Periplaneta americana* of undisclosed age compared with the results of Cook and Eddington (1967) of 2.39 mg per ml using young adult females (see earlier). Clearly many factors are involved in determining blood lipid levels and constitution, and comparisons of analyses of blood from different insects and even the same insect of different ages or sex may not be very meaningful. Other determinations of haemolymph lipid content are included for completeness, despite the fact that the data for *Periplaneta americana* make one sceptical as to value of some determinations carried out for one sex and at one age only. The work of Bailey and Horne (unpublished observations) indicates that in *Locusta migratoria* at least, the lipid content and composition of the haemolymph is very dependent upon age, sex, and nutritional and hormonal status.

Beenakkers and Gilbert (1968) determined the lipid composition of haemolymph from pupae and pharate adults of *Hyalophora cecropia*. In both cases diglyceride is the major component (87 percent of total lipid for pupae, 91 percent for pharate adults) with small amounts of triglycerides, monoglycerides and free fatty acids. Martin (1969a) using adult female *Pyrrhocoris apterus* in the five days after larval-adult ecdysis found that the haemolymph tri-glyceride concentration changed from 14.9 to 16.7 mg per ml between day 1 and day 5 whereas the change in diglyceride concentration was 4.9 to 13.2 mg per ml and the free fatty acid concentration 0.8 to 1.3 mg per ml in the same period of time. The total lipid content of the haemolymph of the cricket, *Acheta domesticus*, ranges from 15 mg per ml in the mature larvae to 27 mg per ml in the young adults (Nowosielski and Patton, 1964). Cholesterol is the principal sterol and phosphatidyl ethanolamine is the most abundant of the phospholipids (Wang and Patton, 1969). Species differences in haemolymph lipids were demonstrated by Cmelik (1969), who found that diglycerides were completely absent from the haemolymph of *Macrotermes goliath* but present in that of *Macrotermes natalensis.* In other respects the species were similar, with the haemolymph containing hydrocarbons, cholesterol esters, triglycerides, free fatty acids, phospholipids and free and esterified alcohols. Mayer and Candy (1969b) detected 1.7 to 2.8 mg free fatty acid, 8.7 to 9.2 mg diglyceride and 3.8 to 4.3 mg triglyceride per ml of haemolymph in adult *Schistocerca gregaria.* The total lipid concentration in *Schistocerca gregaria* haemolymph varies between 6.1 and 16.5 mg per ml during adult male development (Walker *et al.,* 1970). The haemolymph of the tobacco hornworm, *Manduca sexta,* larvae contained 1.45 mg total lipid per ml (10.3 percent triglyceride, 53.1 percent diglyceride, 10.3 percent monoglyceride and 26.3 percent free fatty acid), whereas the haemolymph of the pupa contains 10.4 mg total lipid per ml, consisting of 28.8 percent triglyceride, 50.8 percent diglyceride, 0.1 percent monoglyceride and 20.3 percent free fatty acids (Chang and Friedman, 1971). Values of 5, 5.4 to 8.1, 10.2 to 14.2 and 2 mg total lipid per ml of adult male *Schistocerca gregaria, Locusta migratoria, Gromphadorhina portentosa* and *Tenebrio molitor* haemolymph, respectively have been reported by Goldsworthy *et al.* (1972). Stevenson (1972) has shown that the haemolymph of newly emerged adult *Prodenia eridania* contains 3.3 μmol (approx. 2.8 mg) triglyceride, 49.0 μmol (approx. 29 mg) diglyceride, 4.1 μmol (appox. 1.4 mg) monoglycerides and 3.0 μmol (approx. 0.8 mg) free fatty acids per ml. Finally Chippendale (1973) has shown that the haemolymph lipid composition of fully grown female larval *Sitotrosa cerealella* is 38.2 percent triglycerides, 14.6 percent 1,2 diglycerides, 9.3 percent free fatty acids, 10.6 percent sterols, 7.0 percent sterol esters and 20.3 percent hydrocarbons.

This survey illustrates the many factors which may influence haemolymph lipid content and composition. Changes in haemolymph lipid content in relation to flight will be discussed in a later section. However the data previously described do show that in many cases the lipid concentration in the haemolymph is very high, e.g. a diglyceride concentration of 49 mM (approx. 29 mg/ml) in *Prodenia eridania* haemolymph (Stevenson 1972). These high concentrations are probably related to the supply of fuel to the flight muscles from the haemolymph (Weis-Fogh, 1964). In general either diglycerides or triglycerides are the major haemolymph lipids but large variations exist even in the relatively small number of species so far investigated. As stated at the beginning of this section the total haemolymph lipid content is usually considerably less than the fat body lipid content, e.g. *Sitotroga cerealella* larvae contain 0.74 mg lipid in the fat body and 0.07 mg in the haemolymph (Chippendale, 1973). However Walker *et al.* (1970) have shown that the haemolymph and fat body lipid contents vary during the adult development of male *Schistocerca gregaria*, and although at most stages there is far less lipid in the haemolymph than the fat body, at specific ages similar amounts are found in both tissues, e.g. at 14 days of age 4.5 mg in haemolymph and 7 mg in fat body.

Some workers have analysed the fatty acid composition of haemolymph lipids. Beenakkers (1963a, b; 1965) reported on the fatty acid content of fat body, haemolymph and flight muscles of *Locusta migratoria* before and after flight. Flight causes a 30 percent increase in lipid content in the flight muscles indicating transfer from haemolymph to muscles. The fat body contains 42 percent of its fatty acids as palmitate whereas haemolymph fatty acids are 37 percent palmitate and flight muscle fatty acids 24 percent. Palmitate (C16) and oleate (18:1) appear to be preferentially utilized for flight. Nelson *et al.* (1967) investigated the fatty acid compositions of haemolymph glycerides of adult male and female *Periplaneta americana* at 2, 4 and 6 months of age. Major fatty acids were at all times palmitate, oleate and linoleate (C18:2). Beenakkers and Gilbert (1968) described the fatty acid composition glycerides and total free fatty acids of haemolymph from the pupal and pharate adult silkmoth, *Hyalophora cecropia.* Major fatty acids were usually palmitate, oleate and linolenate (C18:3). Palmitate, oleate and linoleate were the predominant fatty acids in the total lipids of *Acheta domesticus* haemolymph (Wang and Patton, 1969). Cmelik (1969) found that oleate and palmitate predominated in the total lipids in the haemolymph of *Macrotermes natalensis.* The fatty acid composition of total neutral and phospholipids from haemolymph of *Hyalophora cecropia* during adult development have been reported by Stephen and Gilbert (1970). Fatty acid compositions of the neutral and phospholipids are different and both are age dependent. As usual the major fatty acids are C16 and C18 saturated and unsaturated.

Gut

The gut will certainly contain considerable quantities of lipid after a lipid-containing meal. Such lipids may be transported via the haemolymph to the flight muscles to be used as a respiratory fuel for flight, or transported to the fat body for storage. As previously discussed, the fatty acid composition of the diet may have a profound effect on the fatty acid composition of fat body and haemolymph lipids, and hence on the type of fatty acid available as an energy source for flight. Flight muscle and other tissues must have the enzymic capacity to oxidize a range of saturated and unsaturated fatty acids. Although some insects have a predominantly lipid diet e.g. the waxworm *Galleria mellonella* whose normal food, old beescomb, consists largely of waxy esters of palmitic acid, the majority of insects probably ingest a predominantly carbohydrate diet (e.g. nectar, vegetable matter) in which case the lipids have to be synthesized (primarily in the fat body but also in other tissues such as mid-gut – see later) from the ingested carbohydrates. The nature of the fatty acids produced by the insect itself will be governed by the chain length specificity and desaturating ability of its own fatty acid synthesizing systems. Insects, like higher animals, seem incapable of synthesizing the polyunsaturated 'essential fatty acids' linoleate (C18:2) and linoleate (C18:3), which therefore must be obtained from the diet. They are plentiful in plant lipids and as such are available to most insects (Dadd, 1973).

Absorption of dietary lipids and incorporation into fat body lipids

Eisner (1955) studied the absorption of lipids by the gut of the cockroach, *Periplaneta americana*, and concluded that fat absorption occurred in the fore-gut and mid-gut. Although a digestive lipase in the lumen of the mid-gut could bring about hydrolysis of lipids, lipolysis was not a prerequisite for absorption. Treherne (1958a) fed tri [^{14}C]palmitate to *Periplaneta americana* and concluded that absorption occurred in the mid-gut region only. A recent, elegant study by Weintraub and Tietz (1973) using the locust, *Locusta migratoria*, has shed more light on the form in which dietary lipids are absorbed and transported into the haemolymph. [^{14}C]-Triolein was readily hydrolysed to monoglyceride, diglyceride and free fatty acids in the intestinal lumen, but [^{14}C]tripalmitate was excreted unchanged (this result is in contrast to those of Treherne (1958a) and may indicate a species difference). The glycerol and fatty acids formed during digestion were incorporated mainly into the phospholipids of the intestinal wall, the diglycerides of the haemolymph and the neutral lipids (mainly triglyceride) of the fat body. The haemolymph is probably devoid of fatty acid esterifying activity

(Chino and Gilbert, 1965a), although limited esterification of fatty acid does take place in wax moth haemolymph (Wlodawer *et al.*, 1966), and hence the haemolymph diglycerides must have been synthesized elsewhere. The likely possibilities are that either the diglycerides are formed in the fat body from fatty acids transported in the haemolymph from the gut, or else they are formed from digestion products in the intestine itself. The former possibility is supported by the fact that the fat body is known to be able to incorporate fatty acids into the glycerides both *in vivo* and *in vitro* and to release diglycerides into the haemolymph (see pp. 137–145). Further, Weintraub and Tietz (1973) have shown that haemolymph lipoproteins can incorporate and transport free fatty acids and that these fatty acids are readily available to the fat body. Injection of bovine serum albumen to act as a fatty acid trap causes free fatty acid accumulation in the haemolymph during fat absorption.

Evidence for the synthesis of diglycerides from digestion products in the intestine itself, i.e. a situation analagous to that in mammals (in which case however triglycerides are formed) comes from the feeding of $[^3H]$ glyceryl tri $[^{14}C]$ oleate (Weintraub and Tietz, 1973). The $^3H/^{14}C$ ratio of the haemolymph diglycerides was always different from that of fat body triglycerides, indicating that fat body is not the immediate source of haemolymph diglycerides. Further haemolymph diglycerides were readily taken up by the fat body. Thus during fat absorption it is possible that fatty acids enter the intestinal cells and are converted into diglycerides which are released into the haemolymph (probably as lipoproteins) and transported to the fat body for conversion to triglycerides for storage. Irrespective of whether free fatty acids or diglycerides are transported from the intestine to the fat body, the lipid transported will be available to the flight muscles as a fuel for flight. If diglyceride formation does occur in the intestinal cells it is possible that phospholipids are involved as intermediates since Weintraub and Tietz (1973) have shown that much of the radioactivity of labelled, ingested lipid is incorporated into the intestinal phospholipids and that when $[^3H]$ glyceryl tri$[^{14}C]$ oleate is fed, the $^3H/^{14}C$ ratio is very similar in haemolymph diglycerides and intestinal phospholipids. It is possible that phospholipids are formed and then split by a phospholipase to yield diglycerides which are then released into the haemolymph. Phospholipids have also been implicated in lipid absorption by Wlodawer (1956), Niemerko (1959) and Young (1964a, b). It is possible that species differences occur with regard to lipid transport from the intestine, and it would be interesting if studies similar to those of

Weintraub and Tietz (1973) could be carried out on insects which do not have diglyceride as the major haemolymph lipid.

The work of Downer and Steele (1972) indicates that in the cockroach, *Periplaneta americana*, lipids are carried in the haemolymph in the form of triglycerides and are converted to diglycerides in the haemolymph before their transport into the fat body. It appears that the corpus cardiacum-allatum complex contains a factor which stimulates the uptake of lipid by the fat body. The conversion of triglyceride to diglycerides in the haemolymph requires the presence of a lipase and such an enzyme has been found (Downer and Steele 1973). The enzyme activity is unaffected by the corpus cardiacum hypolipaemic factor but is decreased by prolonged periods of starvation. The authors speculate that the enzyme fulfills a similar function in the insect to that of mammalian lipoprotein lipase (Korn, 1955; Robinson, 1965).

The uptake of fatty acids into the fat body lipids has been demonstrated on many occasions both *in vivo* and *in vitro*. Tietz and colleagues have shown that incubation of the fat body of the locust, *Locusta migratoria*, with a medium containing [^{14}C] palmitate, or an injection of radioactive palmitate into the haemocoel results in incorporation of the labelled fatty acid into fat body glycerides (Tietz, 1962, 1967a; Peled and Tietz, 1973). Similar results have been obtained by Gilbert and colleagues using the moth *Hyalophora cecropia* (Chino and Gilbert, 1964, 1965a; Bhakthan and Gilbert, 1968; Beenakkers and Gilbert, 1968), the cockroach, *Periplaneta americana*, (Chino and Gilbert, 1965a; Bhakthan and Gilbert, 1968) and the grasshopper, *Melanoplus differentialis* (Chino and Gilbert, 1965a). Incorporation of [^{14}C] palmitate into fat body glycerides has also been shown by Cook and Eddington (1967) using *Periplaneta americana*, Wlodawer and Lagwinska (1967) using larvae of the wax moth *Galleria melonella*; Beenakkers (1969b) using *Locusta migratoria*; Martin (1969b) using *Pyrrhocoris apterus*; Walker and Bailey (1970a, 1971b) using *Schistocerca gregaria*; Chang and Friedman (1971) using *Manduca sexta* and *Leucophaea maderae*; and Dutkowski and Ziajka (1972) using *Galleria melonella*.

Incorporation of glycerides as well as fatty acids into fat body lipids has also been demonstrated *in vitro*. Martin (1969b) using *Pyrrhocoris apterus* showed that considerably less triglyceride than free fatty acid was taken up by fat body incubated with haemolymph, whereas more diglyceride than free fatty acid was taken up. Weintraub and Tietz (1973) have shown that [^{14}C] diglycerides were taken up from the haemolymph into fat body of *Locusta migratoria*, most of the radioactivity remaining in diglyceride but some being incorporated into free fatty acids and triglycerides. These results suggest that diglycerides continuously enter and are released from the fat body in the locust.

Lipid degradation by the fat body

The discussion in the previous section shows that insects are capable of depositing dietary lipids in the lipid stores of the fat body. As will be seen later, the fat body can also synthesize lipids from non-lipid precursors and the fat body triglycerides whatever their origin can be mobilized and transported to the flight muscles to act as a respiratory fuel. The fat body also oxidizes its own lipids in order to obtain energy to carry out functions such as gluconeogenesis.

Fatty acid oxidation

There is little data available concerning the degradation of lipids by the fat body. Initially fat body glycerides have to be hydrolysed to glycerol and fatty acids by the action of lipases. A discussion of these enzymes is included on pp. 139—141. The fatty acids formed by lipase action can then be oxidized by fat body enzymes. That fat body can degrade fatty acids was indicated by the work of D'Costa and Birt (1966) and Crompton and Birt (1967) who, on the basis of studies on changes in lipid, carbohydrate and amino acid components of the blowfly, *Lucilia cuprina*, concluded that fatty acid degradation provides the principal source of energy for early adult development. D'Costa and Birt (1969) showed that homogenates of the abdomen of *Lucilia cuprina* would readily oxidize butyrate but at a much lower rate than would thorax homogenates. Unfortunately butyrate is not a normal substrate for oxidation since long chain fatty acids predominate in the fat body. However, Martin (1969b) showed that $[1-^{14}C]$palmitate was readily oxidized by fat bodies from adult female *Pyrrhocoris apterus*, although it appears that fatty acids are preferentially stored and glucose is preferentially oxidized.

In contrast, work on *Schistocerca gregaria* (Walker *et al.*, 1970) indicates that fatty acids are the preferred substrates at all ages in the fat body of the adult male of this locust. As can be seen from Fig. 2.3 the respiratory quotients of isolated fat bodies indicate lipid oxidation at all ages and incomplete lipid oxidation at day 8 when the fat body has its maximum lipid content (see Fig. 2.2). $[U-^{14}C]$palmitate degradation to acid soluble material and CO_2 is also highest at about 8 days after emergence (Fig. 2.3). The rates of

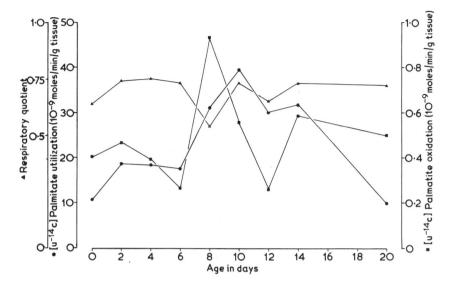

Fig. 2.3. Changes in fat body respiratory quotient (R.Q.) and fatty acid degradation during development of the adult male desert locust. Palmitate oxidation refers to palmitate conversion to CO_2 whereas palmitate utilization refers to palmitate conversion to acid-soluble products. (Data derived from Walker *et al.*, 1970 and Walker and Bailey, 1970a.)

palmitate degradation are simi¹ar to those obtained with flight muscle preparations. It is possible that fatty acid degradation in the fat body is under hormonal control since allatectomy greatly increases the total capacity of fat bodies from adult male *Schistocerca gregaria* to oxidize [U–¹⁴C]palmitate to ¹⁴CO_2, but had little effect on the formation of acid soluble products (Walker and Bailey, 1971b). It is also quite conceivable that the increased fatty acid oxidation in the allatectomized insects is a response to the high levels of lipid in the fat bodies of such insects (Walker and Bailey, 1971a).

Although fat bodies oxidize long chain fatty acids, little is known of the mechanisms involved. Most of the work on fatty acid oxidation has been carried out with flight muscle preparations (see Chapter 1) and nothing is known about the oxidation by fat body of fatty acids of different chain lengths and different degrees of saturation or whether or not carnitine is required in the fatty acid oxidation process. It is assumed that the β-oxidation pathway (Fig. 2.4) operates in fat body, but the activities of the various enzymes involved have been little investigated. Zebe (1959) detected 3-ketoacyl-CoA thiolase in *Locusta migratoria* and Young (1959) obtained some evidence for the presence of 3-hydroxyacyl-CoA

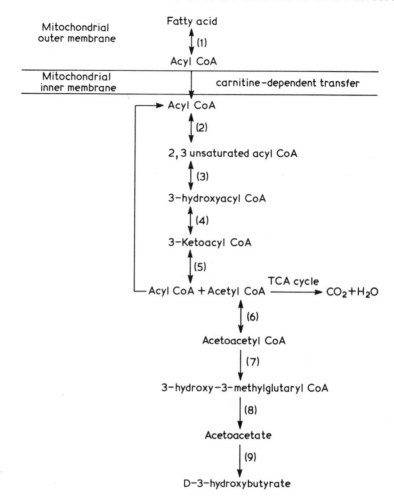

Fig. 2.4. Scheme for fat body β-oxidation of fatty acids and for ketone body formation. The enzymes involved are:

(1) Acyl-CoA synthetase
(2) Acyl-CoA dehydrogenase
(3) Enoyl-CoA hydratase
(4) 3-hydroxyacyl-CoA dehydrogenase
(5) 3-ketoacyl CoA thiolase
(6) Acetoacetyl-CoA thiolase
(7) Hydroxymethylglutaryl-CoA synthase
(8) Hydroxymethylglutaryl-CoA lyase
(9) 3-hydroxybutyrate dehydrogenase

dehydrogenase in *Periplaneta americana* (although it is possible that he was measuring 3-hydroxybutyrate dehydrogenase activity). It is assumed that in the fat body the oxidation process occurs in the mitochondria as it does in well investigated mammalian systems.

Ketone body formation and utilization

The work of Walker and Bailey (1970a) indicated that considerable quantities of long chain fatty acids can be degraded to acid soluble material in the desert locust fat body. By analogy with mammalian liver it seemed possible that some of the soluble material would be in the form of ketone bodies (Lockwood and Bailey, 1970). The acetyl-CoA formed by the β-oxidation of fatty acids in mammalian liver can either be oxidized to CO_2 and H_2O by the tricarboxylic acid cycle or be converted to the ketone bodies acetoacetate and D-3-hydroxybutyrate. The major site of ketone body formation in liver is the mitochondria and the major route of formation is via the hydroxymethyl-glutaryl-CoA pathway illustrated in Fig. 2.4, although some acetoacetate may be formed by deacylation of acetoacetyl-CoA (for recent reviews see Williamson and Hems, 1970; Bailey and Lockwood, 1973). The work of Bailey *et al.* (1971, 1972) and Hill *et al.* (1972) has demonstrated the formation of ketone bodies in *Schistocerca gregaria*, and Hill *et al.* (1972) have shown that these ketone bodies are formed in the fat body and oxidized by fat body, flight muscle and testes. All the tissues oxidize acetoacetate much more rapidly than D-3-hydroxybutyrate and the flight muscles of fed locusts oxidize acetoacetate much more readily than the fat body or testes. The ability of fat body and flight muscles to oxidize ketone bodies in starved locusts is greatly reduced but utilization by the testes remains normal. Thus the flight muscles appear to be major consumers of ketone bodies in fed locusts and the testes the major consumers in starved locusts. The production of ketone bodies is increased under conditions of increased mobilization and degradation of fat body lipids such as starvation (Hill and Goldsworthy, 1970), during flight (Beenakkers, 1965) and after the injection of corpus cardiacum extracts (Mayer and Candy, 1969b). Hill *et al.* (1972) and Bailey *et al.* (1972) have suggested that ketone bodies are formed in the fat body during the mobilization of triglyceride lipid reserves, and are either oxidized by the fat body or transported (mainly as acetoacetate) by the haemolymph to be used as a respiratory fuel (see Fig. 2.5). The importance of ketone bodies as a fuel for flight is doubtful considering the low concentrations in the haemolymph (0.3 mM during flight) as compared with other fuels. A high

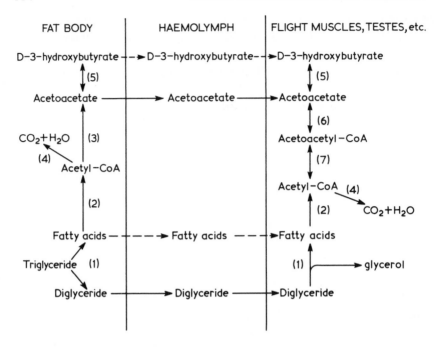

Fig. 2.5. Scheme for the transport and utilization of ketone bodies. The enzymes involved are:

(1) Lipase
(2) β-oxidation sequence
(3) Hydroxymethylglutaryl-CoA pathway
(4) Tricarboxylic acid cycle
(5) 3-hydroxybutyrate dehydrogenase
(6) 3-oxo-acid CoA transferase
(7) Acetoacetyl-CoA thiolase

concentration of fuel in the haemolymph is thought necessary to supply an adequate amount of fuel to the flight muscles to support the high metabolic rate during flight (Weis-Fogh, 1964). Candy (1970) has reported that the optimal concentration of glucose required for oxidation by a perfused working thoracic muscle preparation from the desert locust is about 80 mM. However the rates of oxidation of acetoacetate by flight muscle observed by Hill *et al.* (1972) are very similar to those for trehalose oxidation by similar preparations obtained by Walker and Bailey (1970a). It is possible that ketone bodies provide some key tissues with a guaranteed supply of respiratory fuel in times of starvation as has been suggested for pigeon brain by Bailey and Horne (1972a,b).

Tricarboxylic acid cycle

The acetyl-CoA derived from the β-oxidation of fatty acids or from the degradation of carbohydrates and proteins can be oxidized to CO_2 and H_2O by the tricarboxylic acid (TCA) cycle which is the common terminal pathway of oxidation of all foodstuffs (Krebs and Kornberg, 1957). All the enzymes of the cycle have been demonstrated in acetone powders, extracts, mitochondrial preparations and homogenates prepared from whole *Prodenia eridania* larvae (Levenbook, 1961). Most of the work on the TCA cycle in insects has been carried out with flight muscle preparations. However, as pointed out by Kilby (1963), although all the enzymes of the cycle have not been demonstrated in the fat body of any single species, the findings of various workers on fat bodies of different insects indicate that the cycle operates in the fat body. The evidence for the presence of the individual enzymes of the cycle in fat body has been reviewed by Kilby (1963) and Chefurka (1965a) and few observations have since been made. Measurements on respiration of the fat body of *Schistocerca gregaria* in the presence of tricarboxylic acid cycle intermediates (Bellamy, 1958), and the oxidation of [^{14}C] labelled substrates by such fat bodies (e.g. Clements, 1959; Hines and Smith, 1963; Walker and Bailey, 1970a) support the view that the TCA cycle operates in insect fat bodies.

Respiratory chain and oxidative phosphorylation

Mitochondrial oxidation of substrates by enzymes of the tricarboxylic acid cycle and β-oxidation pathway, is coupled to reduction of coenzymes. Mitochondria also contain the respiratory chain and oxidative phosphorylation systems by which the reduced coenzymes are reoxidized in a process which couples the reoxidation with the phosphorylation of ADP to ATP. Insect fat body respiratory chain and oxidative phosphorylation have not been investigated in any great detail and most of the experiments carried out on these systems have been reviewed by Kilby (1963) and Chefurka (1965a) and will not be elaborated upon here. It is assumed that fat body has the normal respiratory chain and oxidative phosphorylation mechanisms studied in greater detail in mammalian and bacterial systems.

Some experiments have been carried out, in particular by Keeley and colleagues, on the control of respiration in fat body. In vertebrates it is known that thyroid hormone levels affect the oxygen consumption of live animals, tissues and mitochondria (Bronk, 1966; Tata *et al.*, 1963). Early experiments to determine the presence of hormones regulating the basal metabolism of insects were concerned with the role of the corpora allata. Allatectomy depressed oxygen consumption of female insects whereas corpora allata implants had the opposite effect (Thompson, 1949; Sagesser, 1960; Roussel, 1963; Lüscher, 1968). The experiments indicated that the corpora allata stimulate metabolism associated with oocyte development and not the basal metabolism *per se*. The work of Slama (1964) indicated that the corpora allata affected reproductive metabolism whereas the corpora cardiaca affected the basal metabolism. However Clarke and Baldwin (1960) reported a stimulation of O_2 uptake following the addition of corpora allata to mitochondria isolated from *Locusta migratoria*. In contrast Minks (1967) found no effect on oxygen consumption using isolated muscle or fat body mitochondria from allatectomized *Locusta migratoria*, although the addition of corpora allata preparations did appear to increase the coupling of oxidation and phosphorylation.

Several reports have established a relationship between the corpora cardiaca and the oxidative metabolism of the fat body. Lüscher and Leuthold (1965) using *Leucophaea maderae*, have reported the stimulation *in vitro* of fat body respiration following the addition of corpora cardiaca. These observations were supported by Wiens and Gilbert (1965) who suggested that the increased O_2 uptake may be due to a change from the oxidation of carbohydrate to the oxidation of lipids. Müller and Engelmann (1968) demonstrated that a hormonal agent having an effect on fat body respiration in *Leucophaea maderae* was being produced by the corpora cardiaca *per se*.

Extirpation studies on adult males of the cockroach *Blaberus discoidalis* demonstrated a regulatory effect by the corpora cardiaca on basal metabolism. Long term (30 days) cardiacectomy-allatectomy of *Blaberus discoidalis* depressed whole body respiration by 24 percent and *in vitro* fat body respiration by 47 percent (Keeley and Friedman, 1967). Respiration of isolated fat body mitochondria was also decreased 30—40 percent by similar treatment (Keeley & Friedman, 1969), as was electron transport activity as measured by succinate-cytochome C reductase and cytochrome C oxidase activities (Keeley, 1970). Allatectomized animals showed results similar to the controls, indicating that the source of the factor affecting respiratory metabolism was the neurosecretory brain — corpus cardiaca complex. Keeley and Waddill (1971) obtained results which suggest that the respiratory regulating factor in *Blaberus discoidalis* originates from brain neurosecretory cells.

Extracts of the neurosecretory complex return the depressed respiration rates of fat body mitochondria from gland-deficient insects to normal when administered *in vivo* but are ineffective when added to the mitochondria *in vitro.*

Keeley (1971) found that in adult male *Blaberus discoidalis*, the brain — corpora cardiaca neurosecretory complex affects the capacity of fat body mitochondria for respiratory and electron transport activity although respiratory control by ADP, and the efficiency of oxidative phosphorylation remain normal. Keeley (1972) studied neuroendocrine effects in the development of respiratory function in fat body mitochondria of male *Blaberus discoidalis.* The fat body underwent a hormone-independent respiratory development during the first 5 days of adult life followed by a neurohormone-sensitive development during the next 5 days. Increases of 2-3 fold were shown for oxygen uptake and for succinate-cytochrome C reductase and cytochrome C oxidase activities during the 5-10 day period, and an increased capacity for respiratory control by ADP was also evident. No further respiratory changes occurred between 10 and 30 days of age. Respiratory enzyme activities and respiratory control would not develop beyond the 5 day level after disruption of neurosecretory function by cardiacectomy. Conversely, daily administration of corpus cardiacum extracts stimulated precocious 10 day levels of respiration by 5 days of age. Experiments in which 2,4 dinitrophenol was used as uncoupler showed that electron transport was the rate-limiting process of respiratory activity. The incorporation of $[^{14}C]$ amino acids into mitochondria was unaffected by removal of corpora cardiaca indicating that mitochondrial formation was unaffected by the treatment.

Clearly the interesting observations on hormonal control of fat body respiration previously discussed, should be borne in mind when considering the role of such hormones in the control of the mobilization of fat body energy reserves (see pp. 143–145, 161–164).

Biosynthesis of lipids

The pathway and control of fatty acid biosynthesis have been widely studied in vertebrates and micro-organisms and excellent reviews of the field have been published (see e.g. Wakil, 1961, 1963; Lynen, 1967; Vagelos, 1971). Although some chain elongation of fatty acids probably occurs in the mitochondria the major route of *de novo* synthesis of fatty acids from acetyl-CoA takes place in the cell cytoplasm. Fatty acid synthesis from acetyl-CoA is brought about by two enzyme complexes, acetyl-CoA carboxylase and fatty acid synthetase (which contains acyl carrier protein —ACP). The overall conversion of acetyl-CoA to palmitate involves the following stoichiometry

$$8CH_3 COSCoA + 7ATP + 14NADPH + 14H^+ \longrightarrow$$

$$CH_3 (CH_2)_{14} COOH + 8CoASH + 7ADP + 7P_i + 14NADP^+ + 6H_2 O$$

The details of fatty acid synthesis will not be elaborated upon here and the reader is referred to one of the above-mentioned reviews on the subject. An outline of the reactions involved in fatty acid biosynthesis as deduced from studies with yeast is shown in Fig. 2.6. Although palmitate is the major fatty acid produced by most biosynthetic systems studied, the chain length of the fatty acid produced varies in different biological materials and under different experimental conditions. When compared with our knowledge of fatty acid biosynthesis in vertebrates and micro-organisms, our knowledge of the details of fatty acid biosynthesis in insect tissues is, at best, fragmentary.

Whole insects

That insects are able to synthesize lipids from non-lipid precursors is well documented. Wigglesworth (1942) starved larvae of the mosquito, *Aëdes aegypti*, until the fat reserves in the fat body were exhausted, and then demonstrated the reappearance of fat droplets after feeding with protein, amino acids and sugars. Beall (1948) showed that the monarch butterfly, *Danaus plexipus* lays down lipid before migration and since its food (nectar) does not contain any lipid, the insect must have synthesized its lipid reserves from the non-lipid constituents of the diet. Many insects develop well with no dietary lipid other than sterol, and synthesize their fat largely from high levels of dietary carbohydrate (Fraenkel and Blewett, 1946; Pant and Gabrani, 1963; Taylor and Medici, 1966). The waxworm, *Galleria mellonella*, which normally has a diet with a high lipid content (old beescomb) grows and develops well on synthetic diets containing only minor amounts of essential fats (Dadd, 1964, 1966). Van Handel (1965) showed that starved females of the mosquito *Aëdes sollicitans* would carry out an almost quantitative conversion of a fed 1:1 glucose-fructose mixture to lipid, and Van Handel and Lea (1970) have demonstrated that when the mosquitoes are refed a high protein content blood meal, synthesis of lipids also occurs.

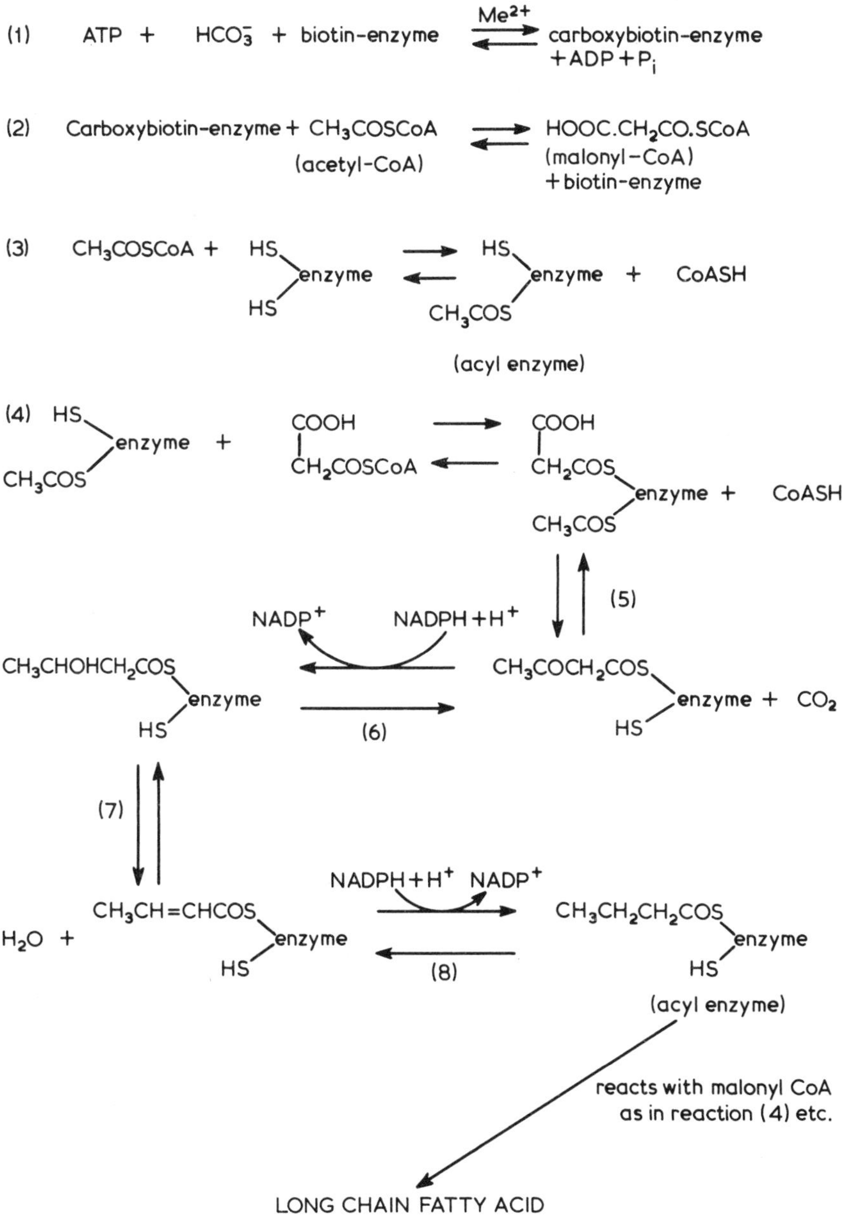

Fig. 2.6. Outline scheme for fatty acid biosynthesis. Reactions (1) and (2) are catalysed by acetyl-CoA carboxylase and reactions (3)–(8) are catalysed by the fatty acid synthetase complex.

Many dipterous larvae are evidently able to synthesize lipids from amino acids and proteins since they grow readily on diets deficient in lipid and carbohydrate. Kon and Monroe (1971) have shown the incorporation of radioactivity from five amino acids in an aseptic diet into the lipids of *Musca domestica*. The work of Walker *et al.* (1970) shows that in the adult male desert locust, *Schistocerca gregaria*, the fat body lipids are primarily derived from the carbohydrate of the diet.

The polyunsaturated fatty acids linoleate (C18:2) and linolenate (C18:3) are essential nutritional requirements for most insects and it seems likely that all insects are incapable of synthesizing them (Dadd, 1973). It is probable that the biosynthesis of polyunsaturated fatty acid by insects, as reported by Louloudes *et al.* (1961) is due to bacterial action. Even Dipterous insects which do not seem to require polyunsaturated fatty acids in the diet are unable to synthesize them (Barlow, 1966; Bridges, 1971; Keith, 1967; Wright and Oehler, 1971). It seems likely that the function of the 'essential fatty acids' is related to membrane permeability, as it is in mammals, since these acids tend to constitute a higher and more stable proportion of the fatty acids of phospholipids than of reserve triglycerides (Bridges, 1971; Fast 1966; Lamb and Monroe, 1968; Moore and Taft, 1970; Terriere and Grau, 1972).

The biosynthesis of lipids in whole insects has been investigated by feeding or injecting [^{14}C]acetate and in some cases [^{14}C]glucose into insects. [^{14}C]Acetate conversion to fatty acids has been investigated in the adult housefly, *Musca domestica* (Robbins *et al.*, 1960) the cockroach, *Periplaneta americana* (Louloudes *et al.*, 1961) the blowfly, *Calliphora erythrocephala* larva (Sedee, 1961), the green peach aphid, *Myzus persicae* (Strong, 1963), the silkworm, *Bombyx mori*, (Sridhara and Bhat, 1964, 1965a), the cockroach, *Eurycotis floridana* (Bade, 1964), adult boll weevils, *Anthonomus grandis* (Lambremont, 1965), larval boll weevils (Lambremont, *et al.*, 1965) and the silkmoth, *Hyalophora cecropia* (Chino and Gilbert 1965b). From these, and other studies it is apparent that insects tend to synthesize predominantly palmitate (C16) palmitoleate (C16:1) stearate (C18) and oleate (C18:1) with [^{14}C]acetate being incorporated to different extents into these acids by various insects. The conversion of [^{14}C]glucose to fatty acids has been shown for the green peach aphid, *Myzus persicae* (Strong, 1963), the silkmoth, *Hyalophora cecropia* (Chino and Gilbert, 1965b) and the silkworm, *Bombyx mori* (Horie *et al.*, 1968). In contrast to the results of Horie *et al.* (1968), Sridhara and Bhat (1965b) found very little conversion·

of [^{14}C]glucose to lipid in *Bombyx mori* larvae. The results of Horie *et al.* (1968) indicate that synthesis of fatty acids from glucose proceeds via pyruvate, decarboxylation of pyruvate and incorporation of 2-carbon units into lipid (see later discussion on conversion of carbohydrate to lipid by fat body). Radioactivity from [^{14}C]glucose is incorporated into the same fatty acids as is label from [^{14}C]acetate (Strong, 1963). Mathur and Yurkiewicz (1969) have compared rates of conversion of [U–^{14}C]glucose to lipids in the blowfly, *Phaenicia sericata*, during flight and at rest. They found that flight decreased the rate of glucose conversion to lipid.

The results of [^{14}C]acetate incorporation studies in *Bombyx mori* (Sridhara and Bhat, 1964, 1965a) the adult boll weevil, *Anthonomus grandis* (Lambremont, 1965), the blowfly, *Calliphora erythrocephala* (Brak, *et al.*, 1966), the cabbage looper, *Trichoplusia ni* (Nelson and Sukkestad, 1968) and the silkmoth, *Hyalophora cecropia* (Stephen and Gilbert, 1969) and [^{14}C]glucose incorporation studies in *Bombyx mori* (Horie *et al.*, 1968) indicate that insects cannot synthesize the polyunsaturated fatty acids. Louloudes *et al.* (1961) did demonstrate the synthesis of [^{14}C]linolenate (C18:3) from [^{14}C]acetate by the cockroach, *Blattella germanica*, but this was probably caused by symbiotic bacteria in the gut. Although insects do not synthesize the essential fatty acids they do synthesize the mono-unsaturated fatty acids, in particular palmitoleate and oleate. Sedee (1961) and Miura *et al.* (1965) studied the *in vivo* metabolism of [^{14}C]acetate by the blowfly, *Calliphora erythrocephala* larva and concluded that saturated and unsaturated fatty acid synthesis takes place by two separate pathways without interconversion by dehydrogenation. However no other workers have reported evidence to support this conclusion and indeed dehydrogenation of saturated fatty acids has been shown in *Eurycotis floridana* (Bade, 1964), *Calliphora erythrocephala* (Brak *et al.*, 1966) and the blowfly, *Aldrichina grahami* (Takaya and Miura, 1968). The work of Bade (1964) on the cockroach *Eurycotis floridana*, Sridhara *et al.* (1966) on the silkworm *Bombyx mori* and Stephen and Gilbert (1969) on the silkmoth, *Hyalophora cecropia*, indicates that to some degree saturated fatty acids can be elongated and desaturated by insects but that unsaturated fatty acids cannot be elongated or saturated.

Recently the biosynthesis of fatty acids by *Ceratitis capitata* has been investigated in great detail *in vivo* and also in whole insect homogenates by Municio and colleagues. Municio *et al.* (1970) studied developmental changes in the biosynthesis of fatty acids from [^{14}C]acetate by homogenates of whole insects. Pupal and larval stages showed greater incorporation of radioactivity into fatty acids than did egg and adult stages. Municio *et al.* (1971) further studied fatty acid and lipid biosynthesis *in vitro* during development. Biosynthesis of fatty acids from [^{14}C] acetate showed marked differences between larval and pharate adult homogenates. Acetate was incorporated by larval homogenates initially into decanoate (C10), laurate (C12) and myristate (C14) and then into palmitate (C16) and palmitoleate (C16:1). The decanoate had the highest specific activity of all fatty acids. Pharate adult homogenates incorporated acetate into palmitate and palmitoleate. The [^{14}C] acetate was predominantly incorporated by larval homogenates into the decanoate contained in triglyceride, whereas the radioactivity was incorporated by pharate adult homogenates into phospholipid fatty acids and not into triglycerides. Municio *et al.* (1972a) studied *in vitro* elongation and desaturation of fatty acids during development. Larval homogenates converted decanoate, laurate, myristate and palmitate into their monounsaturated derivatives and also elongated these acids. The amount of desaturation increased with chain length, but the opposite held for chain elongation. Pharate adult homogenates did not carry out elongation or desaturation. Larval homogenates incorporated more fatty acids into triglycerides than into phospholipids, but a pharate adult homogenate did not incorporate fatty acids into its lipids. Goldin and Keith (1968) have shown that mitochondria prepared from homogenates of whole third instar *Drosophila melanogaster* will readily synthesize palmitate, palmitoleate and oleate from [^{3}H] acetate, and are also capable of desaturating stearate to oleate.

This work of Municio and his colleagues clearly indicates that the type of fatty acid synthesized by insects varies during development and that changes might occur in the pattern of enzymes synthesizing fatty acids, leading to differences in chain length specificity and desaturation ability. Municio *et al.* (1972a) have also shown that the nature of the lipids into which the newly synthesized fatty acids are incorporated also varies with the developmental stage. That the fatty acids formed are incorporated into phospholipids and glycerides has been frequently demonstrated. Thus fatty acids synthesized from [^{14}C] acetate were incorporated into neutral and phospholipids in mosquitoes and green peach aphids (Van Handel and Lum, 1961; Strong, 1963). A study of the incorporation of radioactivity from palmitate into the lipids of *Phaenicia sericata* blowflies showed a slow transfer of label from free fatty acids to triglycerides and the activity of the triglycerides increased with time, both at rest and during flight, whereas that of the free fatty acids decreased. The label recovered in the phospholipids was considerably lower than that of triglycerides (Yurkiewicz and Mathur, 1969). Chino and Gilbert (1965b) studied the conversion of labelled glucose, palmitate and acetate into lipids of *Hyalophora cecropia* pupae and adults. Radioactive material was rapidly incorporated into the diglycerides. It appears that newly synthesized fatty acids are incorporated into various lipid classes according to different patterns which depend upon the developmental stage of the insect and the experimental conditions. This is clearly shown in the results of Municio *et al.* (1973) who studied *in vivo* and *in vitro* [^{14}C] acetate incorporation during development of *Ceratitis capitata*. The pattern of labelled lipids obtained *in vitro* depends on the time of incubation and the stage of development. Larval and pharate adult homogenates incorporate [^{14}C] acetate during the first 60 min mainly into phospholipids; by contrast the major incorporation of label *in vivo* is into triglycerides.

The control of lipid biosynthesis has been studied in work with whole insects. Van Handel and Lum (1961) have shown that while the female mosquito *Aëdes aegypti* can synthesize large amounts of triglyceride when fed on glucose, male mosquitoes and both sexes of houseflies are unable to do so. Both male and female mosquitoes contained about 60 µg fat per insect at emergence but after being fed glucose for 7 days the females contained about 700 µg but the males only 10 to 20 µg lipid. However O'Meara and Van Handel (1971) have shown that when male mosquitoes were feminized by thermal stress, the storage capacity for lipid was not greater than in controls. Van Handel and Lea (1965) found using females of the mosquitoes, *Aëdes taeniorynchus, A. sollicitans* and *A. aegypti* that the medial neurosecretory cells may restrict synthesis of glycogen and stimulate triglyceride synthesis since removal of these cells increases the storage capacity for glycogen at the expense of triglyceride storage. Such reversal of synthesis did not occur in blood-fed insects (Van Handel and Lea, 1970). Lea and Van Handel (1970) found that reimplantation of medial neurosecretory cells into female mosquitoes from which they had been removed, restricted glycogen synthesis but did not restore the interrupted fat synthesis; the authors concluded that the neurosecretory cells exerted an influence on glycogen synthesis (see pp. 161–164). Fat synthesis in female mosquitoes fed on glucose was not affected by removal of the corpora allata, and utilization of glycogen and fat was unaffected by either allatectomy or medial neurosecretory cell removal. The relationship between juvenile hormone titre and the fatty acid composition and synthesis in male and female *Hyalophora cecropia* during development has been discussed by Stephen and Gilbert (1970).

Horie and Nakasone (1971) have shown that in larvae of the silkworm *Bombyx mori* the rate of fatty acid biosynthesis is influenced by the levels of fatty acids and carbohydrates in the diet. The addition of fatty acids to the diet inhibited fatty acid synthesis from [^{14}C]glucose and prolonged feeding with diets containing a high concentration of fatty acids intensified the depression of synthesis. Such negative feedback inhibition of fatty acid synthesis by fatty acids is known in vertebrate systems both in intact animals and tissue slices (Whitney and Roberts, 1955; Brice and Okey, 1956; Hill *et al.*, 1958; Bortz *et al.*, 1963) and in cell free enzyme systems (Porter and Lang, 1958; Bortz and Lynen, 1963). A feedback inhibition of amino acid biosynthesis by dietary amino acids was noted in the silkworm by Inokuchi *et al.* (1969). Horie and Nakasone (1971) also showed that fatty acid synthesis from [^{14}C]glucose was accelerated by an increase in the diet of sucrose but not of starch. The influence of the quantity and type of carbohydrate in the diet on lipogenesis in mammalian tissues is well documented (McDonald, 1966, Bailey *et al.*, 1968) and it is interesting that similar control systems probably exist in insects.

Municio *et al.* (1970) showed that [^{14}C]acetate incorporation into fatty acids was enhanced by NADPH in larval homogenates of *Ceratitis capitata* but decreased in pupal homogenates. Citrate addition decreased acetate incorporation into fatty acids by homogenates of all stages of development whereas ATP-Mg^{2+} had the opposite effect. Municio *et al.* (1972b) have further studied the regulation of fatty acid biosynthesis by NADPH in homogenates of larval and adult *Ceratitis capitata*. Acetyl CoA carboxylase and fatty acid synthetase activities were measured and the effect of NADPH studied. NADPH, 0.1 to 1.0 μmol per ml activates acetyl CoA carboxylase whereas 1.0 to 5.0 μmol per ml produces a marked decrease in activity with either acetate or acetyl-CoA as substrate. This influence of NADPH cannot be interpreted through a concomitant activation of the synthesis of fatty acids from malonyl-CoA since the incorporation of [^{14}C]acetate and acetyl-CoA into fatty acids becomes negatively regulated by NADPH in the same concentration range of 1.0 to 5.0 μmol per ml. The authors discuss the NADPH effect in terms of its competitive influence with respect to malonyl-CoA in the fatty acid synthetase.

Although the above observations on homogenates of whole insects are interesting, as pointed out in the introduction, their value may be very limited. The study of the control of an enzyme in an homogenate of a whole insect is equivalent to a mammalian biochemist working with homogenates of whole rats. In order to make progress with understanding the details of lipid biosynthesis in insects, work must be carried out with individual tissues and purified enzymes from such tissues. Fortunately some work has been carried out on fatty acid biosynthesis in the major lipogenic tissue in insects, the fat body and the results are discussed in the next section. The results on fatty acid biosynthesis in whole insects probably in general reflect fatty acid biosynthesis in the fat body.

Fat body
Synthesis of fatty acids from acetate
Zebe and McShan (1959) found that intact fat body of *Prodenia eridania* would readily incorporate [^{14}C] acetate into long chain fatty acids. The major fatty acid produced was palmitate although small amounts of stearate, oleate, myristate and laurate were also identified. Homogenates of fat body would also synthesise fatty acids in the presence of malonate, ATP, CoASH and glutathione or cysteine. Tietz (1961) obtained similar results using fat body of *Locusta migratoria* and also showed that KHCO$_3$, α-oxoglutarate and NADP were also required. Tietz-Devir (1963) compared the ability of fat body homogenates and sub-cellular fractions to synthesize fatty acids. The supernatant fraction was as effective as the homogenate in fatty acid synthesis. The fatty acids formed did not accumulate but were readily esterified. In the homogenates 18 per cent of the newly synthesized fatty acids were in the form of phospholipids and 81 per cent as glycerides, whereas in the supernatant 10 per cent were in the phospholipid fraction and 90 per cent in the form of glycerides. Tietz-Devir (1963) also demonstrated ^{14}CO$_2$ fixation into malonate by a combined particulate and supernatant system from the locust fat body. The optimum conditions for fatty acid biosynthesis by fat body are similar to those for mammalian and avian systems (Popjak and Tietz, 1954; Porter *et al.*, 1957) and the major product (palmitate) is the same. When [^{14}C] acetate is used as a substrate for fatty acid synthesis experiments, it is assumed that there is ample acetate thiokinase in the preparation to convert the acetate to acetyl-CoA, the true substrate for fatty acid synthesis.

Acetate + ATP + CoASH → acetyl CoA + AMP + PP$_i$

These experiments indicate that the system operating in the fat body for fatty acid synthesis is similar to those occurring in other forms of life, although no studies have been carried out on purified enzymes. It appears that the enzymes involved are mainly in the cytoplasm although as mentioned earlier Goldin and Keith (1968) have obtained evidence that mitochondria of *Drosophila melanogaster* larvae can also synthesize fatty acids. It is interesting that in insects as with other forms of life the products *in vitro* of fatty acid synthesizing systems are saturated fatty acids despite the fact that *in vivo* 65–70 per cent of the acids are unsaturated. However, as indicated earlier experiments with whole insects have demonstrated that they are capable of desaturating fatty acids. Tietz and Stern (1969) have shown that microsomes from the fat body of *Locusta migratoria* are able to convert stearate to oleate, whereas fat body mitochondria had no desaturating ability. In contrast Goldin and Keith (1968) found that the mitochondria of *Drosophila melanogaster* larvae could convert stearate to oleate but not palmitate to palmitoleate.

Synthesis from non-lipid precursors

Zebe and McShan (1959) showed that label from [^{14}C]glucose was incorporated into fatty acids by fat body preparations from *Prodenia eridania*. Clements (1959) showed that amino acids as well as carbohydrates could be converted to fatty acids by intact fat bodies of *Schistocerca gregaria*. [U-^{14}C] Glycine was largely oxidized by fat body but considerable quantities of radioactivity were also incorporated into lipid fractions, particularly the neutral fat which had 50 per cent of its radioactivity in the fatty acid moieties. Radioactivity from [U-^{14}C]leucine and [U-^{14}C]glucose was similarly incorporated into fatty acids. The details of amino acid conversion to fatty acids have not been elucidated, but provided the amino acid can be converted to acetyl-CoA by the fat body transaminases and other enzymes involved in amino acid metabolism (see Kilby, 1963, 1965; Chefurka, 1965b; Chen, 1966), they can be converted to fatty acids.

The conversion of carbohydrate to lipid in the fat body and its control, has been investigated in detail by Walker and Bailey (1969a, b; 1970a, c; 1971b, c). The overall pathway thought to be involved is outlined in Fig. 2.7 and evidence for the pathway has been obtained from studies on changes in [^{14}C]substrate conversion to lipid and the activities of key enzymes of lipogenesis during development of the adult male desert locust, *Schistocerca gregaria*. Carbohydrates are converted in the cytoplasm to pyruvate by the glycolytic sequence of reactions and then the pyruvate is converted

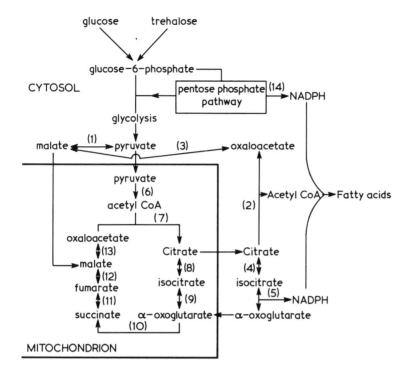

Fig. 2.7. Scheme for the conversion of carbohydrate to lipid in the fat body. The enzymes involved are:

(1) Malic enzyme
(2) ATP citrate lyase
(3) NAD-malate dehydrogenase
(4) Aconitase
(5) NADP-isocitrate dehydrogenase
(6) Pyruvate dehydrogenase
(7)–(13) Tricarboxylic acid cycle enzymes
(14) Glucose 6-phosphate dehydrogenase.

into acetyl-CoA in the mitochondria of the cells. It is known from studies with mammalian systems that the inner mitochondrial membrane is impermeable to acetyl-CoA, and it is thought that the majority of the acetyl-CoA required for lipogenesis in the cytoplasm is transferred from its site of formation in the mitochondria to its site of utilization, by conversion to citrate (Daikuhara *et al.*, 1968) The citrate is formed by the action of citrate synthase on acetyl CoA and oxaloacetate, and then passes into the cytoplasm to be acted upon by ATP citrate lyase to yield oxaloacetate and acetyl-CoA. The oxaloacetate formed in this reaction can be returned to the mitochondria in the form of malate (as illustrated in Fig. 2.7) or alternatively as aspartate after transamination. Such a cyclic scheme for cytoplasmic acetyl-CoA production from carbohydrate had previously been proposed for mammalian tissues by Kornacker and Ball (1965). Support for such a pathway in the insect fat body comes from both radioisotope and enzyme studies. Thus Walker and Bailey (1970a) obtained a developmental pattern for [14C] acetate incorporation into lipid in the adult male desert locust that corresponds to the pattern of accumulation of lipid in the fat body (Walker *et al.*, 1970, see Fig. 2.2). The patterns of glucose and trehalose incorporation into lipid are similar to that of acetate incorporation, indicating that the fat body can readily convert carbohydrate to lipid. Citrate incorporation into lipid also parallels the acetate incorporation as do changes in ATP-citrate lyase activity (Walker and Bailey, 1970a, b) supporting the view that citrate is an important intermediate in carbohydrate conversion to lipid.

Although the pathway for cytoplasmic acetyl CoA formation is similar in mammals and insects, other aspects of carbohydrate conversion to lipid appear to differ. The scheme outlined in Fig. 2.7 requires the formation of mitochondrial oxaloacetate to condense with acetyl-CoA to yield citrate. Such formation of oxaloacetate is thought to occur in rat adipose tissue (Kornacker and Ball, 1965; Ballard and Hanson, 1967) and rat and rabbit mammary gland (Gul and Dils, 1969) by the action of mitochondrial pyruvate carboxylase. However studies on developmental changes in the activity of pyruvate carboxylase in the adult male desert locust (Walker and Bailey, 1970c) suggest that this enzyme cannot fulfill a similar role in the fat body. However, parallel developmental changes in the key glycolytic enzymes phosphofructokinase and pyruvate kinase and in the 'malic enzyme' suggest the formation of mitochondrial oxaloacetate from cytoplasmic pyruvate by the action of cytoplasmic 'malic enzyme' and mitochondrial malate dehydrogenase as indicated in Fig. 2.7. Such a scheme would lead to increased malate concentrations in the cell and thus would facilitate the transfer of citrate into the cytoplasm (Chappell, 1968). Since the oxaloacetate is recycled (see Fig. 2.7) the activity of malic enzyme can decrease while lipogenesis is proceeding (Walker and Bailey, 1969b).

Kornacker and Ball (1965) suggested that in mammalian systems the NADPH required for fatty acid synthesis is provided by the combined activities of the pentose phosphate pathway and of malic enzyme working in the direction of pyruvate formation. As discussed above, it is probable that in the fat body the malic enzyme functions in the direction of malate formation and is therefore an NADPH user. Studies on developmental changes in the activities of glucose-6 -phosphate dehydrogenase (a key enzyme in the pentose phosphate pathway) and cytoplasmic NADP-dependent isocitrate dehydrogenase suggest that these enzymes provide the NADPH for fatty acid biosynthesis in the fat body of the desert locust (Walker and Bailey, 1969b, 1970b) as outlined in Fig. 2.7. Thus citrate metabolism in the cytoplasm provides both acetyl-CoA and NADPH for fatty acid biosynthesis. The extramitochondrial metabolism of citrate, via isocitrate dehydrogenase is interesting in that the diversion of citrate away from the TCA cycle in the mitochondria provides the cell with NADPH for biosynthesis, rather than NADH to enter the respiratory chain. The utilization of NADPH for fatty acid biosynthesis has been studied by Horie (1968), who showed that NADPH oxidation by the cytoplasm of the fat body of *Bombyx mori* is dependent upon malonyl-CoA and acetyl-CoA.

Since the pentose phosphate pathway provides some of the NADPH for lipid biosynthesis this is a convenient point at which to briefly discuss the pathway. The functioning of the pathway in insect tissues has been well established by experiments with labelled substrates and by determination of enzyme activities (see Kilby, 1963; Cherfurka, 1965a; Horie, 1967). The work which has previously been reviewed will not be discussed further here. Although no reports have been made concerning the relative contributions of the pentose phosphate pathway and glycolysis to glucose catabolism in the fat body, a few attempts have been made to assess the relative importance of these two pathways in whole insects *in vivo*. Silva *et al.* (1958) from the results of radiorespirometry experiments have suggested that only 4 to 9 per cent of glucose is metabolized via the pentose cycle in the cockroach, *Periplaneta americana.* Using the same technique, Lambremont and Bennett (1966) and Horie *et al* (1968) obtained figures of 45 and 35 per cent for boll weevils and silkworm larvae respectively. Agosin *et al* (1963) calculated from the specific activities of fatty acids formed from [U-^{14}C] and [6-^{14}C] glucose that in *Triatoma infestans*, 22 per cent of glucose was metabolized via the pentose pathway and Agosin *et al.* (1966) obtained figures of 4 to 17 per cent pentose cycle activity for various strains of house flies. Detailed studies of pentose phosphate pathway activity in the cockroach, *Periplaneta americana*, the milkweed bug, *Oncopeltus fasciatus* and the grasshopper *Melanoplus bivittalus* have been carried out by Chefurka (1966) and Chefurka *et al.* (1970). Figures of 18 per cent pentose cycle activity in the milkweed bug and 38 per cent in the grasshopper were reported while in the cockroach there was a sex difference, 21 per cent of glucose being metabolized by the pentose cycle in the female but only 3 per cent in the male.

The initial part of the pathway of conversion of carbohydrate to lipid involves glycolysis. Although no comprehensive and detailed study of glycolysis has been carried out on the fat body of any one insect, work with whole insects and also fat bodies from a variety of insects has established that the glycolytic pathway functions in insects (see Kilby, 1963; Chefurka, 1965b; Sacktor, 1965, 1970). The work covered by the reviews quoted will not be discussed further here. The two key regulatory enzymes in fat body glycolysis are probably phosphofructokinase and pyruvate kinase as in other forms of life. Partially purified preparations of these enzymes from desert locust fat bodies have been studied by Walker and Bailey (1969b) and Bailey and Walker (1969) respectively. Phosphofructokinase gives sigmoidal hexose monophosphate concentration-activity curves (Fig. 2.8a), which is characteristic of regulatory enzymes. At low ATP concentrations the activity rises with increasing ATP concentration but above an optimum concentration (0.2 mM) ATP becomes inhibitory. AMP activates the enzyme with half-maximal activation occurring at $10\,\mu M$. Unlike their action on the flight muscle enzyme, $3',5'$-(cyclic)-AMP and orthophosphate do not activate the enzyme from fat body. Fructose, 1,6-diphosphate inhibits the enzyme and citrate, phosphoenolpyruvate and α-glycerophosphate had no effect. Fat body pyruvate kinase is allosterically activated (see Fig. 2.8b) by very low concentrations of fructose 1,6-diphosphate $(1\,\mu M)$ but not by a variety of other metabolites at any concentration. The enzyme requires 1-2 mM ADP for maximal activity and is inhibited at higher concentrations. The enzyme is inhibited by ATP with a half maximal inhibition at about 5 mM ATP. Fructose 1,6-diphosphate reverses the inhibition by ATP and high concentrations of ADP.

Fig. 2.9 summarizes the action of effectors on phosphofructokinase and pyruvate kinase. In fat body, ATP inhibits both enzymes and AMP activates phosphofructokinase, thus these two enzymes may be regulated by the adenine nucleotides so that the rate of glycolysis is matched to the energy requirements of the cell. Under energy-requiring conditions ATP is low and because of the presence of adenylate kinase AMP accumulates — this will lead to de-inhibition of both enzymes and allosteric activation of phosphofructokinase giving an increase in glycolytic flux. As the demand for

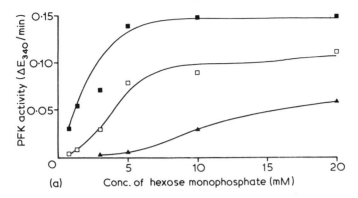

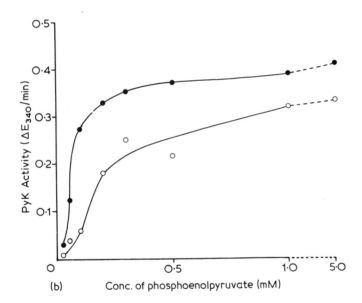

Fig. 2.8. Substrate concentration/activity curves for A. phosphofructokinase (PKF) and B. pyruvate kinase (PyK) of desert locust fat body.

A.　▲　assay contains 2mM ATP
　　　□　assay contains 1mM ATP
　　　■　assay contains 1mM ATP + 2mM AMP
B.　○　assay contains no fructose 1, 6-diphosphate
　　　●　assay contains 0.5mM fructose 1, 6-diphosphate

(Data derived from Bailey and Walker, 1969 and Walker and Bailey, 1969c.)

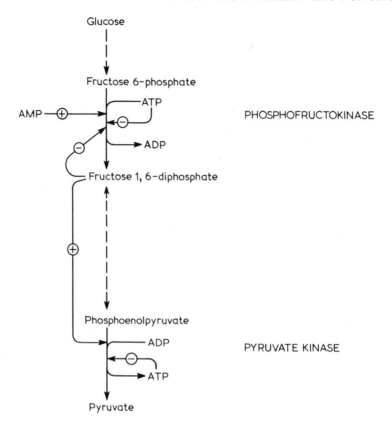

Fig. 2.9. Regulation of glycolysis in fat body
——(+)→   enzyme activator
——(−)→   enzyme inhibitor

energy is met, ATP accumulates and AMP decreases, thus decreasing the activity of both enzymes and decreasing the glycolytic flux. The regulation of glycolysis in vertebrates has been discussed extensively by Newsholme and Gevers (1967) and Scrutton and Utter (1968), particularly with regard to adenylate control of the pathway. The control of glycolysis in insect flight muscle has been discussed by Sacktor (Chapter 1 and 1970). The action of fructose 1,6-diphosphate on phosphofructokinase and pyruvate kinase of fat body is interesting in that it could modulate the activities of the two regulatory enzymes so that they act synchronously to control the overall rate of glycolysis. Increased fructose 1,6-diphosphate production due to increased activity of phosphofructokinase will lead to activation of pyruvate kinase, particularly at low phospho-enolpyruvate concentrations, and then fructose 1,6-diphosphate regulates the activity of phosphofructokinase by feedback inhibition. Thus, even though the absolute activity of pyruvate kinase is much higher than that of phosphofructokinase at all stages of development (Walker and Bailey, 1970b), the activity of both enzymes is regulated by a common effector and synchronized to the overall rate of glycolysis.

Triglyceride biosynthesis

The major storage lipid in all fat bodies is triglyceride and so the fatty acids formed in the fat body have to be esterified with glycerol to form the storage material. As discussed earlier for whole insects, the fatty acids formed from $[^{14}C]$ acetate and $[^{14}C]$ glucose are always incorporated into glycerides and to a limited extent phospho-lipids. Similarly with fat body preparations the label from $[^{14}C]$ acetate is found mainly in triglyceride fatty acids, as shown by Walker and Bailey (1970a) who found that 70 per cent of the label was in triglyceride with smaller amounts in phospholipid, sterol ester and diglyceride fractions and only 2 per cent in the free fatty acid fraction. As discussed on pp. 109–110 fat body preparations will readily incorporate $[^{14}C]$ fatty acids into triglycerides and other neutral- and phospho-lipids. The mechanism by which fatty acids are incorporated into glycerides by insects is not very well understood. The work of Tietz (1967a, b) indicates that the biosynthesis of glycerides proceeds by the pathway previously shown for mammalian tissues (Fig. 2.10). Microsomes from the fat body of *Locusta migratoria* would form $[^{14}C]$ phosphatidic acid when incubated with ATP, $MgCl_2$, glutathione, CoASH, α-glycerophosphate and $[^{14}C]$ palmitate. Addition of a supernatant from centrifugation at 140 000 g caused the phosphatidic acid to disappear and di- and triglycerides to be formed. When the soluble enzyme was incubated with phosphatidic acid, equimolar amounts of phosphate and diglycerides were found. These results indicate the microsomes contain the glycerophosphate and diglyceride acyltransferases and the supernatant contains the phosphatidate phosphohydrolase. Homogenates of the fat body could utilize glycerol as well as α-glycerophosphate for glyceride synthesis, and in the mito-chondrial fraction a glycerol kinase enzyme was found which catalyses the following reaction:

$$\text{glycerol} + \text{ATP} \longrightarrow \alpha\text{-glycerophosphate} + \text{ADP}$$

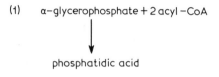

(1)   α–glycerophosphate + 2 acyl –CoA

phosphatidic acid

(2)   phosphatidic acid + H$_2$O

D–α,β–diglyceride + Pi

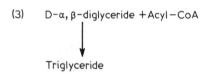

(3)   D–α, β–diglyceride + Acyl –CoA

Triglyceride

Fig. 2.10. Scheme for the biosynthesis of triglycerides. The enzymes involved are:

(1) glycerophosphate acyltransferase
(2) phosphatidate phosphohydrolase
(3) diglyceride acyltransferase

Hormonal control of lipid biosynthesis

There are some indications that lipid synthesis in the fat body is under hormonal control. It is established that removal of the corpora allata in the adult insect is followed by an increase in the lipid content of the fat body (Gilbert, 1964; Bodenstein, 1953; Orr, 1964; Odhiambo, 1966a; Strong, 1968a, b; Walker and Bailey, 1971a). The lipid composition of fat bodies of allatectomized desert locusts is similar to that of controls although the proportion of phospholipid present is somewhat higher (Walker and Bailey, 1971a). Vroman *et al.* (1965) found that in female *Periplaneta americana* [^{14}C] acetate incorporation into triglyceride was increased by allatectomy but triglyceride turnover was reduced. It was suggested that lipid accumulation was due to decreased utilization since the failure of ovarian development which occurs after allatectomy results in lipid not being required by developing oocytes. This explanation is unlikely since the quantity of lipid which accumulates in the fat body of allatectomized female *Schistocerca gregaria* (Hill, 1972) is considerably more than is required for oocyte development and further the explanation cannot be used to account for lipid accumulation in allatectomized male insects (Walker and Bailey, 1971a). Odhiambo (1966b) suggested that in the desert locust lipid accumulation following allatectomy is due to decreased locomotor activity and therefore decreased lipid utilization, but this theory has been questioned by Strong (1968b) who found that in *Locusta migratoria* allatectomy had no effect on locomotor activity. Lipid accumulation following allatectomy is not likely to be due to decreased fatty acid oxidation since this is similar in allatectomized and control insects (Walker and Bailey, 1971b). It is also unlikely that inhibition of lipid release from the fat body contributes to lipid accumulation in allatectomized insects since haemolymph lipid concentrations are normal in allatectomized locusts (Beenakkers, 1969b; Walker and Bailey, 1971a). This view is supported by the work of Doane (1961) who showed that starvation of allatectomized female *Drosophila melanogaster* caused a decrease in fat body lipid level and also by the work of Strong (1968a), who showed that enforced locomotor activity of allatectomized locusts caused a decrease in accumulated lipid of the fat body.

A period of intense feeding by male and female desert locusts coincides with the period of somatic growth which occurs during the first ten days of adult life (see Fig. 2.2, Hill *et al.,* 1968; Walker *et al.,* 1970). Lipids accumulate in the fat body during this period but when feeding declines, the quantity of lipid in the fat body decreases. The corpora allata of laboratory bred locusts normally become active towards the end of the period of somatic growth (Johnson and Hill, 1972). When the desert locust is allatectomized during the first few days of adult life, lipid accumulation in the fat body continues after the period of intense feeding and is accompanied by a higher than normal food intake (Walker and Bailey, 1971a; Hill, 1972). Thus lipid accumulation is not initiated by the absence of corpora allata but rather continues beyond the normal time when the corpora allata are missing. Allatectomy of male *Schistocerca gregaria* after the period of intense feeding causes no lipid accumulation in the fat body (Hill, 1972). Although the period of lipid accumulation in normal locusts is usually short in laboratory bred animals, the period before the onset of corpus allatum activity and sexual maturation can be very much larger in the field so that large reserves can be built up as a potential energy source for migration. The role of juvenile hormone in migration has been discussed by Caldwell and Rankin (1972). A small proportion of laboratory bred locusts show delayed sexual maturation, although they grow normally. These animals have small corpora allata and the fat body lipid reserves resemble those of allatectomized locusts (Hill, 1972).

The capacity for lipogenesis from $[^{14}C]$acetate and $[^{14}C]$glucose in fat bodies of allatectomized desert locusts has been measured by Walker and Bailey (1971b). Although the lipogenic capacity of the fat body as measured per unit of fat body fresh weight is unaltered by allatectomy, the total lipogenic capacity of allatectomized insects is much increased since the fat bodies are much larger than in controls (Walker and Bailey 1971a). These results suggest that accumulation of lipid following allatectomy is due to increased synthetic activity of the fat body. This has been further confirmed by measurements of the activities of key enzymes involved in fat body lipogenesis (Walker and Bailey, 1971c). The total fat body content of pyruvate kinase, phosphofructokinase, 'malic enzyme', ATP citrate lyase, glucose-6-phosphate dehydrogenase and cytoplasmic NADP-isocitrate dehydrogenase (Fig. 2.7) is considerably higher in allatectomized than sham operated male desert locusts although only for ATP citrate lyase is the specific activity increased by allatectomy.

The work discussed above suggests that the fat body of the allatectomized locust has a greater capacity to synthesize lipids than normal fat body. However Walker and Bailey (1971a) have suggested that the primary function of the corpus allatum in the male desert locust is, to regulate the degree of growth and differentiation of the tissues. It is considered that the effects of allatectomy on lipogenesis are secondary to the increased growth of the fat body which occurs and that the corpus allatum hormone probably has no direct effect on lipogenesis. In contrast to this view Gilbert (1967b) obtained a 30 per cent suppression of fat body lipogenesis by the addition of juvenile hormone to incubations of *Leucophaea maderae* fat body. In *Schistocerca gregaria* the addition of corpera allata homogenate to an incubation of fat bodies taken from insects during somatic growth, resulted in an 81 per cent suppression of lipogenesis. The use of tissue from older insects, in which juvenile hormone is already present and lipogenesis is low, resulted in only a 24 per cent suppression of lipogenesis (Hill, 1972). Further, allatectomy also causes an increase in specific activity of the enzyme ATP citrate lyase in the fat body and it is conceivable that the synthesis of this key enzyme of lipogenesis is suppressed by juvenile hormone (Walker and Bailey 1971c).

Other tissues

Although fat body is undoubtedly the major site of lipogenesis in the insect body, a certain amount of lipid synthesis is carried out in other tissues. In *Prodenia eridania* glucose is incorporated into lipids by muscle at 20 per cent the rate of that in fat body. (Zebe and McShane 1959). The mid gut cells of the cockroach *Eurycotis floridana* synthesise many classes of lipid from simple precursors which are believed to be ingested (Bade, 1964).

Lipid release from the fat body

It has been calculated that during flight a locust consumes large quantities of lipid (Weis-Fogh, 1952; Beenakers, 1965) and since the major lipid reserves are in the fat body of the insect it is clear that an efficient system must exist for the mobilization of the fat body triglyceride reserves and the transport of lipid respiratory fuels to the flight muscles.

Nature of the lipid released

Although Gilbert (1967a) suggested that diglyceride is specifically released from the fat body of all insects, it is now clear that the lipid mobilized can be in the form of triglyceride, diglyceride and free fatty acids depending on the insect species. Tietz (1962, 1967) studied lipid release from the fat body of the locust *Locusta migratoria*. When fat body tissue was incubated *in vitro* with [1-^{14}C] palmitate, the radioactivity was incorporated into di- and triglycerides, and the specific activity of the diglycerides was 50-fold that of the triglycerides. When prelabelled tissue was transferred to a medium containing haemolymph, diglycerides were released into the medium and the amount released was dependent upon the time of incubation and the quantity of haemolymph in the medium. The concentration of diglyceride in the fat body did not change but the specific activity was markedly reduced. When the fat body lipid was prelabelled *in vivo* the specific activities of the fat body di- and triglycerides were identical and the specific activity of the diglyceride was not reduced during diglyceride release. These results indicate that the diglyceride released from the fat body is replaced by lipid formed from hydrolysis of fat body triglycerides. The effectiveness of the haemolymph in the medium was destroyed by heating indicating the presence of an essential protein factor.

Similar experiments to those of Tietz (1962, 1967) have been carried out by Gilbert and colleagues (Chino and Gilbert, 1965a;

Thomas and Gilbert, 1968, 1969). Chino and Gilbert (1965a) concluded that in *Hyalophora cecropia, Periplaneta americana* and *Melanoplus differentialis,* diglyceride is released from pre-labelled fat bodies. Diglyceride was only released into media containing haemolymph and the process was inhibited by the presence of azide or dinitrophenol, indicating that the release is energy-dependent. Little triglyceride was released together with some free fatty acids, although release of the latter is not specific for insect haemolymph but also occurs with bovine serum albumin. It has been suggested that free fatty acid release is a passive process which follows a concentration gradient, dependent upon the level of free fatty acids in the fat body (Gilbert 1967a). Diglyceride release has been shown to occur *in vivo* as well as *in vitro* (Chino and Gilbert, 1964) in *Hyalophora cecropia* and it is also possible that in this insect two different pools of glycerides exist in the fat body only one of which releases lipids into the haemolymph (Beenakkers and Gilbert, 1968). The work of Downer and Steele (1972) indicates that in *Periplaneta americana* lipid release also occurs as diglyceride.

The concentration of diglyceride in the haemolymph of locusts increases several fold during flight (Beenakkers, 1965; Mayer and Candy, 1967, 1969a), but little change occurs in other haemolymph lipids. An increase in haemolymph diglyceride content, following injection of corpus cardiacum extracts has also been observed in locusts (Beenakkers, 1969b; Mayer and Candy, 1969b) but, in contrast, such injections in *Periplaneta americana* cause a decrease in haemolymph triglyceride and diglyceride content (Downer and Steele, 1969, 1972; Downer, 1972). Van Handel and Nayar (1972b) have studied the turnover of diglycerides during flight and rest in the moth *Spodoptera frugipenda* and found that the diglyceride pool turns over in one hour during flight and in 6 hours during rest. The turnover is sufficient to account for the energy used during both flight and rest and is consistant with the hypothesis that diglycerides function in lipid transport. It was also concluded that flight stimulates the *de novo* synthesis of lipid from non-lipid precursors.

The results of some workers indicate that free fatty acids rather than diglycerides are released from the fat body into the haemolymph. Wlodawer *et al.* (1966) and Wlodawer and Lagwinska (1967) have suggested that prelabelled fat bodies of the larvae of the wax moth *Galleria mellonella* release free fatty acids into the haemolymph-containing medium, where they are then esterified to triglycerides by a lipase in the haemolymph. Little radioactivity was found in the haemolymph diglyceride even though this is the major haemolymph glyceride. Bhakthan and Gilbert (1968) also found that free fatty acids were released in four genera of cockroach. Chang and Friedman (1971) have studied lipid release from pre-labelled fat bodies of the tobacco hornworm *Manduca sexta,* the cockroach *Leucophaea maderae* and the locust, *Schistocerca gregaria.* When pre-labelled fat body from pupal, or pharate adult hornworms is incubated with haemolymph or buffered bovine serum albumin, there is considerable release of lipids (mainly in the form of free fatty acid) into the medium. No lipid release was obtained in buffered saline. Dilution of haemolymph causes a decrease in free fatty acid release which is also decreased by azide and arsenate but not cyanide. There was little radioactivity release from pre-labelled larval fat body. Chang and Friedman (1971) also confirmed the results of Bhakthan and Gilbert (1968) that free fatty acids are released from cockroach fat body and the results of Tietz (1967) that diglycerides are released from locust fat body. The work of Stevenson (1969, 1972) on flight muscle and fat body lipases and haemolymph lipids suggests that in *Prodenia eridania* free fatty acids are transported in the haemolymph from the fat body to the flight muscles even though diglycerides are the major haemolymph lipids.

The findings of Cook and Eddington (1967) using *Periplaneta americana* indicate that in this insect both triglycerides and free fatty acids are released from the fat body into the haemolymph despite the fact that diglycerides are the major haemolymph lipids. Haemolymph was required in the incubation medium and buffer or buffer containing bovine serum albumin was ineffective. The Authors point out that the difference between their results and those of Chino and Gilbert (1965a) could be due to a difference in experimental technique. Thus Cook and Eddington measured actual amounts of lipid released whereas Chino and Gilbert measured radioactivity released from pre-labelled fat body. Martin (1969b) has also obtained results using *Pyrrhocoris apterus* which indicate that triglyceride is the major lipid released from the fat body.

Lipase activity in haemolymph and fat body
Clearly if lipid is stored in the fat body in the form of triglyceride and transported to the flight muscles by the haemolymph in the form of diglyceride or free fatty acids, then the fat body must contain lipase activity. As Gilbert *et al.* (1965) and Gilbert (1967a) have pointed out, most studies purporting to have demonstrated lipase activity in extra-digestive tissues have demonstrated the occurrence of esterase activity only. The literature on esterases has

been reviewed by Gilbert (1967a) and will not be elaborated upon here. Although Gilbert *et al.* (1965) were able to demonstrate considerable esterase activity in the fat body of *Hyalophora cecropia* which would hydrolyse ester bonds of water-soluble tributyrin and ethyl acetate, true lipase activity towards water-insoluble glycerides which are the *natural* substrates could not be demonstrated by the normal assays. However using [^{14}C] triolein as substrate the fat body was shown to possess slight lipase activity.

Unlike Gilbert *et al.* (1965), other workers have been able to detect and assay reasonable lipase activities in fat body. Wlodawer and Baranaska (1965) used Ediol (commercially prepared coconut oil emulsion) as a substrate and demonstrated by a titrimetric method two different lipase activities in the fat body of wax moth larvae. A possible criticism of the use of Ediol is that it contains emulsifying agents such as Tweens which are known to be hydrolysed by some esterases in mammalian tissues (Lech and Calvert, 1968). The presence of lipases in the fat body of *Manduca sexta* has been indicated by Chang and Friedman (1971). Stevenson (1972) found that the fat body of *Prodenia eridania* contains lipases capable of hydrolysing tri-, di- and monoglycerides. The lipolytic activity of the fat body of *Galleria mellonella* has been studied during development by Dutkowski and Ziajka (1972) and Dutkowski and Sarzola-Drabikowska (1973). The ability to hydrolyse [^{14}C] triolein is low in the female fat body after the larval-pupal ecdysis but rises during the next 6 days. In contrast, the lipolytic activity of the fat body of male insects is high at all stages of development and does not undergo changes. In females ovariotectomy causes a fall in lipolytic activity and pre-incubation of fat body of ovariotectomized females in a medium containing ovaries of pharate adults 6 days after the larval-pupal ecdysis brings about a rise in the lipolytic activity of the tissue. Sex differences in the lipolytic activity of *Galleria mellonella* fat body have been described by Dutkowski (1973).

As discussed in the previous section, it appears that frequently lipid is transported into the haemolymph from the fat body in the form of free fatty acids or triglycerides in insects in which the major haemolymph lipid is diglyceride. This might imply that diglycerides are the transport form of lipids in the haemolymph of many insects and that the triglycerides and free fatty acids released from the fat body may have to be converted to diglycerides by the haemolymph itself. It might be expected therefore that the haemolymph will contain lipase activity. As discussed earlier Wlodawer *et al.* (1966) have suggested that free fatty acids released from the fat body of

Galleria mellonella are incorporated into glycerides by a haemo-lymph lipase. The ability of *Periplaneta americana* haemolymph to hydrolyse triglycerides is indicated by the work of Cook and Eddington (1967) and lipolytic activity in the haemolymph of *Menduca sexta* has been reported by Chang and Friedman (1971). A 'true' lipase in the haemolymph of *Periplaneta americana* has recently been studied (Downer and Steele, 1973, see p. 109).

Transport of lipid in the haemolymph

There is now very good evidence that lipids are transported in the haemolymph in the form of lipoproteins. The work of Siakotos (1960a, b) resulted in the finding that three of the five protein fractions in the haemolymph of *Periplaneta americana* nymphs are lipoproteins. Tietz (1962) has shown that a lipoprotein in the haemolymph of *Locusta migratoria* becomes radioactive after incubation with pre-labelled fat body. Chino and Gilbert (1965a) have identified three lipoproteins in the blood of *Hyalophora cecropia*. Incubation studies with pre-labelled fat body indicates, however, that only one of these lipoproteins appears to carry the diglyceride. Chino *et al.* (1967) isolated and purified two lipo-proteins from the pupal haemolymph of the silkmoth *Philosamia cynthia* by ammonium sulphate precipitation and chromatography on DEAE cellulose columns and showed that the diglycerides released *in vitro* from pre-labelled fat body were incorporated into one of the lipoproteins. Thomas and Gilbert (1968, 1969) isolated and purified by preparative ultracentrifugation three classes of lipoproteins from the haemolymph of pupae and adult *Hyalophora cecropia* and *Hyalophora gloveri*. A high density lipoprotein con-tained about 75 per cent of the total lipid and electrophoresis demonstrated that each class contained several lipoprotein species. Similar lipoproteins were also found in the fat body and the yolk of the developing oocytes of these insects.

Mayer and Candy (1967) have studied changes in haemolymph lipoproteins during flight of the desert locust. Electrophoresis of haemolymph of resting locusts on cellulose acetate demonstrated the presence of two lipoproteins (Group A). During flight lipids are also associated with two other proteins (Group B) and after 2 hours of flight, 70 per cent of lipid is in Group A and 30 per cent in Group B. Locust fat body lipids were labelled by feeding a diet containing [^{14}C] palmitate, and electrophoresis of haemolymph of resting locusts indicated radioactivity only in Group A lipoproteins. During flight the radioactivity in Group A increased and activity was also found in Group B. The total radioactivity in the haemolymph increases four-fold during flight due to increased diglyceride content. Group A lipoproteins contain triglyceride and diglyceride but Group B lipoproteins only contain diglyceride. Lipid in Group B is also present in Vth instar hoppers just before moulting and in adult females during egg production and it seems likely that these lipoproteins are concerned with rapid mobilization of lipid from the fat body.

Thomas (1972) studied the synthesis of lipoproteins during larval and pupal development of *Hyalophora cecropia*. Fat body of larvae, pharate pupae and newly ecdysed pupae incorporated [^{14}C] palmitate and [^{3}H]leucine into lipoproteins and on subsequent incubation of the fat bodies with medium containing haemolymph, the lipoproteins were released into the medium. Puromycin inhibited [^{3}H]leucine incorporation into lipoproteins but did not affect fatty acid incorporation. Fat body extracts (from which the lipids had been removed) incorporated labelled haemolymph lipids into lipoproteins when incubated *in vitro*. Its is suggested that the fat body may contain apo-lipoproteins that are able to bind haemolymph lipids to form lipoproteins.

Lipoprotein synthesis in *Locusta migratoria* has been investigated by Peled and Tietz (1973). To determine the possible correlation between protein and lipid synthesis and lipoprotein release from the fat body, the incorporation of [^{14}C]amino acids into fat body proteins, the release of [^{14}C]proteins, the incorporation of [^{3}H] palmitate into tissue glycerides and the release of [^{3}H]diglycerides were measured. Although equal amounts of [^{14}C]proteins were released from fat body into buffer or haemolymph, diglyceride release was dependent upon the addition of haemolymph. Cyclo-heximide blocked protein synthesis but did not inhibit the formation of diglycerides and their release in the presence of haemolymph. It was concluded that although lipoproteins are synthesized in the fat body, the process of diglyceride release is not dependent upon lipoprotein synthesis but seems to be stimulated by a system which leads to the addition of the diglycerides onto a circulating protein. This view is further supported by the fact that, although the lipoproteins contain the phospholipids phosphatidyl choline and phosphatidyl ethanolamine as major components as well as protein and diglyceride the *in vitro* or *in vivo* incorporation of [^{14}C] palmitate into phospholipids and the release and incorporation of phospholipids into lipoproteins could not be demonstrated. In contrast to these results Thomas and Gilbert (1967) have shown that

[^{14}C] palmitate is incorporated *in vitro* into glycerides and phospholipids of *Hyalophora cecropia* fat body and that [^{14}C] fatty acids, glycerides and phospholipids are released into the medium when the fat body is then incubated with haemolymph. The findings of Peled and Tietz (1973) were in agreement with earlier results of Walker and Bailey (1970a) who had also obtained very little incorporation of [^{14}C] palmitate into phospholipids of desert locust fat body *in vitro*. Peled and Tietz (1973) showed that when haemolymph is fractionated into lipoprotein-rich and lipoprotein-poor fractions, only the former promotes diglyceride release from fat body. However, the specific activity of the purified lipoprotein was lower than intact haemolymph and it appears either that part of the lipoprotein is lost during purification or that additional proteins are necessary for the process of diglyceride release. As discussed in the next section such an additional factor may be a peptide hormone released from the corpus cardiacum.

Hormonal control of lipid release

Bhakthan and Gilbert (1968) studied the effect of several vertebrate hormones on the *in vitro* release of lipids from insect fat bodies. Growth hormone, gonadotrophin and thyroxine exerted a stimulatory effect on diglyceride and free fatty acid release from *Hyalophora cecropia* fat body, i.e. they had an adipokinetic effect. Insulin, on the other hand, inhibited the release of diglycerides and free fatty acids. Adrenaline greatly stimulated the release of free fatty acids from the fat body of four species of cockroaches and also from the fat body of *Hyalophora cecropia* but, in contrast, inhibited the release of diglycerides in *Hyalophora cecropia*. It was suggested that lipolysis in cockroach fat body in the presence of adrenaline is mediated by cyclic 3',5' AMP.

The work of Beenakkers (1969b) and Mayer and Candy (1969b) suggests that the corpora cardiaca of locusts contain an adipokinetic hormone. Mayer and Candy (1969b) showed that injection of extracts of corpora cardiaca into adult *Schistocerca gregaria* produced a two-fold increase in the lipid concentration of the haemolymph within 30 min followed by a return to normal values within 2 hours. Analysis of haemolymph showed that the increase in haemolymph lipid was due to an increase in the diglyceride fraction. A change also occurred in the lipoproteins of the haemolymph which was similar to the change which occurred during flight (see Mayer and Candy 1967, and earlier discussion). The site of action of the adipokinetic hormone was shown to be the fat body, since addition of corpus cardiacum extract to the fat body *in vitro* increased the release of diglycerides into the incubation medium. The active factor from the corpus cardiacum appeared to be stable to boiling, but was inactivated by proteolytic enzymes, suggesting that it was a peptide. Haemolymph from flown locusts contained significant amounts of the active factor but haemolymph from unflown locusts did not. It was proposed that the control of the diglyceride concentration of locust haemolymph during flight is mediated, at least in part, by a peptide hormone from the corpora cardiaca and that the hormone acts by stimulating the release of diglyceride from the fat body.

Goldsworthy, Mordue and Guthkelch (1972) have confirmed and extended the findings of Mayer and Candy (1969b). An adipokinetic factor was shown to be present in the corpora cardiaca of the locust, *Locusta migratoria* and the mealworm *Tenebrio molitor* as well as *Schistocerca gregaria*. In both species of locust used the amount of corpus cardiacum extract needed to produce an elevation of blood lipid *in vivo* was considerably less than that used by Mayer and Candy (1969b) and the duration of the effect was longer. The return of blood lipids to normal appeared to be due to excretion or inactivation of hormone since a second injection of gland extract would elicit a further response. Glandular lobe extracts in *Locusta migratoria* are about 50 times more potent in elevating blood lipids than storage lobe extracts. The relationship between the neuro-secretory storage lobes and the glandular lobes of the corpora cardiaca has been discussed by Hill (1972).

Goldsworthy, Johnson and Mordue (1972) have shown that sectioning of the afferent nerves, the nervi corporis cardiaci interni and externi (NCC I and NCC II), to the locust corpus cardiacum prevents the *in vivo* release of adipokinetic hormone from the glandular lobes. The failure to release the hormone during flight and the consequent lack of lipid mobilization brings about an impairment of flight which can be corrected by injections of corpus cardiacum extracts. The results of individual sectioning of the NCC I and NCC II suggest that a double innervation of the glandular lobes functions to control adipokinetic hormone release. The authors discuss the possibility that the adipokinetic hormone (or some other corpus cardiacum factor) may facilitate the uptake of haemolymph glyceride by flight muscles.

The action of the adipokinetic hormone is not simply that of a releasing agent. The lipid contained in the fat body of the locust is mainly triglyceride whereas that in the haemolymph is diglyceride.

Thus a fatty acid moiety must be split off each triglyceride molecule before transport can occur. It seems likely that the adipokinetic hormone stimulates lipase action in the fat body and it will be interesting to see if future work indicates that similar hormonal controls of lipase action occur in insect fat body as do in mammalian adipose tissue.

In contrast to the results obtained using the locust, Downer and Steele (1969, 1972) have shown that the corpus cardiacum of the cockroach *Periplaneta americana* contains a factor which decreases levels of triglyceride and diglyceride in the haemolymph. Coincident with these changes is an increase in the concentration of glycerides within the fat body. The authors suggest that triglycerides are converted to diglycerides in the haemolymph prior to their transport into the fat body and that the corpus cardiacum contains a hypolipaemic factor which stimulates lipid uptake by the fat body, and may also be involved in dietary lipid uptake by this tissue. Downer (1972) has shown that extracts of corpus cardiacum from *Periplaneta americana* elicit the characteristic locust response of hyperlipaemia when injected into male *Locusta migratoria*. Conversely, gland extracts of *Locusta migratoria* cause the typical cockroach response of hypolipaemia when injected into adult male *Periplaneta americana*. It is proposed that similar factors are present in both species with specificity being determined at the site of action. Goldsworthy, Mordue and Guthkelch (1972) also demonstrated that extracts of *Periplaneta americana* had a hyperlipaemic (adipokinetic) effect in the locust but found that they had little effect in *Periplaneta americana* itself. The hypolipaemic affect also could not be demonstrated in similar experiments with another cockroach *Gromphadorhina portentosa*.

Goldsworthy, Mordue and Guthkelch (1972) have discussed the possibility that the adipokinetic hormone and the corpus cardiacum hyperglycaemic hormone (see p. 163) are the same hormone. They are both peptides and are predominantly found in the glandular region of the corpus cardiacum. Further similar amounts of corpus cardiacum extract are required to produce an adipokinetic or a hyperglycaemic response. However the fact that Downer and Steele (1972) have found that purified hyperglycaemic factor from corpora cardiaca of *Periplaneta americana* has no effect on haemolymph lipid levels in the same insect indicates that the adipokinetic and the hyperglycaemic hormones may be different. The different response of haemolymph lipids and carbohydrates to stress (Hill, 1972) supports this view.

CARBOHYDRATES AND THEIR METABOLISM

Some of the topics dealt with in this section have been the subjects of previous reviews by Kilby (1963), Chefurka (1965a), Sacktor (1965, 1970) and Wyatt (1967) and it is clear that the major sources of carbohydrate fuels for flight, apart from flight muscle glycogen, are glycogen deposits in the fat body and trehalose and other sugars in the haemolymph and gastro-intestinal tract. The glycogen reserves of the fat body are mobilized not as glucose but as trehalose, which pools with the blood trehalose to act as a major energy supply for flight.

Tissue content of carbohydrates

Whole insects

The literature on the carbohydrate content of whole insects has been reviewed by Wyatt (1967) and indicates that considerable amounts of glycogen and trehalose are found in most insects and that variable amounts of glucose and other sugars also occur. Much of the literature is involved with developmental changes in insect carbohydrate content and recent reports which are not included in Wyatt's review include those of Crompton and Birt (1967), Lindh (1967), Pant and Morris (1969), Tate and Wimer (1971), Wright and Rushing (1973) and Stafford (1973). These and earlier studies indicate that glycogen is an important nutrient reserve supplying energy for adult differentiation and flight in many insect species.

Fat body

The fat body is the major site of glycogen deposition in the insect body (Kilby, 1963; Wyatt, 1967). As discussed in the aforementioned reviews, considerable developmental changes occur in fat body glycogen content and more recent reports of such developmental changes include those of Lipke, Grainger and Siakotos (1965), Hill and Goldsworthy (1968, 1970), Wimer (1969) and Chippendale (1973). The glycogen content of adult locust fat body has been determined by Goldsworthy (1969), Mayer and Candy (1969a) and Walker et al. (1970). Apart from its glycogen stores, the fat body also contains small amounts of other carbohydrates such as glucose and trehalose (Egorova, 1963; Lenartowitz et al., 1967; Wyatt, 1967; Wimer, 1969; Mayer and Candy 1969a).

Haemolymph

An excellent summary of the carbohydrate content of the haemolymph of a wide variety of insects is given by Wyatt (1967). Insect haemolymph usually contains more that 0.5 per cent of total carbohydrate and a value as high as 8.1 per cent has been obtained for *Megoura viciae* (Ehrhardt, 1962). The most characteristic sugar of insect haemolymph is the disaccharide trehalose (α-D-glucopyranosyl-α-D-glucopyranoside) and usually haemolymph contains a high concentration of trehalose (up to 6.5 per cent) and a relatively low concentration of glucose. However, trehalose is absent from the haemolymph of some insects at various stages of development (Evans and Dethier, 1957; Mochnacka and Petryszyn, 1959; Barlow and House, 1960; Dutrieu, 1961; Crompton and Birt, 1967; Wimer, 1969; Leader and Bedford, 1972).

Glycogen has been detected in small amounts in haemolymph but as Wyatt (1967) points out, since the glycogen was not isolated and characterized it is likely that in some cases, other polysaccharides of glycoproteins have been mistakenly included in these measurements. Often hexoses are found in the blood which reflect the diet of the insect. Thus locusts fed on pears (high fructose content) contain fructose in the haemolymph, but insects fed wheat do not (Hansen, 1964). Bees fed fructose, glucose or galactose also have the appropriate hexose as a major haemolymph constituent (Maurizio, 1965). Occasionally unusual carbohydrates occur, e.g. *scyllo*-inositol has been identified in desert locust haemolymph (Candy, 1967).

Gut and other tissues

As will be seen later, the lumen of the gastro-intestinal tract after a meal can act as a major source of carbohydrate fuels for flight. The carbohydrates present in the gut will vary enormously with the diet of particular insects, but a large number of different carbohydrates may be ingested, e.g. nectar and honeydew which are principal foods for many insects contain fructose, glucose, sucrose, melizitose and dextrin as major constituents and maltose, trehalose, mannitol, dulcitol, melibiose and raffinose as minor components (Wykes, 1952; Auclair, 1963). Apart from the carbohydrate foodstuffs in the lumen of the gut, the mid-gut cells themselves are known to contain glycogen stores (Wigglesworth, 1949; Mayer and Candy, 1969a).

Wigglesworth (1949) showed that apart from flight muscles, fat body and mid-gut cells of *Drosophila melanogaster*, glycogen was also associated with the halteres and proventriculus. In larvae of *Gastrophilus intestinalis* glycogen can account for 40 per cent of the dry weight of the trachael organ (Levenbook, 1951). Recently D'Costa *et al.* (1973) and D'Costa and Rice (1973) have measured the glycogen content of the proventriculus of the tsetse fly, *Glossina morsitans* and the blowfly *Chryomozia putoria* and suggested that it supplies the energy required for secretory processes.

Utilization of carbohydrate reserves during flight

The utilization of fat body glycogen, blood sugars and gastro-intestinal tract carbohydrates during flight has frequently been observed. Changes in locust tissue carbohydrates during flight are shown in Fig. 2.11 and other examples are discussed below.

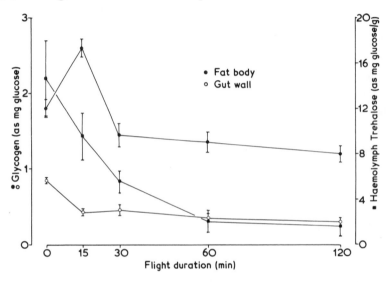

Fig. 2.11. Changes in carbohydrate reserves of the desert locust during flight. (Data derived from Mayer and Candy, 1969a.)

Wigglesworth (1949) found that continuous flight of *Drosophila melanogaster* led to a depletion of glycogen in the fat body and haltere. Similarly in *Culex pipiens* (Clements, 1955) flight to exhaustion led to the disappearance of glycogen from the fat body. Clegg and Evans (1961) studied the decrease in abdominal (presumably mainly fat body) glycogen during flight of *Phormia regina* and showed that the mobilized glycogen is transported in the haemolymph as trehalose. Depletion of abdominal glycogen during flight was also shown for the cockroach *Periplaneta americana* (Polacek and Kubista, 1960). The amount of glycogen in the fat body of *Schistocerca gregaria* increases during the first 15 min of flight but later falls to below the resting value (Mayer and Candy, 1969a)

Blood glucose and trehalose are depleted during the flight of blowflies (Evans and Dethier, 1957; Clegg and Evans, 1961) as is total blood sugar in the honey-bee during flight (Beutler, 1936). Flight causes a considerable fall in the concentration of blood trehalose in cockroaches (Polacek and Kubista, 1960) and in the desert locust (Weis-Fogh, 1964). Mayer and Candy (1969a) found that the amount of trehalose in the haemolymph of the desert locust falls during flight, whereas the glucose level increases in the early part of flight, falling later.

Mayer and Candy (1969a) found that the amount of glycogen in locust gut wall decreased during flight. Further it has long been recognized that the dietary carbohydrates in the gut lumen can be used to provide the energy for flight. Beutler (1936) observed that sugar from the honey-bladder region of the gut of the honey-bee was used during flight, and Hocking (1953) concluded that crop sugar of *Tabanus* was an energy source for flight. Use of gastro-intestinal sugars during flight of blowflies has been studied by Hudson (1958) and Clegg and Evans (1961) and it is clear that if the blowflies are made to fly for long periods of time after being fed, then the gut supplies a large proportion of the energy for flight.

Hudson (1958) has suggested that during flight, utilization of tissue glycogen takes place before the utilization of sugars from the gastro-intestinal tract since total body glycogen disappeared rapidly during the flight of blowflies despite the fact that the crop contained much glucose. However, most of the available evidence suggests that the sugars in the intestinal lumen are rapidly, and perhaps preferentially used as a source of flight energy. Thus Wigglesworth (1949) and Hudson (1958), using *Drosophila melanogaster* and *Phormia regina* respectively, showed that dietary glucose could restore flight in exhausted insects within 30 s of ingestion. In sugar-fed mosquitoes *Aëdes sollicitans* and *Aëdes taeniorhynchus*, glycogen was not utilized as a flight substrate as long as sugar was available (Nayar and Van Handel, 1971). Starved mosquitoes and insects flown to exhaustion could resume vigorous flight immediately after a sugar meal. Trehalose did not change during vigorous flight and made a negligible contribution to exhaustive flight. Flight was sustained by glycogen stores in unfed mosquitoes.

It seems unlikely that gut carbohydrates are absorbed mainly as monosaccharides which can then be metabolized by the other tissues. Some of the absorbed sugars are probably oxidized directly by the flight muscles during exercise, while most of the remaining sugars are converted to glycogen and trehalose by the fat body, a process which occurs mostly during periods of rest. Flight muscles can oxidise a variety of hexoses (Sacktor, 1965, 1970), and various sugars can restore flight in exhausted insects. Thus Wigglesworth (1949) found that whereas glucose restored continuous flight of *Drosophila melanogaster* within 30 s of ingestion, fructose, mannose, sucrose, maltose and trehalose required several minutes. Flight was supported in exhausted mosquitoes by feeding glucose, sucrose, raffinose, mannose, dextrin, maltose, melizitose or trehalose (Nayar and Sauerman, 1971). Many carbohydrates such as lactose and sorbose did not support flight at all, but galactose, mellibiose, α-methylglucoside, glycerol, mannitol and sorbitol supported delayed flight after feeding; however 24 h after feeding these carbohydrates, flight was supported immediately suggesting that they had to be converted to storage compounds such as glycogen before acting as respiratory fuels. Glycogen formation from a wide variety of sugars was demonstrated in *Aëdes taeniorhynchus* and *A. aegypti* but a species difference was apparent because although *A. aegypti* had the capacity to convert xylose, α-methylglucoside and glycerol to glycogen, *A. taeniorhynchus* did not. Nayar and Van Handel (1971) have shown that glycogen accumulates at rest or during flight in mosquitoes fed sugar, while in *Spodoptera frugiperda*, glycogen and also trehalose are formed both at rest and during flight (Van Handel and Nayar, 1972a) after feeding glucose.

Metabolism of sugars other than glucose

Dietary glucose can readily be converted in the fat body to glucose 6-phosphate and glucose 1-phosphate by the enzymes hexokinase and phosphoglucomutase and then the phosphorylated glucose can either be converted to trehalose or glycogen or be metabolized via glycolysis and the pentose phosphate pathway (see Fig. 2.12; pp. 129–133, Sacktor, 1965; Chefurka, 1965a). As indicated in the previous section, some insects live on a diet which contains a wide variety of sugars many of which have shown to be oxidized and/or converted to glycogen and trehalose. The small amount of work that has been carried out on the metabolism of sugars other than glucose has usually involved whole insect preparations and little is known of the metabolism of such sugars by a tissue such as the fat body.

Studies of insect hexokinases (Chefurka, 1954; Kerly and Leaback, 1957; Shigematsu, 1958; Ito and Horie, 1959; Sols *et al.*, 1960) indicate that the enzymes are not specific for glucose but will also phosphorylate fructose and mannose. The mannose 6-phosphate formed by hexokinase activity can be converted to fructose 6-phosphate by the enzyme phosphomannoseisomerase (Sols *et al.*,

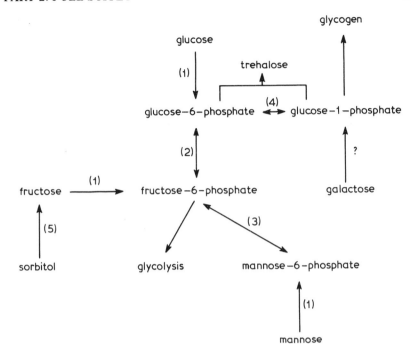

Fig. 2.12. Possible interconversions of dietary sugars. The enzymes involved are:

(1) Hexokinase
(2) Phosphoglucoseisomerase
(3) Phosphomannoseisomerase
(4) Phosphoglucomutase
(5) Polyol: NAD oxidoreductase

1960) and the fructose 6-phosphate derived from this reaction, or from the action of hexokinase on fructose can be converted to glucose 6-phosphate by phosphoglucoseisomerase and thus be metabolized in the same manner as glucose (Chefurka, 1965a). Sols *et al.* (1960) have suggested that the mannose toxicity in honey-bees (Von Frisch, 1927) is due to the fact that phosphomannoseisomerase has a much lower activity than phosphoglucoseisomerase and so mannose 6-phosphate accumulates and blocks glycolysis by inhibiting phosphoglucoseisomerase. However this explanation of mannose toxicity has been challenged by Saunders *et al.* (1969) and Van Handel (1971). Saunders *et al.* (1969) reported considerably higher activities of phosphomannoseisomerase in the honey-bee than had Sols *et al.* (1960) and Van Handel (1971) showed that the activity of the enzyme was more than tenfold greater than that required to account for the oxidation rates of mannose observed *in vivo*. Van Handel (1971) found that mannose was metabolized by the mosquito at the same rate as glucose but at a much slower rate by the bee and that mannose did not inhibit the metabolism of glucose. In both insects the activity of the phosphoglucoseisomerase is 100 times that of the phosphomannoseisomerase.

Van Handel (1969a) has shown that sorbitol, which is present in the diet of some insects, is readily converted to fructose in the mosquito *Aëdes sollicitans* by the enzyme polyol:NAD oxido-reductase, but is not converted to glucose. The fructose formed can then be metabolized as indicated above. The *in vivo* metabolism of fructose, mannose and galactose by the mosquito has been studied by Van Handel (1969b). The results suggest that glucose is not an intermediate between the hexoses and glycogen or trehalose and that trehalose is not an intermediate between the hexoses and glycogen. The ready conversion of fructose and mannose and the slow conversion of galactose to trehalose by the fat body of *Phormia regina* has been demonstrated by Clegg and Evans (1961).

The possible pathways of interconversion of various sugars is outlined in Fig. 2.12. However much further work is required to see, for example, if special enzymes exist for fructose and galactose metabolism as is known to be the case in mammalian tissues.

Glycogen biosynthesis

It is probable that in insect tissues the synthesis of glycogen proceeds via the same pathway as that demonstrated for mammalian tissues and outlined below:

(1) Glucose 1-phosphate + UTP $\rightleftharpoons$ UDP-glucose + PP_i

(2) UDP-glucose + glycogen$_n$ $\longrightarrow$ UDP + glycogen$_{(n+1)}$

UDP-glucose formed in reaction (1) by the action of UDP-glucose pyrophosphorylase reacts with pre-existing polysaccharide in the presence of glycogen synthetase (UDP-glucose-glycogen trans-glycosylase) as shown in reaction (2) to extend the polysaccharide by one glucose unit.

Although the synthesis of glycogen by locust muscle has been demonstrated by incorporation of radioactivity from UDP-[^{14}C] glucose into glycogen and the release of UDP (Trivelloni, 1960) and by histochemical methods (Hess and Pearse, 1961) most studies have been carried out on fat body. Thus Vardanis (1963) showed that fat body of *Periplaneta americana* would incorporate [^{14}C]glucose from

UDP-[^{14}C] glucose into glycogen and that the tissue possessed UDP-glucose pyrophosphorylase activity. Glucose 6-phosphate had little effect on glycogen synthesis but glucose 1-phosphate stimulated it, probably due to the fact that it prevented glycogen breakdown by inhibiting glycogen phosphorylase which was present in high activity.

Murphy and Wyatt (1965) studied glycogen synthetase from fat bodies of larval *Hyalophora cecropia* and found that the enzyme could be sedimented largely free of phosphorylase and trehalose 6-phosphate synthetase (see later) by centrifugation at 37 000 g. As in mammalian tissues the synthetase is bound to particulate glycogen and is activated by glucose 6-phosphate, galactose 6-phosphate and glucosamine 6-phosphate (Rosell-Perez and Larner, 1964a). The Km for UDP-glucose is 1.6 mM and the Km for glucose 6-phosphate is 0.6 mM. Glucose 6-phosphate activates without altering the Km for UDP-glucose as is the case with the D form of the enzyme in mammalian tissues (Rosell-Perez and Larner, 1964b).

The glycogen synthetase of whole honey-bee larvae has been studied by Vardanis (1967). The enzyme is similar to mammalian enzymes in many ways and since the enzyme is activated by glucose 6-phosphate which greatly increases the affinity of the enzyme for UDP-glucose it seems that the enzyme has the properties of both the D and the I forms of the mammalian enzyme (for review of mammalian glycogen synthetases see Larner and Villar-Palasi, 1971). However in contrast to the mammalian enzyme, the insect enzyme bound with particulate glycogen is an active enzyme-substrate complex. Vardanis (1967) has suggested that the main factor limiting the synthesis of glycogen is the length of the outer branch chains in the primer and when this limit is reached, glucose incorporation stops. Addition of glucose 6-phosphate causes incorporation to resume until a longer limit is reached. It is possible that glucose 6-phosphate either changes the specificity of the enzyme for primer outer chains or, alternatively, activates a glucose 6-phosphate dependent form of the enzyme that can utilize longer outer chains as effective primary units.

The control of fat body glycogen synthesis is, as in the well-studied mammalian systems, intimately linked to glycogen breakdown. Glucose 6-phosphate which stimulates glycogen synthesis concomitantly inhibits glycogen breakdown by inhibiting glycogen phosphorylase activity (see later). The control of fat body glycogen synthesis in relation to glycogen breakdown and trehalose synthesis is outlined in Fig. 2.13.

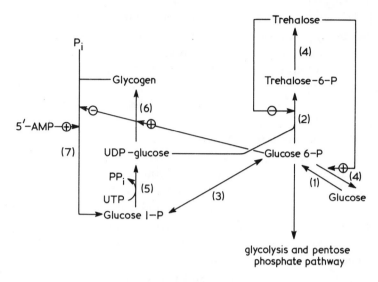

Fig. 2.13. Control of trehalose and glycogen synthesis and glycogen breakdown in the fat body. The enzymes involved are:

(1) Hexokinase
(2) Trehalose 6-phosphate synthetase
(3) Phosphoglucomutase
(4) Trehalose 6-phosphatase
(5) UDP-glucose pyrophosphorylase
(6) Glycogen synthetase
(7) Phosphorylase
⊕— activates
⊖— inhibits

Trehalose biosynthesis

The *in vivo* conversion of [^{14}C]glucose to trehalose has been demonstrated in a number of insects (Treherne, 1958b; Winteringham, 1959; Clegg and Evans, 1961; Lipke, Graves and Leto, 1965; Van Handel and Nayar, 1972a) and [^{14}C]pyruvate and [^{14}C]glucose 1-phosphate incorporation into trehalose by silkworms has been studied by Bricteux-Gregoire *et al.* (1964, 1965). Rapid conversion of fructose and mannose and slow conversion of galactose to trehalose by *Phormia regina* was shown by Clegg and Evans (1961).

The ability of fat body to synthesise trehalose has been shown by incubating fat bodies *in vitro* with [^{14}C]glucose and measuring its conversion to radioactive trehalose. Such experiments have demonstrated trehalose synthesis in fat bodies of *Schistocerca gregaria*

(Clements, 1959; Candy and Kilby, 1959), *Phormia regina* and *Leucophaea maderae* (Clegg and Evans, 1961), *Bombyx mori* (Saito, 1963) and *Hyalophora cecropia* (Murphy and Wyatt, 1965). Although the fat body is the major site of trehalose synthesis, other tissues may also have limited ability to synthesise the disaccharide. Thus Trivelloni (1960) and Hines and Smith (1963) demonstrated $[^{14}C]$ glucose conversion to trehalose by locust muscle and Shyamala and Bhat (1965) detected trehalose synthesis in the isolated mid-gut of the silkworm. In contrast to these results Candy and Kilby (1959) and Clements (1959) using locusts and Clegg and Evans (1961) using *Phormia regina* observed little or no synthesis of trehalose in muscle or gut.

Candy and Kilby (1959, 1961) and Mikolashek and Zebe (1967), demonstrated that the pathway of trehalose synthesis in locust fat body was the same as that discovered in yeast by Cabib and Leloir (1958):

(1) UDP-glucose + glucose 6-phosphate $\longrightarrow$

trehalose 6-phosphate + UDP

(2) trehalose 6-phosphate + H_2O $\longrightarrow$ trehalose + P_i

As well as the enzymes involved in reactions 1 (trehalose 6-phosphate synthetase, trehalose phosphate-UDP-glucosyltransferase) and 2 (trehalose 6-phosphatase) fat body extracts also contained essential accessory enzymes such as phosphoglucomutase, nucleoside diphosphate kinase and UDP-glucose pyrophosphorylase. Trehalose 6-phosphatase has been partially purified from *Phormia regina* and acts upon glucose 6-phosphate as well as trehalose 6-phosphate (Friedman, 1971).

The kinetics of trehalose 6-phosphate synthetase of fat bodies of *Hyalophora cecropia* larvae has been studied by Murphy and Wyatt (1965). The Km for UDP-glucose is 0.3 mM and half-maximal velocity is obtained at about 5 mM glucose 6-phosphate. The velocity/glucose 6-phosphate concentration curve is sigmoidal indicating that the enzyme has allosteric properties (Monod *et al.*, 1963, 1965). Partially purified enzyme shows Michaelis-Menten kinetics with a Km for glucose 6-phosphate of 5 mM. Trehalose 6-phosphate synthetase is inhibited by trehalose (i.e. feedback inhibition) and the extent of the inhibition depends on the trehalose, glucose 6-phosphate and Mg^{2+} concentrations. Trehalose decreases the affinity of the enzyme for glucose 6-phosphate and is non-competitive with respect to UDP-glucose.

The inhibition of trehalose 6-phosphate synthetase by trehalose is probably important in the control of haemolymph trehalose levels. Murphy and Wyatt (1965) have suggested that elevation of blood glucose after ingestion of carbohydrate causes a rise in fat body glucose 6-phosphate which activates both glycogen (see pp. 152–154) and trehalose 6-phosphate synthetases and therefore stimulates both glycogen and trehalose biosynthesis. At the same time glucose 6-phosphate inhibits glycogen phosphorylase and therefore glycogen breakdown (see later). When the blood trehalose level builds up, the sugar inhibits its own synthesis via its action on trehalose 6-phosphate synthetase leading to an accumulation of UDP-glucose which is directed into glycogen synthesis (at low UDP-glucose concentration trehalose synthesis is favoured since the Km for trehalose 6-phosphate synthetase (0.3 mM) is much lower than for glycogen synthetase (1.6 mM)). When haemolymph trehalose levels fall during flight or starvation, the inhibition of trehalose 6-phosphate synthetase is relieved and trehalose synthesis may lead to a lowering of glucose 6-phosphate levels in the fat body and this may inhibit glycogen synthesis and stimulate glycogen breakdown and thus allow continued trehalose synthesis from stored glycogen in the absence of dietary carbohydrate.

The work of Friedman (1968, 1971) suggests that the control of trehalose synthesis may be accomplished without involving simultaneous control of glycogen synthesis. Friedman (1968) has shown that trehalose specifically increases the activity of glucose 6-phosphatase considerably in *Phormia regina*, and thus removes one of the substrates of trehalose 6-phosphate synthetase. Friedman (1971) has purified the enzyme capable of carrying out a trehalose activated hydrolysis of glucose 6-phosphate and it has been shown that the same enzyme has trehalose 6-phosphatase activity but that this is not activated by trehalose. The hydrolytic sites overlap at the phosphate binding site. The enzyme has a molecular weight of 25–26 000 and requires Mg^{2+} for activity towards either substrate. The pH optimum for trehalose 6-phosphate hydrolysis is 6–6.5 whereas for glucose 6-phosphate it is 7.0 in the absence of trehalose and 6.0 in its presence.

The factors involved in trehalose synthesis in the fat body are summarized in Fig. 2.13. Utilization of trehalose involves hydrolysis to glucose catalysed by the enzyme trehalase. Probably only flight muscle trehalase which has been discussed in the previous article, is directly involved in the supply of fuel for flight. The literature on trehalase activity in other tissues has been ably reviewed by Wyatt (1967) and more recent work includes that by Yanagawa (1971)

Dahlman (1971) and Duve (1972). Since these enzymes are not directly involved with the supply of fuel for flight they will not be discussed further here.

Carbohydrate biosynthesis from non-carbohydrate precursors

The fact that insects can convert certain amino acids into carbohydrate has been appreciated for a long time. Thus Wigglesworth (1942) showed that glycogen stores of *Aëdes aegypti* depleted by starvation could be restored by feeding casein, alanine and glutamic acid but not by feeding olive oil, i.e. the insect can carry out gluconeogenesis from some amino acids but not from lipid. Similarly Van Handel (1965) found that *Aëdes sollicitans* could synthesize glycogen from dietary protein albeit at only one tenth the rate of synthesis from dietary sugar. Nayar and Sauerman (1971) have shown that glycerol, which can be obtained in the diet or during lipid degradation, can be converted to glycogen in mosquitoes.

The enzymes involved in insect gluconeogenesis have been little investigated. Gluconeogenesis involves reversal of the glycolytic sequence of reactions and work with mammalian systems has shown that extra enzymes are required to convert pyruvate to phosphoenolpyruvate and fructose 1,6-diphosphate to fructose 6-phosphate since the pyruvate kinase and phosphofructokinase reactions are essentially irreversible. As shown in Fig. 2.14 conversion of pyruvate to phosphoenolpyruvate involves two reactions catalysed by the enzymes pyruvate carboxylase and phosphoenolpyruvate carboxykinase and conversion of fructose 1,6-diphosphate to fructose 6-phosphate involves the enzyme fructose 1,6-diphosphatase.

Glycerol can enter the glycolytic sequence at the triose level via the action of the enzymes glycerokinase and α-glycerophosphate dehydrogenase which are both found in insect tissues (Chefurka, 1965a; Crabtree and Newsholme, 1972a,b). Amino acids can be degraded via transamination and deamination to pyruvate, acetyl-CoA or tricarboxylic acid cycle intermediates [which can be converted to oxaloacetate by the reactions of the cycle – Kilby (1963), Chefurka (1965b) Chen (1966)]. Amino acids giving rise to pyruvate or oxaloacetate can be converted to carbohydrate but those giving rise to acetyl-CoA cannot do so because of the irreversibility of the pyruvate dehydrogenase catalysed reaction (hence fatty acids cannot be converted to carbohydrate).

Changes in the activities of fat body mitochondrial and cytoplasmic pyruvate carboxylase and cytoplasmic fructose 1,6-diphosphatase during development of the adult male desert locust

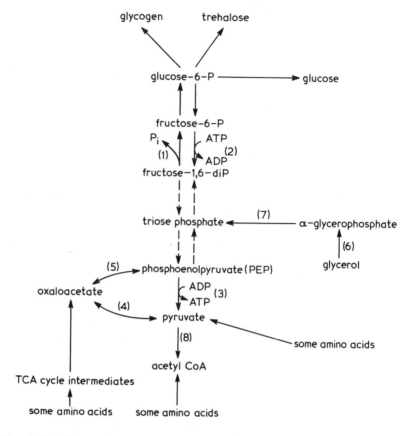

Fig. 2.14. Gluconeogenesis from amino acids and glycerol. The enzymes involved are:

(1) Fructose 1, 6-diphosphatase
(2) Phosphofructokinase
(3) Pyruvate kinase
(4) Pyruvate carboxylase
(5) Phosphoenolpyruvate carboxykinase
(6) Glycerokinase
(7) α-Glycerophosphate dehydrogenase
(8) Pyruvate dehydrogenase

have been measured by Walker and Bailey (1970b, c). Most of the pyruvate carboxylase activity is in the mitochondria and the developmental changes can be correlated with the synthesis of carbohydrate in the fat body (Walker and Bailey, 1970b). However the developmental changes in the activity of fructose 1,6-diphosphatase correlate with changes in glycolysis and not carbohydrate formation (Walker and Bailey, 1970b). Fructose 1,6-

diphosphatase, pyruvate carboxylase and phosphoenolpyruvate carboxykinase activities have been measured in the muscles of a number of insects (Crabtree *et al.*, 1972).

There have been suggestions that insects can convert lipid to carbohydrate (for review see Gilbert, 1967a; Wyatt, 1967) but such a conversion has not been unequivocally demonstrated. It appears that only those microorganisms and plants equipped with the enzymes of the glyoxylate cycle are capable of bringing about conversion of lipid to carbohydrate.

Glycogen breakdown

The degradation of glycogen in mammalian tissues by the enzyme phosphorylase has been much studied (for review of phosphorylase and the control of glycogen degradation see Fischer *et al.*, 1971). The enzyme catalyses the following reaction

$$\text{glycogen}_n + P \rightleftharpoons \text{glycogen}_{(n-1)} + \text{glucose } 1-P$$

The skeletal muscle enzyme exists in active (phosphorylase *a*) and relatively inactive (phosphorylase *b*) forms. Phosphorylase *b* is converted to phosphorylase *a* by phosphorylation which involves the enzyme phosphorylase kinase activated by cyclic $3',5'-$AMP. The production of cyclic $3'5'-$AMP is accomplished by the enzyme adenyl cyclase the activity of which is stimulated by certain hormones. The conversion of phosphorylase *a* to phosphorylase *b* is brought about by phosphorylase phosphatase.

Much of the detailed work on insect phosphorylase has been carried out on flight muscle preparations (Sacktor, 1970; Chapter 1) and knowledge of phosphorylases in other insect tissue is fragmentary. However it is clear that when fat body glycogen is broken down it is usually converted to trehalose.

Phosphorylase activity in the fat body of *Bombyx mori* was demonstrated by Shigematsu (1956) and in the mid-gut of the silkworm by Ito and Horie (1959). Changes in the activity of the enzyme during development of the silkworm have been reported by Yamashita (1965). Developmental changes in phosphorylase activity of whole *Philosamia ricini* have been measured by Pant and Morris (1969). Takehara (1962) assayed the enzyme activity in prepupae and pupae of *Monema fluorescens*. The enzyme had an optimum pH of 6.6, the Km for glucose 1-phosphate was 11 mM and AMP greatly increased the activity.

Steele (1964) showed that the phosphorylase activity of *Periplaneta americana* fat body is stimulated by cyclic 3', 5'—AMP. Glycogen phosphorylase was assayed in homogenates of fat body from *Samia cynthia* and *Hyalophora cecropia* pupae by Stevenson and Wyatt (1964). Activity without AMP (active enzyme) was much less than in its presence (total enzyme). Apparent Km's for active and total enzymes were 1.4 and 0.6 mg/ml for glycogen and 20 and 30 mM for glucose 1-phosphate. Incubation of fat body tissue *in vitro* caused an increase in active enzyme activity without a change in total enzyme activity suggesting conversion of an inactive to an active form. The proportion of active enzyme present changed during development. It is of interest that the activity of adenyl cyclase in the fly *Ceratitis capitata* also changes during development (Castillon *et al.*, 1973) and as with the active phosphorylase in *Samia cynthia* (Stevenson and Wyatt, 1964) is elevated in the later stages of development to the adult. Wiens and Gilbert (1967a) have also studied the phosphorylase system of *Hyalophora cecropia*. In the presence of ATP, fat body, muscle and gut could transform inactive phosphorylase to the active form. Phosphorylase in the fat body was activated by phosphorylase kinase and evidence was also obtained to suggest the presence of phosphorylase phosphatase in the tissue. Developmental changes in phosphorylase activity were also described.

Recently the control of phosphorylase in the fat body of the locust, *Locusta migratoria*, has been studied by Applebaum and Schlesinger (1973). The enzyme was partially purified by differential centrifugation and dissociation from glycogen particles at two pH values. Optimum activity was obtained at pH 6.6—6.7. The calculated apparent Km values for glycogen and glucose 1-phosphate were 0.08 per cent and 10—13 mM respectively. 5'—AMP activated in the range 5 mM—1 mM. Glucose 6-phosphate was a competitive inhibitor for the substrate glucose 1-phosphate (Ki = 1.7 mM) and 5'—AMP abolished this inhibition. Glucose weakly inhibited (Ki = 25—30 mM), but trehalose did not inhibit even at 100 mM. Since glucose concentrations in the fat body are extremely low the effect of glucose is probably physiologically unimportant.

Applebaum and Schlesinger (1973) suggest that glucose 6-phosphate is a major regulator of glycogen phosphorylase activity in locust fat body. The concomitant control that this metabolite exerts on trehalose and glycogen synthesis has been discussed previously and is outlined in Fig. 2.13 (p. 154) along with the control of phosphorylase. The activation of glycogen breakdown by AMP is interesting in view of the role of this nucleotide in controlling the rate of fat body glycolysis (see pp. 130–133). Clearly in times of low cell ATP concentrations, the accompanying high AMP concentration stimulates glycogen breakdown to glucose 1-phosphate and the metabolism of this compound via the glycolytic sequence.

Hormonal control of carbohydrate metabolism

Glycogen synthesis

After removal of cerebral neurosecretory cells, glycogen has been shown to accumulate in the tissues of *Calliphora erythrocephala* (Thomsen, 1952), *Aëdes taeniorhynchus* (Van Handel and Lea, 1965, 1970; Lea and Van Handel, 1970) and *Locusta migratoria* (Goldsworthy, 1971). Such accumulation of glycogen is not likely to be due to lack of carbohydrate mobilizing factors from the corpus cardiacum (see later) since corpus cardiacum extracts do not produce hyperglycaemia in cauterized *Locusta migratoria* (Goldsworthy, 1971) and removal of cerebral neurosecretory cells does not affect fat body phosphorylase levels in this species (Goldsworthy, 1970), nor does it alter the rate of glycogen utilization in the fat body of the mosquito (Lea and Van Handel, 1970). It has been suggested that in the mosquito a hormone from the neurosecretory cells may suppress glycogen synthesis in the fat body (Lea and Van Handel, 1970; Van Handel and Lea, 1970). However this explanation of glycogen accumulation after removal of neurosecretory cells may not be valid for the locust since the level of glycogen synthetase does not increase in the fat body of cauterized locusts (Hill, 1972).

Removal of the corpora allata also causes glycogen accumulation in insects such as *Pyrrhocoris apterus* (Janda and Slama, 1965) and *Phormia regina* (Orr, 1964). However a direct effect on carbohydrate metabolism is unlikely (Wyatt, 1967) although Liu (1973) has reported that allatectomy of female *Musca domestica* results in a large decrease in fat body glycogen synthetase activity.

Glycogen breakdown

Steele (1961) was the first person to demonstrate the presence of a hyperglycaemic factor in the corpus cardiacum of an insect. In *Periplaneta americana* marked elevation of blood trehalose content occurred within 30 min of injection of corpus cardiacum extracts. Ralph (1962) and McCarthy and Ralph (1962) showed that the corpus cardiacum extract acted by causing a breakdown of fat body glycogen and also found that gland extracts caused a release of trehalose from fat body incubated *in vitro*. Extracts of corpus cardiacum injected into *Blaberus discoidalis* cause a rise in blood trehalose and glucose and a concomitant decrease in fat body glycogen (Bowers and Friedman, 1963).

Steele (1963) showed that injection of corpora cardiaca extracts in the cockroach causes a depletion of nerve cord glycogen as well as of fat body glycogen but had little effect on the glycogen content of muscle and gut. The glycogenolytic effect of the extracts was shown to be due to their ability to increase the activity of glycogen phosphorylase. Hart and Steele (1973) have shown that glyco-genolysis in cockroach nerve cord is accelerated in the presence of extracts of corpus cardiacum because of increased levels of cyclic 3', 5'−AMP which causes activation of phosphorylase.

The effect of corpus cardiacum extracts on blood trehalose and fat body glycogen and phosphorylase in the cockroach *Leucophaea maderae* has been studied by Wiens and Gilbert (1965, 1967b). Fat body phosphorylase is rapidly activated by low concentrations of gland extract and the concentration of phosphorylase-activating factor in the gland appears to be related to changes in glycogen content during oogenesis. The hormone elicits an accelerated release of trehalose from the fat body and a switch of fat body metabolism towards lipid oxidation. It is noteworthy that corpus cardiacum extracts from the silkmoth which will activate the fat body phosphorylase in the cockroach had no such effect on the silkmoth itself (Wiens and Gilbert, 1967a). Friedman (1967) showed that corpus cardiacum extracts injected into *Phormia regina* would only cause a rise in blood trehalose in 24 h starved insects and not in fed insects, although the extract stimulated trehalose synthesis in isolated fat bodies *in vitro*. It appears that in the fed fly, synthesis of blood trehalose proceeds at a maximum rate from dietary sugar and that gland extracts only affect trehalose synthesis when breakdown of fat body glycogen is a limiting factor, (i.e. in 24 h starved insects).

Natalizi and Frontali (1966) separated and partially purified heart-accelerating and hyperglycaemic factors from *Periplaneta americana* corpora cardiaca. Hyperglycaemic factor is a peptide and was purified by the use of Sephadex and Biogel columns. It is of interest that Matthews and Downer (1973) have shown that certain anaesthetics cause hyperglycaemia in cockroaches and it is likely that this is due to the release of corpus cardiacum hyperglycaemic factor since stress is known to cause the release of neurosecretion from the storage lobes of the corpus cardiacum and to increase the level of active phosphorylase in the fat body of locusts (Hill, 1972).

The role of hyperglycaemic factors from the corpora cardiaca of locusts has been reviewed by Hill (1972). Two chromatographically distinct carbohydrate mobilizing factors are present in locust corpora cardiaca, a potent hyperglycaemic factor being present in the glandular lobes and a less potent factor in the storage lobes (Goldsworthy, 1969; Mordue and Goldsworthy, 1969). Although both of the locust hyperglycaemic factors increase the level of active phosphorylase in the fat body of *Locusta migratoria*, the glandular lobes exhibit more activity than the neurosecretory lobes (Goldsworthy, 1970). Flying is known to cause release of neurosecretion from the storage lobes of the corpus cardiacum (Highnam and Haskell, 1964). The hyperglycaemic factors are only effective in locusts containing reasonable amounts of fat body glycogen (Goldsworthy, 1969) and in the field a build up of such energy stores probably takes place in preparation for migration. Highnam and Goldsworthy (1972) have shown that the hyperglycaemic factor in the neurosecretory storage lobes originates from the cerebral neurosecretory cells. Since brain extracts do not have a hyperglycaemic action, it has been suggested that modification of neurosecretion occurs at the axon terminals or during passage down the axons. The relationship between the two hyperglycaemic factors in the lobes of the locust corpora cardiaca is not clear although it is possible that the neurosecretory factor has a tropic effect on the glandular lobe (Hill, 1972).

It will be interesting to see if the hormones of the corpora cardiaca exert their influence on fat body glycogen breakdown in a manner similar to the effect of adrenaline on muscle glycogen in mammals, i.e. by stimulating adenyl cyclase activity leading to production of cyclic $3', 5'-$AMP, which can activate phosphorylase kinase thus leading to the conversion of inactive phosphorylase to the active form. The action of insect hormones on adenyl cyclase is clearly a very important area of future study. The only measurements of adenyl cyclase activity so far reported are for whole fly homogenates (Sutherland *et al.*, 1962; Castillon *et al.*, 1973) and moth pupal wing epidermis (Applebaum and Gilbert, 1972) with no values yet recorded for the fat body enzyme. It is also possible that corpus cardiacum adipokinetic hormone (see pp. 143–145) acts upon fat body lipases via cyclic $3', 5'-$AMP as a second messenger in the same way that mammalian hormones act upon adipose tissue lipases. The purification and characterization of the various corpus cardiacum hormones and reports of studies of the effect of purified hormones on enzymes of individual tissues are awaited with interest.

The effect of integumentary injury on fat body glycogen mobilization has been reviewed by Wyatt (1967) and will not be discussed further here. Juvenile hormone has also been implicated in glycogen breakdown. Liu (1973) has shown that allectectomy of the female housefly leads to a decrease in fat body phosphorylase activity which might suggest that juvenile hormone stimulates glycogen breakdown. In contrast, Wright and Rushing (1973) showed that application of a juvenile hormone analogue to pupae of the stable fly *Stomoxys calcitrans* decreased the utilization of glycogen by the pupae. However Wright *et al.* (1973) have shown that the effect is not due to a direct action of juvenile hormone on phosphorylase activity.

THE SUPPLY OF AMINO ACID FUELS FOR FLIGHT

Although carbohydrates and lipids are the major fuels for flight it has been suggested that a role for proline as an additional energy reserve may be widespread in insects (Gilmour, 1965). High concentrations of proline are found in the flight muscles of a number of insects such as the housefly (Price, 1961), the locust (Kirsten *et al.*, 1963), the tsetse fly (Bursell, 1963), the blowfly (Sacktor and Wormser-Shavit, 1966) and the Colorado beetle (De Kort *et al.*, 1973) and the concentration of proline in the flight muscles of the tsetse fly (Bursell, 1963, 1966) the locust (Kirsten *et al.*, 1963) and the blowfly (Sacktor and Wormser-Shavit, 1966) decreases during flight. Furthermore flight muscle mitochondria of the tsetse fly (Bursell, 1963, 1966), the locust (Brosemer and Veerabhadrappa, 1965), the blowfly (Sacktor and Childress, 1967) and the Colorado beetle (De Kort *et al.*, 1973) readily oxidize proline. The activity of proline dehydrogenase, the initial enzyme in proline oxidation, has been measured in the flight muscles of a number of insects and particularly high activities are found in the cockchafer, *Melolantha melolantha*, and the tsetse fly, *Glossina austeni* (Crabtree and Newsholme, 1970). The pathway of proline oxidation in flight muscle has been elucidated by Bursell (1967). The amino acid glutamate is also readily oxidized by flight muscle mitochondria of locusts (Van den Bergh, 1964), moths (Stevenson, 1968), cockroaches (Cochran, 1963) and the periodical Cicida *Magicada septendecim* (Hansford, 1971).

The source of proline and glutamate and other amino acids for oxidation by the flight muscles is the haemolymph. The concentration of free amino acids in insect haemolymph is usually very high and correspondingly high and often higher concentrations are found in the tissues (see reviews by Florkin and Jeuniaux, 1964; Sacktor, 1965; Chen, 1966). Winteringham (1959) suggested that in the adult insect, haemolymph amino acids may provide a soluble and readily available substrate reserve for the tricarboxylic acid cycle since many amino acids are readily degraded to acetyl CoA, pyruvate or tricarboxylic acid cycle intermediates (Chefurka, 1965b). Proline is the major amino acid in the haemolymph of insects such as the housefly, *Musca domestica* (Price, 1961) and the concentration of proline in locust haemolymph decreases during the first two hours of flight but then returns to its resting level (Mayer and Candy, 1969a).

Haemolymph amino acids are probably derived from three sources, firstly from the gut during digestion and absorption of dietary proteins, secondly from the breakdown of proteins in the fat body and other tissues and finally by synthesis in the fat body. Clearly an insect receiving a high protein diet such as the tsetse fly or the mosquito will obtain a plentiful supply of amino acids from the diet and these may become major sources of fuels for flight (Clements, 1955; Bursell, 1963, 1966). However the physiological importance of a pathway providing energy from amino acids such as proline in other insects is not known. It is possible that the dietary amino acids are converted into carbohydrate or lipid before being used as a source of flight energy (see e.g. Van Handel, 1965). There are few reports in the literature of studies on protein degradation in tissues such as the fat body although Maynard-Smith *et al.* (1970) have measured the rate of turnover of proteins in tissues of adult *Drosophila melanogaster* and have shown that 20 percent of the body protein is replaced with a half life of approximately 10 days but the other 80 per cent of the proteins are not replaced. However, there is no indication in the literature of a specifically controlled breakdown of tissue proteins to provide amino acids as a respiratory fuel for flight muscles.

Neither dietary amino acids nor protein degradation would lead specifically to large amounts of a particular amino acid such as proline in the flight muscles or haemolymph and if this amino acid is truly a major source of flight energy in some insects then it is either specifically removed by the flight muscles from the general amino acid pool or alternatively the insects specifically synthesize proline for use as a fuel for flight. The biosynthesis of proline in mammalian and bacterial systems (Shemin and Rittenburg, 1945; Vogel and Davis, 1952) proceeds from glutamate which, of course, is formed from α-oxoglutarate by the action of glutamic dehydrogenase,

(1) α-oxoglutarate + NH_3 + NADH + H^+ $\longrightarrow$

$$\text{glutamate} + NAD^+$$

(2) glutamate + NADH + H^+ + ATP $\longrightarrow$

$$\text{glutamic-}\gamma\text{-semialdehyde} + NAD^+ + ADP + P_i$$

(3) glutamic-γ-semialdehyde + NADPH + H^+ $\longrightarrow$

$$\text{proline} + NADP^+$$

The conversion of [^{14}C]acetate *in vivo* to proline, alanine, aspartate, glutamate and glutamine was demonstrated by Winteringham and Harrison (1956) using *Musca domestica*, and the conversion of [^{14}C]glucose to amino acids *in vivo* was demonstrated by Kasting and McGinnis (1958) using *Phormia regina*. The time course of incorporation of [^{14}C]acetate into amino acids in the housefly *in vivo* indicates that proline is formed from glutamate (Price, 1961). The site of amino acid formation in insects is likely to be the fat body. Clements (1959) showed that fat body of the desert locust would readily incorporate radioactivity from [^{14}C]acetate into aspartate, glutamate and proline. It appears likely then that the pathway of biosynthesis of proline is similar in insects to other forms of life. However McEnroe and Forgash (1958) showed that radioactivity from [^{14}C]formate was readily incorporated into proline by fat body from *Periplaneta americana* but little labelled glutamate was formed. It is likely that the long period of incubation (5 h) allowed time for all the glutamate formed to be further metabolized to proline and other products. Although insects can clearly synthesize proline, there are no indications in the literature of a specific biosynthesis of proline in the fat body for transport to the flight muscles to be used as a fuel for flight. If such a specific flight-oriented amino acid synthesis does occur in fat body, it will be interesting to know if it is controlled by factors from the corpus cardiacum such as those discussed earlier that control the mobilization of fat body lipid and carbohydrate reserves.

CONCLUSION

Clearly from the literature quoted in this review we now have quite a good understanding of the origins, synthesis, storage and transport of fuels for flight in insects and of the control of fuel supply to the flight muscles. However there are still large gaps in our knowledge, particularly with regard to the supply of amino acids as a source of energy for flight. It is hoped that if this review achieves little else it will help to highlight deficiencies in our knowledge and help stimulate the work required to give us a more complete understanding of the subject.

REFERENCES

Agosin, M., Scaramelli, N., Dinamarca, M. L. and Aravera, L. (1963) *Comp. Biochem. Physiol.*, **8**, 311–320.

Agosin, M., Fine, B. C., Scaramelli, I. N., Ilivicky, J. and Aravera, L. (1966) *Comp. Biochem. Physiol.*, **19**, 339–349.

Applebaum, S. W. and Gilbert, L. I. (1972) *Dev. Biol.*, **27**, 165–173.

Applebaum, S. W. and Schlesinger, H. M. (1973) *Biochem. J.*, **135**, 37–41.

Atkins, M. D. (1966) *Can. Ent.*, **98**, 953–991.

Atkins, M. D. (1969) *Can. Ent.*, **101**, 164–165.

Auclair, J. L. (1963) *Ann. Rev. Ent.*, **8**, 439–490.

Bade, M. L. (1964) *J. Insect Physiol.*, **10**, 333–341.

Bailey, E. and Horne, J. A. (1972a) *Comp. Biochem. Physiol.*, **42**, 659–667.

Bailey, E. and Horne, J. A. (1972b) *Biochem J.*, **128**, 108–109 P.

Bailey, E. and Horne, J. A., Izatt, M. E. G. and Hill, L. (1971) *Life Sci.*, **10**, 1415–1419.

Bailey, E., Horne, J. A., Izatt, M. E. G. and Hill, L. (1972) *Biochem. J.*, **128**, 79P.

Bailey, E. and Lockwood, E. A. (1973) *Enzyme*, **15**, 239–253.

Bailey, E., Taylor, C. B. and Bartley, W. (1968) *Nature* (London), **217**, 471–472.

Bailey, E. and Walker, P. R. (1969) *Biochem. J.*, **111**, 359–364.

Ballard, F. J. and Hanson, R. W. (1967) *J. Lipid. Res.*, **8**, 73–79.

Barlow, J. S. (1966) *Can. J. Zool.*, **44**, 775–779.

Barlow, J. S. and House, H. L. (1960) *J. Insect Physiol.*, **5**, 181–189.

Barroso, C., Municio, A. M. and Ribera, A. (1969) *Comp. Biochem. Physiol.*, **28**, 239–244.

Bartley, W., Dean, B., Taylor, C. B. and Bailey, E. (1967) *Biochem. J.*, **103**, 550–555.

Beall, G. (1948) *Ecology*, **29**, 80–94.

Beenakkers, A. M. T. (1963a) *Naturwissenshaften*, **50**, 361.

Beenakkers, A. M. T. (1963b) *Biochem. Z.*, **337**, 436–439.

Beenakkers, A. M. T. (1965) *J. Insect Physiol.*, **11**, 879–888.

Beenakkers, A. M. T. (1969a) *J. Insect Physiol.*, **15**, 353–361.

Beenakkers, A. M. T. (1969b) *Gen. Comp. Endocrinol.*, **13**, 492.

Beenakkers, A. M. T. and Gilbert, L. I. (1968) *J. Insect Physiol.*, **14**, 481–494.

Beenakkers, A. M. T. and Scheres, J. M. J. C. (1971) *Insect Biochem.*, **1**, 125–129.

Bellamy, D. (1958) *Biochem. J.*, **70**, 580–589.

Beutler, R. (1936) *Z. Vergleich. Physiol.*, **24**, 71–115.

Bhakthan, N. M. G. and Gilbert, L. I. (1968) *Gen. Comp. Endocrinol.*, **11**, 186–197.

Bodenstein, D. (1953) *J. exp. Zool.*, **124**, 105–116.

Bortz, W., Abrahams, S. and Chaikoff, I. L. (1963) *J. biol. Chem.*, **238**, 1266–1272.

Bortz, W. and Lynen, F. (1963) *Biochem. Z.*, **337**, 505–509.

Bowers, W. S. and Friedman, S. (1963) *Nature* (London), **198**, 685.

Brak, J. A. W., Vonk, H. J. and Daniels, F. J. A. (1966) *Archs. int. Physiol. Biochim.*, 74, 821–829.

Brice, E. G. and Okey, R. (1956) *J. biol. Chem.*, 218, 107–114.

Bricteux-Gregoire, S., Jeuniaux, Ch. and Florkin, M. (1964) *Archs. int. Physiol. Biochim.*, 72, 482–488.

Bricteux-Gregoire, S., Jeuniaux, Ch. and Florkin, M. (1965) *Comp. Biochem. Physiol.*, 16, 333–340.

Bridges, R. G. (1971) *J. Insect. Physiol.*, 17, 881–895.

Bronk, J. R. (1966) *Science*, 153, 638–639.

Brosemer, R. W. and Veerabhadrappa, P. S. (1965) *Biochim. biophys. Acta.*, 110, 102–112.

Bücher, T. and Klingenberg, M. (1958) *Angew Chem.*, 70, 552–570.

Bursell, E. (1963) *J. Insect. Physiol.*, 9, 439–452.

Bursell, E. (1966) *Comp. Biochem. Physiol.*, 19, 809–818.

Bursell, E. (1967) *Comp. Biochem. Physiol.*, 23, 825–829.

Bursell, E. (1970) *An Introduction to Insect Physiology*, Academic Press, London and New York.

Cabib, E. and Leloir, L. F. (1958) *J. biol. Chem.*, 231, 259–275.

Caldwell, R. L. and Rankin, M. A. (1972) *Gen. Comp. Endocrinol.*, 19, 601–605.

Candy, D. J. (1967) *Biochem. J.*, 103, 666–671.

Candy, D. J. (1970) *J. Insect. Physiol.*, 16, 531–543.

Candy, D. J. and Kilby, B. A. (1959) *Nature* (London), 183, 1594–1595.

Candy, D. J. and Kilby, B. A. (1961) *Biochem. J.*, 78, 531–536.

Castillon, M. P., Catalan, R. E. and Municio, A. M. (1973) *Febs. Letters*, 32, 113–115.

Chadwick, A. (1947) *Biol. Bull., Woods Hole*, 93, 229–239.

Chang, F. and Friedman, S. (1971) *Insect Biochem.*, 1, 63–80.

Chappell, J. B. (1968) In: *Biological Structure and Function.* (Goodwin, T. W. and Lindberg, O. eds.), 2, p. 71 Academic Press, London.

Chen, P. S. (1966) In: *Advances in Insect Physiology.* (Beament, J. W. L., Treherne, J. E. and Wigglesworth, V. B. eds.) Vol. 3, pp. 53–132 Academic Press, New York and London.

Chefurka, W. (1954) *Enzymologia*, 17, 73–89.

Chefurka, W. (1965a) In: *Physiology of Insecta.* (Rockstein, M. ed.) Vol. 2, pp. 581–667 Academic Press, New York and London.

Chefurka, W. (1965b) In: *Physiology of Insecta.* (Rockstein, M. ed.) Vol. 2, pp. 669–768 Academic Press, New York and London.

Chefurka, W. (1966) *Proc. entomol. Soc. Ont.*, 96, 17–23.

Chefurka, W., Horie, Y. and Robinson, J. R. (1970) *Comp. Biochem. Physiol.*, 37, 143–165.

Chino, H. and Gilbert, L. I. (1964) *Science*, 143, 359–361.

Chino, H. and Gilbert, L. I. (1965a) *Biochim. biophys. Acta*, 98, 94–110.

Chino, H. and Gilbert, L. I. (1965b) *J. Insect. Physiol.*, 11, 287–295.

Chino, H., Sudo, A. and Harashima, K. (1967) *Biochim. biophys. Acta*, 144, 177–179.

Chippendale, G. M. (1971) *Insect Biochem.*, 1, 39–46.

Chippendale, G. M. (1973) *Insect Biochem.*, 3, 1–10.

Clarke, K. U. and Baldwin, R. W. (1960) *J. Insect. Physiol.*, 5, 37–46,

Clegg, J. S. and Evans, D. R. (1961) *J. Expt. Biol.*, **38**, 771–792.

Clements, A. N. (1955) *J. Expt. Biol.*, **32**, 547–554.

Clements, A. N. (1959) *J. Expt. Biol.*, **36**, 665–675.

Cmelik, S. H. W. (1969) *J. Insect Physiol.*, **16**, 851–864.

Cochran, D. G. (1963) *Biochim. biophys. Acta*, **78**, 393–403.

Cockbain, A. J. (1961) *J. Exp. Biol.*, **38**, 163–174.

Cook, B. J. and Eddington, L. C. (1967) *J. Insect. Physiol.*, **13**, 1361–1372.

Crabtree, B., Higgins, S. J. and Newsholme, E. A. (1972) *Biochem. J.*, **130**, 391–396.

Crabtree, B. and Newsholme, E. A. (1970) *Biochem. J.*, **117**, 1019–1021.

Crabtree, B. and Newsholme, E. A. (1972a) *Biochem. J.*, **126**, 49–58.

Crabtree, B. and Newsholme, E. A. (1972b) *Biochem. J.*, **130**, 697–705.

Crompton, M. and Birt, L. M. (1967) *J. Insect. Physiol.*, **13**, 1575–1592.

Dadd, R. H. (1964) *J. Insect. Physiol.*, **10**, 161–178.

Dadd, R. H. (1966) *J. Insect. Physiol.*, **12**, 1479–1492.

Dadd, R. H. (1973) *Ann. Rev. Ent.*, **18**, 381–420.

Dahlman, D. L. (1971) *J. Insect. Physiol.*, **17**, 1677–1687.

Daikuhara, Y., Tsunemi, T. and Takeda, V. (1968) *Biochim. biophys. Acta.*, **158**, 51–61.

Davis, R. A. and Fraenkel, G. (1940) *J. Expt. Biol.*, **17**, 402–407.

D'Costa, M. A. and Birt, L. M. (1966) *J. Insect Physiol.*, **12**, 1377–1394.

D'Costa, M. A. and Birt, L. M. (1969) *J. Insect. Physiol.*, **15**, 1629–1645.

D'Costa, M. A. and Rice, M. T. (1973) *Comp. Biochem. Physiol.*, **45B**, 483–485.

D'Costa, M. A., Rice, M. T. and Latif, A. (1973) *J. Insect. Physiol.*, **19**, 427–433.

De Kort, C. A. D., Bartelink, A. K. M. and Scuurmans, R. R. (1973) *Insect. Biochem.*, **3**, 11–17.

Demyanovsky, S. Ya and Zubova, U. A. (1957) *Biokhimiya*, **21**, 698–704.

Doane, W. W. (1961) *J. exp. Zool.*, **146**, 275–298.

Downer, R. G H. (1972) *Can. J. Zool.*, **50**, 63–65.

Downer, R. G. H. and Steele, J. E. (1969) *Proc. Entomol. Soc. Ont.* **100**, 113–116.

Downer, R. G. H. and Steele, J. E. (1972) *Gen. Comp. Endocrinol.*, **19**, 259–265.

Downer, R. G. H. and Steele, J.,E. (1973) *J. Insect. Physiol.*, **19**, 523–532.

Dutkowski, A. B. (1973) *J. Insect Physiol.*, **19**, 1721–1726.

Dutkowski, A. B. and Sarzala-Drabikowska, M. G. (1973) *J. Insect. Physiol.*, **19**, 1341–1350.

Dutkowski, A. B. and Ziajka, B. (1972) *J. Insect. Physiol.*, **18**, 1351–1367.

Dutrieu, J. (1961) *C.r. Lebd. Searc. Acad. Sci. Paris*, **252**, 347–349.

Duve, H. (1972) *Insect Biochem.*, **2**, 445–450.

Egorova, T. A. (1963) *Nauch. Dokl. vyssh, Shk. Biol. Nauki.*, No. 1, 88–91.

Ehrhardt, P. (1962) *Z. vergl. Physiol.*, **46**, 169–211.

Eisner, T. (1955) *J. Exp. Zool.*, **130**, 159–181.

Evans, D. R. and Dethier, U. G. (1957) *J. Insect. Physiol.*, **1**, 3–17.

Fast, P. G. (1964) *Mem. ent. Soc. Can.*, No. 37, 1–50.

Fast, P. G. (1966) *Lipids*, **1**, 209–215.

Fast, P. G. (1970) In: *Progress in the Chemistry of Fats and other Lipids.* (Holman, R. T. ed.), vol. **XI** part 2 pp. 181–244, Pergamon Press, Oxford.

Fernandez-Sousa, J. M., Municio, A. M., Ribera, A. (1971) *Biochim. biophys. Acta.*, **248**, 226–233.

Fischer, E. H., Heilmeyer, L. M. G. and Haschke, R. H. (1971) In: *Current topics in cellular regulation.* (Horecker, B. L. and Stadtman, E. R. eds.), vol. 4 pp. 211–251, Academic Press, New York and London.

Florkin, M. and Jeuniaux, Ch. (1964) In: *Physiology of Insecta.* (Rockstein, M. ed.), vol. 3, pp. 109–152, Academic Press, New York and London.

Fraenkel, G. and Blewett M. (1946) *J. Exp. Biol.*, **22**, 172–190.

Friedman, S. (1960) *Arch. Biochem. Biophys.*, **76**, 532–545.

Friedman, S. (1967) *J. Insect. Physiol.*, **13**, 397–405.

Friedman, S. (1968) *Science N.Y.*, **159**, 110–111.

Friedman, S. (1971) *J. biol. Chem.*, **246**, 4122–4130.

Fulton, R. A. and Romney, U. E. (1940) *J. Agric. Res.*, **161**, 737–743.

Gelperin, A. (1971) *Ann. Rev. Ent.*, **16**, 365–378.

Gilbert, L. I. (1964) In: *Physiology of Insecta* (Rockstein, M. ed.), vol. 1 pp. 149–225, Academic Press, New York and London.

Gilbert, L. I. (1967a) In: *Advances in Insect Physiology* (Beament, J. W. L., Treherne, J. E. and Wigglesworth, V. B. eds.), vol. 4, pp. 69–211, Academic Press, New York and London.

Gilbert, L. I. (1967b) *Comp. Biochem. Physiol.*, **21**, 237–257.

Gilbert, L. I., Chino, H. and Domroese, K. A. (1965) *J. Insect. Physiol.*, **11**, 1057–1070.

Gilbert, L. I. and Schneiderman, H. A. (1961) *Gen. Comp. Endocrinol.*, **1**, 453–472.

Gilby, A. R. (1965) *Ann. Rev. Ent.*, **10**, 141–160.

Gilmour, D. (1965) *The metabolism of insects*, pp. 16–17, Oliver and Boyd, Edinburgh.

Giral, F. (1946) *J. biol. Chem.*, **162**, 61–63.

Goldin, H. H. and Keith, A. D. (1968) *J. Insect. Physiol.*, **14**, 887–899.

Goldsworthy, G. J. (1969) *J. Insect. Physiol.*, **15**, 2131–2140.

Goldsworthy, G. J. (1970) *Gen. comp. Endocrinol.*, **14**, 78–85.

Goldsworthy, G. J. (1971) *J. Endocrinol.*, **50**, 237–240.

Goldsworthy, G. J., Johnson, R. A. and Mordue, W. (1972) *J. comp. Physiol.*, **79**, 85–96.

Goldsworthy, G. J., Mordue, W. and Guthkelch J. (1972) *Gen. comp. Endocrinol.*, **18**, 545–551.

Gul, B. and Dils, R. (1969) *Biochem. J.*, **111**, 263–271.

Hansen, O. (1964) *Biochem. J.*, **92**, 333–337.

Hansford, R. G. (1971) *Biochem. J.*, **121**, 771–780.

Hart, D. E. and Steele, J. E. (1973) *J. Insect. Physiol.*, **19**, 927–939.

Hess, R. and Pearse, A. G. E. (1961) *Enzymol. biol. clin.*, **1**, 15–33.

Highnam, K. C. and Goldsworthy, G. J. (1972) *Gen. Comp. Endocrinol.*, **18**, 83–88.

Highnam, K. C. and Haskell, P. T. (1964) *J. Insect. Physiol.*, **10**, 849–864.

Hill, L. (1972) *Gen. Comp. Endocrinol.*, Supp. 3, 174–183.

Hill, L. and Goldsworthy, G. J. (1968) *J. Insect. Physiol.*, **14**, 1085–1098.

Hill, L. and Goldsworthy, G. J. (1970) *Comp. Biochem. Physiol.*, **36**, 61–70.

Hill, L., Izatt, M. E. G., Horne, J. A. and Bailey, E. (1972) *J. Insect. Physiol.*, **18**, 1265–1285.

Hill, L., Luntz, A. J. and Steele, P. A. (1968) *J. Insect Physiol.*, **14**, 1–20.

Hill, R., Linazasoro, J. M., Chevallier, F. and Chaikoff, I. L. (1958) *J. biol. Chem.*, **226**, 497–509.

Hines, W. J. W. and Smith, M. J. H. (1963) *J. Insect Physiol.*, **9**, 463–468.

Hocking, B. (1953) *Trans. Roy. Entomol. Soc.* London, **104**, 223–229.

Hopf, H. S. (1940) *Biochem. J.*, **34**, 1396–1403.

Horie, Y. (1967) *J. Insect. Physiol.*, **13**, 1163–1175.

Horie, Y. (1968) *J. Insect. Physiol.*, **14**, 417–425.

Horie, Y., and Nakasone, S. (1971) *J. Insect Physiol.*, **17**, 1441–1450.

Horie, Y., Nakasone, S. and Ito, T. (1968) *J. Insect Physiol.*, **14**, 971–981.

Hudson, A. (1958) *J. Insect Physiol.*, **1**, 293–304.

Inokuchi, T., Horie, Y. and Ito, T. (1969) *Biochim. biophys. Res. Commun.*, **35**, 783–787.

Ito, T. and Horie, Y. (1959) *Arch. Biochem. Biophys.*, **80**, 174–186.

Janda, U. V. and Slama, K. (1965) *Zool. Jb. Physiol.*, **71**, 345–358.

Johnson, R. A. and Hill, L. (1972) *Gen. Comp. Endocrinol.*, **18**, 598.

Kasting, R. and McGinnis, A. J. (1958) *Nature* (London), **182**, 1380–1381.

Kato, T. and Lowry, O. H. (1973a) *J. Neurochem.*, **20**, 151–163.

Kato, T. and Lowry, O. H. (1973b) *J. biol. Chem.*, **248**, 2044–2048.

Keeley, L. L. (1970) *Life Sci.,*. **9**, 1003–1011.

Keeley, L. L. (1971) *J. Insect. Physiol.*, **17**, 1501–1515.

Keeley, L. L. (1972) *Arch. Biochem. Biophys.*, **153**, 8–15,

Keeley, L. L. and Friedman, S. (1967) *Gen. Comp. Endocrinol.*, **8**, 129–134.

Keeley, L. L. and Friedman, S. (1969) *J. Insect. Physiol.*, **15**, 509–518.

Keeley, L. L. and Waddill, V. H. (1971) *Life Sci.*, **10**, 737–745.

Keith, A. D. (1967) *Life Sci.*, **6**, 213–218.

Keith, A. D. (1967) *Comp. Biochem. Physiol.*, **21**, 587–600.

Kerly, M. and Leaback, D. H. (1957) *Biochem. J.*, **67**, 245–250.

Kerukize, G. R. (1973) *Comp. Biochem. Physiol.*, **43**, 563–569.

Kilby, B. A. (1963) In: *Advances in Insect Physiology.* (Beament, J. W. L., Treherne, J. E. and Wigglesworth, V. B. eds.) vol. 1, pp. 111–174. Academic Press, New York and London.

Kilby, B. A. (1965) In: *Aspects of Insect Biochemistry.* (Goodwin, T. W. ed.), pp. 39–48. Academic Press, New York and London.

Kirsten, E., Kirsten, R. and Arese, P. (1963) *Biochem. Z.*, **337**, 167–178.

Kon, R. T. and Monroe, R. E. (1971) *Ann. Entomol. Soc. An.*, **64**, 247–250.

Korn, E. D. (1955) *J. biol. Chem.*, **215**, 1–14.

Kornacker, M. S. and Ball, E. G. (1965) *Proc. Nat. Acad. Sci. U.S.A.*, **54**, 899–904.

Krebs, H. A. and Kornberg, H. L. (1957) *Energy Transformations in Living Matter.* Springer-Verlag, Berlin, Gottingen, Heidelberg.

Krogh, A. and Weis-Fogh, T. (1951) *J. Expt. Biol.*, **28**, 344–357.

Lamb, N. J. and Monroe, R. E. (1968) *Ann. Entomol. Soc. Am.*, **61**, 1167–1169.

Lambremont, E. N. (1965) *Comp. Biochem. Physiol.*, **14**, 419–424.

Lambremont, E. N. and Bennett, A. F. (1966) *Can. J. Biochem.*, **44**, 1597–1606.

Lambremont, E. N. and Stein, C. I. and Bennett, A. F. (1965) *Comp. Biochem. Physiol.*, **16**, 289–302.

Larner, J. and Villar-Palasi, C. (1971) In: *Current Topics in Cellular Regulation (Horecker, B. L. and Stadtman, E. R. eds.), vol.* 3, pp. 195–236, Academic Press, New York and London.

Lea, A. O. and Van Handel, E. (1970) *J. Insect Physiol.,* 16, 319–321.

Leader, J. P. and Bedford, J. J. (1972) *Comp. Biochem. Physiol.,* 43, 233–235.

Lech, J. J. and Calvert, D. N. (1968) *Can. J. Biochem.,* 46, 707–714.

Lenartowitz, E., Zaluska, H. and Niemerko, S. (1967) *Acta. biochim. Pol.,* 14, 267–275.

Levenbook, L. (1950) *Biochem. J.,* 47, 336–346.

Levenbook, L. (1951) *J. Expt. Biol.,* 28, 173–180.

Levenbook, L. (1961) *Arch. Biochem. Biophys.,* 92, 114–121.

Lindh, N. O. (1967) *Comp. Biochem. Physiol.,* 20, 209–216.

Lipke, H., Grainger, M. M. and Siakotos, A. N. (1965) *J. biol. Chem.,* 240, 594–600.

Lipke, H., Graves, B. and Leto, S. (1965) *J. Biol. Chem.,* 240, 601–608.

Liu, T. P. (1973) *Comp. Biochem. Physiol.,* 46, 109–113.

Lockwood, E. A. and Bailey, E. (1970), *Biochem. J.,* 120, 49–54.

Louloudes, S. J., Kaplanis, J. N., Robbins, W. E. and Munroe, R. E. (1961) *Ann. Entomol. Soc. Am.,* 54, 99–103.

Ludwig, D. and Wugmeister, M. (1953) *Physiol. Zool.,* 26, 254–259.

Lüscher, M. (1968) *J. Insect. Physiol.,* 14, 499–511.

Lüscher, M. and Leuthold, R. (1965) *Rev. suisse Zool.,* 72, 618–623.

Lynen, F. (1967) *Biochem. J.,* 102, 381–400.

Madariago, M. A., Municio, A. M. and Ribera, A. (1970a) *Comp. Biochem. Physiol.,* 35, 57–62.

Madariago, M. A., Municio, A. M. and Ribera, A. (1970b) *Comp. Biochem. Physiol.,* 36, 271–278.

Martin, J. S. (1969a) *J. Insect. Physiol.,* 15, 1025–1045.

Martin, J. S. (1969b) *J. Insect. Physiol.,* 15, 2319–2344.

Mathur, C. F. and Yurkiewicz, W. J. (1969), *J. Insect. Physiol.,* 15, 1567–1571.

Matthews, J. R. and Downer, R. G. H. (1973) *Can. J. Zool.,* 51, 395–397.

Maurizio, A. (1965) *J. Insect Physiol.,* 11, 745–763.

Mayer, R. J. and Candy, D. J. (1967) *Nature* (London), 215, 987.

Mayer, R. J. and Candy, D. J. (1969a) *Comp. Biochem. Physiol.,* 31, 409–418.

Mayer, R. J. and Candy, D. J. (1969b) *J. Insect. Physiol.,* 15, 611–620.

Maynard-Smith, J., Bozcuk, A. N. and Tebbutt, S. (1970) *J. Insect. Physiol.,* 16, 601–613.

McCarthy, R. and Ralph, C. L. (1962) *Am. Zool.,* 2, 429.

McDonald, I. (1966) *Adv. Lipid Res.,* 4, 39–47.

McEnroe, W. O. and Forgash, A. J. (1958) *Ann. Entomol. Soc. Am.,* 51, 126–129.

Mikolaschek, G. and Zebe, E. (1967) *J. Insect Physiol.,* 13, 1483–1485.

Minks, A. K. (1967) *Arch. neerl. Zool.,* 17, 175–258.

Miura, K., Vonk, H. K., Zandee, D. I. and Houx, N. W. H. (1965) *Arch. int. Physiol. Biochim.,* 73, 65–72.

Mochnacka, I. and Petraszyn, C. (1959) *Acta. biochim. pol.,* 6, 307–311.

Monod, J., Changeux, J. P. and Jacob, F. (1963) *J. molec. Biol.,* 6, 306–326.

Monod, J., Wyman, J. and Changeux, J. P. (1965) *J. molec. Biol.,* 12, 88–118.

Moore, R. F. and Taft, H. M. (1970) *Ann. Entomol. Soc. Am.,* 63, 1275–1279.

Mordue, W. and Goldsworthy, G. J. (1969) *Gen. Comp. Endocrinol.,* 12, 360–369.

Müller, H. P. and Engelmann, F. (1968) *Gen. Comp. Endocrinol.,* 11, 43–50.

Municio, A. M., Odriozola, J. M. and Pineiro, A. (1970) *Comp. Biochem. Physiol.,* 37, 387–395.

Municio, A. M., Odriozola, J. M., Pineiro, A. and Ribera, A. (1971) *Biochim. biophys. Acta.,* 248, 212–225.

Municio, A. M., Odriozola, J. M., Pineiro, A. and Ribera, A. (1972a) *Biochim. biophys Acta.,* 280, 248–257.

Municio, A. M., Odriozola, J. M., Pinerio, A. and Ribera, A. (1973) *Insect Biochem.,* 3, 19–24.

Municio, A. M., Odriozola, J. M. and Ramos, J. A. (1972b) *Insect Biochem.,* 2, 353–360.

Murphy, T. A. and Wyatt, G. R. (1965) *J. biol. Chem.,* 240, 1500–1508.

Natalizi, G. M. and Frontali, N. (1966) *J. Insect Physiol.,* 12, 1279–1287.

Nayar, J. K. and Sauerman, D. M. (1971) *J. Insect Physiol.,* 17, 2221–2233.

Nayar, J. K. and Van Handel, E. (1971) *J. Insect Physiol.,* 17, 471–481.

Nelson, D. R. and Sukkestad, D. R. (1968) *J. Insect Physiol.,* 14, 293–300.

Nelson, D. R., Terranova, A. C. and Sukkestad, D. R. (1967) *Comp. Biochem. Physiol.,* 20, 907–917.

Newsholme, E. A. and Gevers, W. (1967) *Vitamins and Hormones,* 25, 1–87.

Niemierko, W. (1959) *Fourth Int. Congr. Biochem.,* 12, 185–197.

Niemierko, S., Wlodawer, P. and Wojtczak, A. F. (1956) *Acta. Biol. exp. Vars.,* 17, 255–276.

Nowosielski, J. W. and Patton, R. L. (1964) *J. Insect Physiol.,* 11, 263–270.

Odhiambo, T. R. (1966a) *J. Expt. Biol.,* 45, 45–50.

Odhiambo, T. R. (1966b) *J. Expt. Biol.,* 45, 51–64.

O'Meara, G. F. and Van Handel, E. (1971) *J. Insect Physiol.,* 17, 1411–1413.

Orr, C. W. M. (1964) *J. Insect Physiol.,* 10, 103–119.

Pant, N. C. and Gabrani, K. (1963) *Indian J. Entomol.,* 25, 110–115.

Pant, R. and Morris, I. D. (1969) *J. Biochem.,* 66, 29–31.

Peled, Y. and Tietz, A. (1973) *Biochim. biophys. Acta.,* 296, 499–509.

Polacek, I. and Kubista, V. (1960) *Physiol. Bohemoslav.,* 9, 228–236.

Popjak, G. and Tietz, A. (1954) *Biochem. J.,* 56, 46–54.

Porter, J. W. and Lang, R. W. (1958) *J. biol. Chem.,* 233, 20–25.

Porter, J. W., Wakil, S. J., Tietz, A., Jacob, M. J. and Gibson, D. M. (1957) *Biochim. biophys. Acta.,* 25, 35–41

Price, G. M. (1961) *Biochem. J.,* 80, 420–428.

Ralph, C. L. (1962) *Am. Zool.,* 2, 550.

Robbins, W. E., Kaplanis, J. E., Louloudes, S. J. and Monroe, R. E. (1960) *Ann. Entomol. Soc. Am.,* 53, 128–129.

Robinson, D. S. (1965) In: *Handbook of Physiology* (Renald, P. E. and Cahill, G. F. J. eds.), Section 5, pp. 295–299, Williams and Wilkins, Maryland.

Rosell-Perez, M. and Larner, J. (1964a) *Biochemistry,* 3, 773–778.

Rosell-Perez, M. and Larner, J. (1964b) *Biochemistry,* 3, 75–81.

Roussell, J. (1963) *J. Insect. Physiol.,* 9, 721–729.

Sacktor, B. (1955) *J. biophys. biochem. Cytol.,* 1, 29–46.

Sacktor, B. (1961) *Ann. Rev. Ent.,* 6, 103–130.

Sacktor, B. (1965) In: *Physiology of Insecta.* (Rockstein, M. ed.), vol. 2, pp. 483–580, Academic Press, New York and London.

Sacktor, B. (1970) In: *Advances in Insect Physiology,* (Beament, J. W. L.,

Treherne, J. E. and Wigglesworth, V. B. eds.), vol. 7, pp. 267–347, Academic Press, New York and London.

Sacktor, B. and Childress, C. C. (1967) *Arch. Biochem. Biophys.*, 120, 583–588.

Sacktor, B. and Wormser-Shavit, E. (1966) *J. biol. Chem.*, 241 624–631.

Sagesser, H. (1960) *J. Insect. Physiol.*, 5, 264–285.

Saito, S. (1963) *J. Insect Physiol.*, 9, 509–519.

Saunders, S. A., Gracy R. W., Schnackery, K. O. and Noltmann, E. A. (1969) *Science*, 164, 858–859.

Scoggin, J. K. and Tauber, O. E. (1950) *Iowa State Coll. J. Sci.*, 25, 99–124.

Scrutton, M. C. and Utter, M. F. (1968) *Ann. Rev. Biochem.*, 37, 249–302.

Sedee, P. D. S. W. (1961) *Arch. int. Physiol. Chem.*, 69, 295–309.

Shemin, D. and Rittenburg, D. (1945) *J. biol. Chem.*, 158, 71–76.

Shigematsu, H. (1956) *J. seric Sci.*. Tokyo, 25, 115–121.

Shigematsu, H. (1958) *Annotnes Zool. Jap.*, 31, 6–12.

Shyamala, M. R. and Bhat, J. V. (1965) *Indian J. Biochem.*, 2, 101–104.

Siakotos, A. N. (1960a) *J. gen. Physiol.*, 43, 999–1013.

Siakotos, A. N. (1960b) *J. gen. Physiol.*, 43, 1015–1030.

Silva, G. M., Doyle, W. P. and Wang, C. H. (1958) *Nature* (London), 182, 102–104.

Slama, K. (1964) *J. Insect. Physiol.*, 10, 283–303.

Sols, A., Cadenas, E. and Alvarado, F. (1960) *Science*, 131, 297–298.

Sridhara, S. and Bhat, J. V. (1964) *Biochem J.*, 91, 120–123.

Sridhara, S. and Bhat, J. V. (1965a) *Biochem. J.*, 94, 700–704.

Sridhara, S.and Bhat, J. V. (1965b) *Life Sci.*, 4, 979–982.

Sridhara, S. and Bhat, J. V. (1965c) *J. Insect Physiol.*, 11, 449–462.

Sridhara, S., Rao, U. R. and Bhat, J. V. (1966) *Biochem. J.*, 98, 260–265.

Stafford, W. L. (1973) *Comp. Biochem. Physiol.*, 45, 763–768.

Steele, J. E. (1961) *Nature* (London), 192, 680–681.

Steele, J. E. (1963) *Gen. Comp. Endocrinol.*, 3, 46–52.

Steele, J. E. (1964) *Am. Zool.*, 4, 328.

Stephen, W. F. and Gilbert, L. I. (1969) *J. Insect. Physiol.*, 15, 1833–1854.

Stephen, W. F. and Gilbert, L. I. (1970) *J. Insect. Physiol.*, 16, 851–864.

Stevenson, E. (1968) *J. Insect Physiol.*, 14, 179–198.

Stevenson, E. (1969) *J. Insect Physiol.*, 15, 1537–1550.

Stevenson, E. (1972) *J. Insect Physiol.*, 18, 1751–1756.

Stevenson, E. and Wyatt, G. R. (1964) *Arch. Biochem. Biophys.*, 108, 420–429.

Strong, F. E. (1963) *Science*, 140, 983–984.

Strong, L. (1968a) *J. Exp. Biol.*, 48, 624–634.

Strong, L. (1968b) *J. Insect Physiol.*, 14, 1685–1691.

Sutherland, E. W., Rall, T. W. and Menon, T. (1962) *J. biol. Chem.*, 237, 1220–1227.

Takaya, T. and Miura, K. (1968) *Arch. int. Physiol. Biochim.*, 76, 603–614.

Takehara, I. (1962) *Low Temp. Sci. Ser. B.*, 20, 36–43.

Tata, J. R., Ernster, L., Lindberg, D., Arrhenius, E., Pedersen, S. and Hedman, R. (1963) *Biochem. J.*, 86, 408–428.

Tate, L. G. and Wimer, L. T. (1971) *Insect Biochem.*, 1, 199–206.

Taylor, M. W. and Medici, J. C. (1966) *J. Nutr.*, 88, 176–180.

Terriere, L. C. and Grau P. A (1972) *J. Insect Physiol.*, 18, 633–647.

Thomas, K. K. (1972) *Insect Biochem.*, 2, 107–118.

Thomas, K. K. and Gilbert, L. I. (1967) *J. Insect Physiol.*, 13, 963–980.

Thomas, K. K. and Gilbert, L. I. (1968) *Arch. Biochem. Biophys.*, 127 512–521.

Thomas, K. K. and Gilbert, L. I. (1969) *Physiol. Chem. Phys.*, 1, 293–311.

Thompson, S. N. and Bennett, R. B. (1971) *J. Insect Physiol.*, 17, 1555–1563.

Thomsen, E. (1949) *J. Exp. Biol.*, 26, 137–144.

Thomsen, E. (1952) *J. Exp. Biol.*, 29, 137–172.

Tietz, A. (1961) *J. Lipid Res.*, 2, 182–187.

Tietz, A. (1962) *J. Lipid Res.*, 3, 421–426.

Tietz, A. (1965) In: *Handbook of Physiology*. (Renald, A. E. and Cahill, G. F. eds.), Section 5, pp. 45–54, Williams and Wilkins, Baltimore.

Tietz, A. (1967a) *Eur. J. Biochem.*, 2, 236–242.

Tietz, A. (1967b) *Israel J. Chem.*, 5, 135p.

Tietz, A. and Stern, N. (1969) *Febs Letters.* 2, 286–288.

Tietz-Devir, A. (1963) *Fifth Int. Congr. Biochem.*, 7, 85–89.

Treherne, J. E. (1958a) *J. Exp. Biol.*, 35, 862–870.

Treherne, J. E. (1958b) *J. Exp. Biol.*, 35, 611–625.

Trivelloni, J. C. (1960) *Archs. Biochem. Biophys.*, 89, 149–150.

Vagelos, P. R. (1971) In: *Current Topics in Cellular Regulation*. (Horecker, B. L. and Stadtman, E. R. eds.), pp. 119–116, Academic Press, New York and London.

Van Den Bergh, S. G. (1964) In: *Methods of Enzymology*. (Estabrook, R. W. ed.), vol. 10, pp. 117–122. Academic Press, New York and London.

Van Handel, E. (1965), *J. Physiol.*, 181, 478–486.

Van Handel, E. (1969a) *Comp. Biochem. Physiol.*, 29, 1023–1030.

Van Handel, E. (1969b) *Comp. Biochem. Physiol.*, 29, 413–421.

Van Handel, E. (1971) *Comp. Biochem. Physiol.*, 38, 141–145.

Van Handel, E. and Lea, A. O. (1965) *Science*, 149, 298–300.

Van Handel, E. and Lea, A. O. (1970) *Gen. Comp. Endocrinol.*, 14, 381–384.

Van Handel, E. and Lum, P. T. M. (1961) *Science*, 134, 1979–1980.

Van Handel, E. and Nayar, J. K. (1972a) *Insect Biochem.*, 2, 203–208.

Van Handel, E. and Nayar, J. K. (1972b) *Insect Biochem.*, 2, 8–12.

Vardanis, A. (1963) Biochim. biophys. Acta., 73, 565–573.

Vardanis, A. (1967) *J. biol. Chem.*, 242, 2306–2311.

Von Frisch, V. K. (1927) *Naturwissenschaften*, 15, 321–327.

Vogell, H. J. and Davies, B. D. (1952) *J. Am. Chem. Soc.*, 74, 109–112.

Vroman, H. E., Kaplanis, J. N. and Robbins, W. E. (1965) *J. Insect. Physiol.*, 11, 897–903.

Wakil, S. J. (1961) *J. Lipid Res.*, 2, 1–24.

Wakil, S. J. (1963) *Fifth Int. Cong. Biochem.*, 7, 3–43.

Walker, P. R. and Bailey, E. (1969a) *Biochem. J.*, 115, 50P.

Walker, P. R. and Bailey, E. (1969b) *Biochim. biophys. Acta.*, 187, 591–593.

Walker, P. R. and Bailey, E. (1969c) *Biochem. J.*, 111, 365–369.

Walker, P. R. and Bailey, E. (1970a) *J. Insect Physiol.*, 16, 499–509.

Walker, P. R. and Bailey, E. (1970b) *J. Insect Physiol.*, 16, 679–690.

Walker, P. R. and Bailey, E. (1970c) *Comp. Biochem. Physiol.*, 36, 623–626.

Walker, P. R. and Bailey, E. (1971a) *J. Insect Physiol.*, 17, 1125–1137.

Walker, P. R. and Bailey, E. (1971b) *J. Insect Physiol.*, 17, 813–821.

Walker, P. R. and Bailey, E. (1971c) *J. Insect Physiol.*, 17, 1359–1369.

Walker, P. R., Hill, L. and Bailey, E. (1970) *J. Insect Physiol.*, 16, 1001–1015.
Wang, C. M. and Patton, R. L. (1969) *J. Insect Physiol.*, 15, 851–860.
Weintraub, H. and Tietz, A. (1973) *Biochim. biophys. Acta.*, 306, 31–41.
Weis-Fogh, T. (1952) *Phil. Trans. R. Soc.B.*, 237, 1–36.
Weis-Fogh, T. (1952) *J. Exp. Biol.*, 41, 229–256.
Whitney, J. E. and Roberts, S. (1955) *Am. J. Physiol.*, 181, 446–450.
Wiens, A. W. and Gilbert, L. I. (1965), *Science*, 150, 614–616.
Wiens, A. W. and Gilbert, L. I. (1967a) *Comp. Biochem. Physiol.*, 21, 145–159.
Wiens, A. W. and Gilbert, L. I. (1967b) *J. Insect Physiol.*, 13, 779–794.
Wigglesworth, V. B. (1942) *J. Exp. Biol.*, 19, 56–77.
Wigglesworth, V. B. (1949) *J. Exp. Biol.*, 26, 150–163.
Williamson, D. H. and Hems, R. (1970) In: *Essays in Cell Metabolissm*, (Bartley, W., Kornberg, H. L. and Quayle, J. R. eds.) pp. 251–281, Wiley-Interscience. London.
Wimer, L. T. (1969) *Comp. Biochem. Physiol.*, 29, 1055–1062.
Winteringham, F. P. W. and Harrison, A. H. (1956), *Nature* (London), 171, p. 81.
Winteringham, F. P. W. (1959) *Fourth Int. Cong. Biochem.*, 12, 201–215.
Wlodawer, P. (1956) *Acta. Biol. exp. Vars.*, 17, 221–230.
Wlodawer, P. and Baranska, J. (1965) *Acta biochem. pol.*, 12, 39–47.
Wlodawer, P. and Lagwinska, E. (1967) *J. Insect Physiol.*, 13, 319–331.
Wlodawer, P., Lagwinska, E. and Baranska, J. (1966) *J. Insect Physiol.*, 12, 547–560.
Wright, J. E., Crookshank, H. R. and Rushing, D. D. (1973) *J. Insect Physiol.*, 19, 1575–1578.
Wright, J. E. and Oehler, D. D. (1971) *J. Insect Physiol.*, 17, 1479–1488.
Wright, J. E. and Rushing, D. D. (1973) *Ann. Entomol. Soc. Am.*, 66, 274–276.
Wyatt, G. R. (1961) *Ann. Rev. Ent.*, 6, 75–102.
Wyatt, G. R. (1967) In: *Advances in Insect Physiology.* (Beament, J. W. L., Treherne, J. E. and Wigglesworth, V. B. eds.), 4, pp. 287–360, Academic Press, New York and London.
Wykes, G. R. (1952) *New Phytol.*, 51, 210–215.
Yamashita, O. (1965) *J. seric Sci.* Tokyo., 34, 1–8.
Yanagawa, H. A. (1971) *Insect Biochem.*, 1, 102–112.
Young, R. C. (1959) *Ann. Entomol. Soc. Am.*, 52, 567–570.
Young, R. C. (1964a) *Ann. Entomol. Soc. Am.*, 57, 321–324.
Young, R. C. (1964b) *Ann. Entomol. Soc. Am.*, 57, 325–327.
Yurkiewicz, W. J. and Mathur, C. F. (1969) *J. Insect Physiol.*, 15, 439–444.
Zebe, E. (1954) *Verh. d. dtsch. Zool in Munster/Westf.*, 309–314.
Zebe, E. (1959) *Z. vergl. Physiol.*, 36, 290–317.
Zebe, E. and McShan, W. H. (1959) *Biochim. biophys. Acta.*, 31, 513–518.

3 Excretion in Insects

D. G. COCHRAN

INTRODUCTION

Certain chemical substances are produced in excessive quantities in animal tissues as a consequence of life processes, and must be eliminated from active metabolic sites before they cause problems for the organisms involved. The concentration at which these substances may become obtrusive varies widely depending upon the chemical nature and state of the specific substance. For example, ammonia is quite toxic at low concentrations and must normally be disposed of rapidly. On the other hand, uric acid or urates can apparently build up to quite high concentrations internally and still not cause death or major metabolic derangement. Among animals, elaborate systems have been developed to accomplish the voiding of metabolic wastes. They are usually referred to as excretory systems, although it should be remembered that the respiratory system often performs an essentially equivalent function in the elimination of carbon dioxide. While the latter is obviously an important function, it will not be considered in this review since it is usually treated as a part of respiratory exchanges (Bursell, 1970).

The Malpighian tubule-rectum complex is the most highly developed and widely distributed type of excretory system in insects although it is by no means the only one found (p. 258). It is a very efficient system and plays a vital role in osmoregulation as well as in excretion. Indeed, it is quite difficult to separate these two functions physiologically. Substances, such as ions and water, which may be excreted under one set of circumstances because they are present in excess, may be differentially reabsorbed under other conditions to help maintain homeostasis. The problem is further complicated by the fact that faecal wastes and excretory products produced by the Malpighian tubules are voided by a common route. In some instances this raises the question of accuracy in distinguishing between true excretory products and substances, sometimes similar in nature, which have merely passed through the gut unabsorbed.

Other problems arise out of considerations of a suitable definition for excretion. Maddrell (1971) has made a distinction between the processes of excretion and secretion. He considered excretion to be the process by which those substances that interfere with the normal functioning of the animal, either by their presence or their toxic action, are removed from the metabolic pool. Secretion, he defined as *removal of substances* from the metabolic pool to achieve some useful purpose outside the metabolic pool. These definitions are satisfactory when one considers such simple cases as the voiding of uric acid via the Malpighian tubule-rectum complex on the one hand, and the secretion of a sex or trail-laying pheromone on the other hand. Unfortunately, other processes occur which cannot be clearly classified under the above definitions. For example, there is the matter of the production and internal storage of uric acid, often referred to as storage excretion. If it can be shown that the storage is a true and permanent sequestration there is no particular problem. In this sense it matters little whether the material is voided or permanently stored. However, evidence is currently accumulating which indicates that in some insects the internal uric acid or urates are in a dynamic state, and can be mobilized particularly in times of dietary stress (p. 212). Under these conditions neither excretion nor secretion are adequate descriptive terms. Rather, the concept of a metabolic reserve seems more appropriate even though the compound involved is normally a waste product. There is the added difficulty that the term secretion is used by renal physiologists to mean substance elaboration by cells, such as those of the Malpighian tubules.

Another variation of this problem is exemplified by the presence of pteridine pigments in the wings of certain Lepidoptera. These compounds resemble uric acid in that they contain a significant amount of nitrogen and are highly oxidized. They could, therefore, be end products of metabolism, while at the same time serving as pigments producing colour in the wings. The compounds are synthesized by the insects and are irreversibly deposited in the wing scales. There is little doubt that they are effectively removed from the general metabolic pool, satisfying that portion of the above definition of excretion. However, their importance as end products of metabolism as opposed to their role in adaptive pigmentation has not been completely resolved (p. 246).

There is also the phenomenon of males of certain cockroach species which possess uricose glands. These glands are part of the male accessory gland apparatus which fill with uric acid shortly after maturity. At mating the male transfers a spermatophore to the female and covers it with uric acid from the glands. This process has been described as being excretory in nature. Again, it fits the definition of excretion given above in the sense that it is not confined to some preconceived organ or system. However, as discussed later (p. 261), there is a strong possibility that uric acid in this case serves a useful purpose in reproduction. At the same time there is little evidence to show that voiding of uric acid via this route is critical to the survival of individual males.

From these examples it is clear that a satisfactory definition of excretion is difficult to achieve except in general terms and in simple, straight-forward cases. Perhaps the best position to assume is one of flexibility. It must be recognized that dual or multiple functions can and do occur for certain processes, and that sometimes one or more of these functions may impinge upon excretion. It is also true that one should not automatically assume excretion to be involved in a primary way merely because a given process incorporates a known insectan excretory compound. Full evaluation of that phenomenon may reveal some unexpected results and render a simple explanation impossible.

In considering the biochemistry of these and other processes it is first necessary to identify the end products of insect metabolism. Subsequently, one may search for the relevant metabolic pathways and the sites where they occur. As a result of the efforts of many people over a long period of time, it is known that several major and many minor insect excretory products occur. From this it becomes immediately apparent that a variety of enzymes and tissues are likely to be involved in the production of such a diversity of compounds. Often more than one tissue will possess a given enzymatic capability. Sometimes all or parts of a synthetic process will occur in one tissue or another which is different from the organ of excretion. Thus, the haemolymph may be concerned with the transport of waste products from the site(s) of production to the site of excretion. Alternatively, the site of excretion may have the enzymatic capability to carry out the final step(s) in the production of end products, and also bring about their excretion at least into the gut. In other instances all that can be done is to describe the formation of the urine and indicate the sites and importance of active transport in the elimination of certain simple compounds and ions.

In the pages that follow an attempt will be made to discuss each of the important excretory products with emphasis on the biochemical aspects of the problem. At the same time, the physiological components of their excretion, including the organ or site of excretion, will be indicated where known. Interpretations of the processes discussed will incorporate the concept of recognizing multiple functions wherever the evidence so indicates.

EXCRETION ASSOCIATED WITH THE MALPIGHIAN TUBULE-RECTUM SYSTEM

The major portion of our current knowledge about normal excretion in insects falls under this heading. This is because the Malpighian tubule-rectum complex is their principal excretory system. In most insects this system is responsible for the voiding of excess ions, water, nitrogen-containing compounds, and other substances which must be eliminated (except CO_2). In some instances the body of relevant biochemical knowledge is quite large and complex. Hopefully, the corresponding sections of this review will reflect that breadth and complexity. Other sections will be extremely brief because of the paucity of facts concerning them, but without intending to indicate their ultimate importance. Excretory processes associated with other organs and tissues will be treated later in the review.

Ions, water and active transport

Among the more intensively studied excretory products of insects are certain inorganic ions and water. Interest in these substances stems not only from their involvement in the excretory process, but also because of their role in osmoregulation. Conceptually, osmoregulation is probably the broader of the two processes because it is concerned with the regulation of the internal environment, whereas excretion has the more limited task of ridding the system of useless or noxious substances. Nevertheless, the two processes are so intertwined that their separation is very difficult. Therefore the approach followed here will be to discuss only that information pertinent to excretion, and not to cover the entire topic of osmoregulation. A number of other references should be consulted for a fuller treatment of this topic (Shaw, 1955; Stobbart and Shaw, 1964; Treherne, 1965; Wigglesworth, 1972; Berridge, 1970a; Bursell, 1970; Maddrell, 1971; Stobbart, 1971; Riegel, 1972). It is apparent from the recent, exhaustive reviews (Maddrell, 1971; Riegel, 1971, 1972), that the bulk of information on excretory mechanisms is physiological in nature. Since this chapter is intended to stress the biochemical aspects of excretion, it will include just enough physiological data to clarify the processes of urine[1] formation by the Malpighian tubules, how it relates to the principal excretory products, and the role of rectal regulation in determining the final composition of the excreta.

[1] While the term urine will be used here for simplicity, it should be recognized that this fluid is only the primary filtrate. It is subjected to considerable modification as it passes through the hindgut. The final filtrate, as found in the rectum, is the fluid that should be referred to as urine *senso stricto.*

The Malpighian tubules of insects are extremely variable in number, length, shape, histology, and position in the body. In general, they should be considered as a series of tubules which lie in the body cavity, and have rather free access to the circulating haemolymph. The distal end of each tubule is closed. It may lie free in the body cavity (Wigglesworth, 1931), or be buried in the walls of the rectum forming a cryptonephridial system (Srivastava, 1962; Saini, 1964; Gouranton, 1968; Irvine, 1969; Phillips, 1970; Koefoed, 1971). Proximally, the tubules are attached to the gut in the region of the midgut-hindgut junction. They empty into the gut, but entry in relation to the pyloric valve is quite variable (Wigglesworth, 1972). The tubules may move in the body cavity in response to body movements or as a result of muscular contractions by the muscles which frequently girdle them in a helical manner (Crowder and Shankland, 1972). A tracheal tube accompanies the muscles, and presumably furnishes oxygen to them. In addition, the outer wall of the Malpighian tubules is abundantly supplied with tracheoles often originating with and binding them to other organs such as the intestine. The account of the Malpighian tubules given by Wigglesworth (1972) should be consulted for greater detail and pertinent references.

The process of urine formation by the Malpighian tubules has been studied intensively. Ramsay laid much of the groundwork for our current understanding of this process in a series of papers (1952, 1953, 1954, 1955a, b, 1956). He found that the urine of several insect species is rich in potassium, particularly in the distal end of the tubules. Sodium and other cations are also present, but at concentrations lower than in the blood. The urine is nearly isotonic with the haemolymph, and the production of urine alters the blood level of these ions only slightly. The flow of urine by isolated Malpighian tubules is greatly stimulated by increasing the concentration of K^+ in the surrounding medium. There is an electrical potential across the tubule wall. Based on these and other facts, Ramsay proposed that K^+ are actively transported into the lumen of the distal part of the tubules. He also recognized the probable involvement of water, Na^+, and phosphate ions in the formation of urine, and suggested that some of these materials may be reabsorbed in the proximal portion of the tubules or in the rectum. A similar suggestion was made by Wigglesworth (1931) many years ago.

Subsequent workers have elaborated upon this base. Berridge (1969) pointed out that K^+ movement alone will not bring about water flow into the Malpighian tubule lumen. Rather anions must accompany K^+ to achieve fluid movement. He suggested that Cl^- are the normal anions, but recognized that phosphate ions support the best rate of urine flow. The action of these two anions seems to be independent of each other. Also Na^+ support a low rate of urine flow and greatly stimulate the flow from isolated tubules in the presence of low concentrations of K^+ in the surrounding medium. Maddrell (1969) found similar results, but reported that NH_4^+ in the external medium will likewise support urine flow. He concluded that water movement follows and is closely linked with solute movement. Other workers have produced results which are in essential agreement with the above findings (Coast, 1969; Pilcher, 1970b; Maddrell and Klunsuwan, 1973). The reviews by Maddrell (1971) and Riegel (1971, 1972) provide much fuller accounts of the details.

From the above discussion it is clear that active transport of K^+ appears to be the key to fluid secretion by the Malpighian tubules. If this is correct, then a detailed examination of the secreting cells should provide evidence in support of this contention. Berridge and Oschman (1969) found that the Malpighian tubules of *Calliphora* have two types of cells. Electron micrographs disclosed that the primary cells have complex infoldings in the basal region and a microvillate lumenal border. These cells are supplied with abundant mitochondria, and their structure is suggestive of a secretory or transport role (Berridge, 1968). The other type of cell present is much simpler in structure and has few mitochondria. Similarly, Taylor (1971a, b) has examined the fine structure of the Malpighian tubules of *Carausius* (*Dixippus*). Two types of cells are again present. Those found in the main urine-secreting tubule are complex with basal infoldings and apical microvilli, which greatly amplify the surface area available for transport of solutes. Mitochondria are arranged in two bands near the basal and apical surfaces, suggesting active processes at both surfaces. A function of mucus production was proposed for the other cell type. An equally complex cellular organization has been reported for the Malpighian tubules of a number of other insect species (Beams *et al.*, 1955; Wessing, 1965, 1966; Dressler, 1968). In addition, Berridge (1966a) has pointed out the similarity between the fine structure of the insect Malpighian tubule and the vertebrate nephron. He demonstrated that isolated tubules utilize metabolic substrates, which indicates that they have an actively functioning glycolytic-Krebs cycle system as might be expected from the abundance of mitochondria. Magnesium-activated adenosine triphosphatases (ATPases) have been localized on both the basal and apical plasma membranes of the tubule cells in *Calliphora*

(Berridge, 1967). The apical membrane also has an active alkaline phosphatase. The metabolic system is sensitive to cyanide, azide and DNP (Berridge, 1966a). At the physiological level it has been demonstrated that certain of these inhibitors bring about the cessation of urine secretion (Maddrell, 1969; Pilcher, 1970b). Cutting the tracheal supply is another method of limiting the rate of secretion (Maddrell, 1971). These several types of evidence all support the contention that active K^+ transport by the distal region of the Malpighian tubules is the driving force in urine production. Of course, variations on this theme must be expected in so diverse a group of animals as insects (Srivastava, 1962; Irvine, 1969; Maddrell, 1971), but the main outline of the process now seems to be clear.

Some interesting information is available indicating that the process of urine formation is under hormonal control. The best documented case concerns the blood-sucking insect *Rhodnius*. Maddrell (1962, 1963) demonstrated a diuretic hormone in this insect which is produced by the neurosecretory cells of the mesothoracic ganglionic mass. There is a rapid onset of diuresis following a blood meal (Maddrell, 1964a) which is triggered by nervous information originating from the abdominal stretch receptors (Maddrell, 1964b). The process can be thought of as a reflex response due to abdominal stretching following feeding. The result is a prompt release of hormone into the circulating haemo-lymph. The Malphighian tubules respond by rapidly increasing the rate of urine production (Maddrell, 1962, 1963). The circulating hormone reaches a titre higher than is required for maximal response, but is quickly destroyed or eliminated by the Malpighian tubules (Maddrell, 1964a). The length of diuresis is mainly controlled by the period of ganglionic release of hormone into the blood (Maddrell, 1964a, 1966). 5-Hydroxytryptamine (serotonin) stimulates a rapid flow of urine in *Rhodnius* (Maddrell, 1969), but not in *Schistocerca* (Maddrell and Klunsuwan, 1973). With the latter insect, cyclic AMP was stimulatory.

A diuretic hormone has also been demonstrated from the median neurosecretory cells of the brain, the corpora cardiaca, and the fused mesothoracic ganglion in *Dysdercus* (Berridge, 1966b). The hormone is capable of making inactive Malpighian tubules actively produce urine. In *Carausius* a diuretic hormone is produced by the brain, corpora cardiaca, and the suboesophageal ganglion (Pilcher, 1970a). The level of hormone in the haemolymph varies with the state of hydration of the insect, and the hormone is inactivated by the Malpighian tubules. Pilcher (1970b) stated that the diuretic hormone changes only the rate of urine production, perhaps by stimulating active K^+ transport.

Several investigators have suggested the existence of an antidiuretic hormone (Wall and Ralph, 1964; Wall, 1965; 1966; 1967; Unger, 1965; Gersch, 1967; Mills, 1967; Mills and Nielson, 1967; Vietinghoff, 1967; Cazel and Girardie, 1968), but usually, a diuretic hormone was also reported from the same insect. The two hormonal actions would presumably be antagonistic. In some instances the antidiuretic role could possibly be explained by the absence of the diuretic hormone or by target tissue insensitivity. Berridge (1966b) failed to find evidence for an antidiuretic hormone in *Dysdercus.* On the other hand, preliminary purification of both hormones has been reported from *Periplaneta* (Goldbard *et al.,* 1970). Obviously, this issue is sufficiently important to receive a high priority for further study.

From this brief account of urine production and its regulation, it is seen that excess water and ions can be eliminated from the insect's internal environment under varying physiological conditions. With the exception of K^+, the movement of these substances into the lumen of the Malpighian tubules is apparently passive. Ramsay (1958) extended this concept to include other chemicals, some of which have excretory significance. He found that six amino acids, three simple sugars and urea are all transferred into the tubule lumen by passive diffusion. Thus, urine production may be regarded as a means of removing all soluble substances of low molecular weight from the blood. The rate of removal is influenced by the blood concentration of individual solutes, and the permeability characteristics of the tubule wall. As will be discussed shortly, this is an energetically expensive operation because many of these substances are useful metabolites that must be actively reabsorbed by the rectum. The manner in which uric acid fits into this process is not known with certainty, but Wigglesworth (1931) proposed that potassium or sodium urates are the important products. In view of the above discussion, it seems logical to presume that the urate ion can serve as an anion balancing the active transport of K^+ and facilitating the flow of water. Urine at the time of formation is reported always to be more alkaline than the haemolymph (Ramsay, 1956; Stobbart and Shaw, 1964), which would favour the ionization of uric acid. The possibility of active urate secretion also exists (Bursell, 1970). Further consideration will be given to the mechanism of urate excretion in the section on uric acid (p. 193).

An additional means of voiding material by cells of Malpighian tubules is formed-body secretion. Riegel (1966) reported the presence of formed bodies, containing digestive enzymes, in the urine of *Carausius*. He proposed that formed bodies are produced by the cells, and are voided into the urine for the purpose of excreting large molecules. This would represent a mechanism whereby molecules not amenable to passive filtration could still be excreted without first undergoing extensive degradation. Wessing (1965) suggested a channelled transport of vesicles across the cells of Malpighian tubules. Formed bodies have also been reported to be a common feature of *Calliphora* tubules (Berridge and Oschman, 1969). Contrarily, Taylor (1971a) described various stages of vesicle formation and liberation at both the basal and apical borders of *Carausius* Malpighian tubule cells, but concluded that they probably do not contribute significantly to transcellular transport. Much additional study needs to be accomplished to clarify the role of formed bodies in insect excretion (Maddrell, 1971).

The primary excretory fluid produced by the Malpighian tubules is subject to considerable modification before it leaves the body with the excreta. Some removal of water and ions, as well as a lowering of the pH, may occur in the proximal portion of the tubule itself. Subsequently, the anterior region of the hindgut probably removes some substances, including water, and continues the acidification process. Other activities, notably those of gut microbes, undoubtedly occur here, and in some species may alter the primary filtrate to a considerable extent. The nature and pervasiveness of these changes remain largely unknown. Perhaps because of this situation, Maddrell (1971) concluded that the fluid entering the rectum is not very different from the primary filtrate produced by Malpighian tubules. Whatever the case may be for the individual species, it is apparently still the function of the rectum to carry out the major modifications which are so important to both excretion and osmoregulation. In this sense, the rectum can be thought of as the second part of the excretory system whose mission is to complete the task begun by the Malpighian tubules.

The rectum is a very complex structure (Hopkins, 1967; Gupta and Berridge, 1966b; Veitinghoff, 1967; Ferreira and Cruz-Landin, 1969; Mordue, 1969; Oschman and Wall, 1969). It usually has a series of rectal glands, pads or papillae which are the most prominent structural features. There are several cell types present in the rectum including neurosecretory components. Extensive tracheation is supplied to the rectum from large tracheal trunks. The fine structure of the rectal papillae and pads indicates that they carry out a secretory or transport role. Finally, the rectal lumen is lined with a thin cuticular sac known as the intima.

Intensive study of the rectum in recent years has demonstrated that several types of regulation occur there. The first of these is conducted by the intima in a very interesting way. It has been shown that the intima, at least in *Schistocerca,* acts like a molecular sieve severely restricting the rate of penetration of large molecules (Phillips, 1965a, b; Phillips and Dockrill, 1968; Phillips and Beaumont, 1971). This has the effect of allowing the rapid exchange of simple substances such as ions, water, amino acids, and sugars, but not proteins, dyes, uric acid and large molecules. The cells of the rectal epithelium, which lie beneath the intima, are accordingly presented with a fluid which has already undergone a preliminary screening. This simplifies their task and at the same time assures that certain types of substances will be voided with the excreta. It may also serve to protect the epithelial cells from toxic substances produced, for example, by gut microbes. The general applicability of this finding to insects is not yet known. The proctodaeum of virtually all insects is lined with cuticle (Wigglesworth, 1972). Presumably, this molecular sieving role could be quite important.

It is the cells of the rectal epithelium which carry out the major role of rectal regulation. In a classical series of experiments on *Schistocerca,* Phillips (1964a–c, 1965a) demonstrated that the rectum can remove water, K^+, Na^+ and Cl^- from the excreta. Furthermore, the rate at which these substances are removed was related to the nutritional state of the insect, and to its need to conserve or eliminate water or ions. Thus, the rectum plays a vital role in osmoregulation (Maddrell, 1971). It was demonstrated that K^+ are reabsorbed 10 times faster than Na^+, but that K^+, Na^+ and Cl^- are all actively transported from the rectal lumen to the blood. As a consequence of rectal regulation of water and individual ions, Phillips (1965a) estimated that less than 5 per cent of the Malpighian tubule output is actually eliminated from the body usually as a hypertonic urine. Similar findings on ion and water reabsorption were reported for *Calliphora* (Phillips, 1969).

As was the case with Malpighian tubule functioning, active transport appears to be important in rectal regulation. The problem was particularly perplexing initially because Phillips (1964c, 1965a) demonstrated the movement of water in the apparent absence of ion transport. He cautiously interpreted this finding to imply only that an energy requiring process is involved, not necessarily a water-molecule carrier type of transport. Subsequent physiological and fine structural studies have provided an alternative explanation, and active water transport is usually discounted (Maddrell, 1971).

Gupta and Berridge (1966b) demonstrated that the rectal epithelium of *Calliphora* is composed of rectal, cortical and junctional cells. The rectal cells are simple and presumed not to participate in active transport, but may foster passive transport (Berridge and Gupta, 1967). Junctional cells are located between the rectal cells and the cortical cells which make up the rectal papillae or cones. Tracheal, nervous and neurosecretory components are closely associated with the cones. Internally, the cones have a series of complex infoldings of lateral plasma membrane, and stacks of repeating sub-units on the internal cytoplasmic surface which connect with intercellular sinuses (Gupta and Berridge, 1966a). This complex arrangement is closely associated with mitochondrial concentrations, and is reported to be the site of active transport of K^+ and Cl^- (Berridge and Gupta, 1967). It corresponds with the localization of ATPase activity (Berridge and Gupta, 1968). During active transport there is a magnesium-activated secretion of ions into the enclosed intercellular sinus which creates the local osmotic gradient necessary to extract water from the rectum. To explain the transport of water to the haemocoel, a two-membrane theory was proposed (Berridge and Gupta, 1967). The second membrane was suggested to be a 'leaky' connection between the intercellular sinus and the haemocoel. Through this membrane the intercellular sinus fluid passes due to high hydrostatic pressure moving both water and ions into the haemocoel. This model can accommodate the observed physiological facts and the fine structural organization of the rectal papillae.

Several other authors have reported ultrastructural arrangements in rectal pads and papillae which are in essential agreement with the above results (Phillips, 1965a; Hopkins, 1967; Stobbart, 1968; Oschman and Wall, 1969; Wall and Oschman, 1970; Wall *et al.,* 1970). In *Periplaneta* it was reported that the rectal pads are particularly complex as is demonstrated by their intensive tracheation and other features (Oschman and Wall, 1969; Wall and Oschman, 1970). It was suggested that this greater complexity is necessary to explain the cockroach's ability to carry out the movement of water in an ion-free fluid. It was proposed that ion recycling occurs within the cells, and partially accounts for the local osmotic gradient necessary to move water out of the rectum. Along similar lines a recycling of water by *Leucophaea* rectal cells has been reported but not explained (Hopkins *et al.,* 1971; Hopkins and

Srivastava, 1972). Sauer *et al.* (1970) demonstrated that ion reabsorption in *Periplaneta* is sensitive to DNP. In examining the sinus fluid of *Periplaneta,* Wall and Oschman (1970) discovered that it is rich in amino nitrogen. This seems to be in agreement with the report of Balshin and Phillips (1971) that the rectum of *Schistocerca* carries out active transport of certain amino acids from the lumen. Water movement may also thereby be facilitated.

In many species of insects, particularly in the Coleoptera and Lepidoptera, the distal ends of the Malpighian tubules are buried in the rectal wall forming a cryptonephridial system. Ultrastructural examination of this system has revealed a complex, intimate association between the wall of the rectum and the wall of the tubule (Saini, 1964; Gouranton, 1968; Irvine, 1969; Phillips, 1970; Koefoed, 1971). There are convoluted cytoplasmic borders, abundant mitochondria, and intercellular sinuses, again suggesting active processes. It was proposed that this is a system which is especially effective in water reabsorption, resulting in production of the dry excreta characteristic of many beetles (Saini, 1964). In addition, water transfer directly from the rectum into the tubule would have the effect of setting up a fluid circulation in a closed loop system, perhaps thereby aiding in the flow of urine and flushing out the tubule contents (Srivastava, 1962; Gouranton, 1968; Irvine, 1969). Of course, the more typical functioning of the excretory system accomplishes the same objective, but involves the open circulatory system.

The acidification of the primary Malpighian tubule fluid as it moves posteriorly has already been mentioned. This represents yet another regulatory function which is accomplished, or at least completed, by the rectum. It is an especially important function for those insects which excrete uric acid. The transformation of the weakly alkaline primary urine into moderately acid (pH 4-6) excreta assures that soluble, ionized urate will be converted into free uric acid. Since uric acid is highly insoluble it will form a precipitate thereby removing itself as an osmotic component of the excreta, and assuring its elimination from the body. Several authors have proposed that acidification is accomplished through an ion exchange whereby K^+ are removed from the rectal lumen and H^+ are secreted into it (Phillips, 1965a; Gupta and Berridge, 1966a; Berridge and Gupta, 1968). The process could be coupled with electron transport (Gupta and Berridge, 1966a) and would account for the observed change in rectal pH (Maddrell, 1971).

Hormonal control of rectal reabsorption is also known to occur.

Diuretic factors from the corpora cardiaca of *Schistocerca* were reported to increase secretion by the Malpighian tubules and reduce rectal reabsorption (Mordue, 1969). Both diuretic and antidiuretic hormones active on the rectum have been found in several insect species (Wall, 1965, 1966, 1967, Berridge and Gupta, 1967; Vietinghoff, 1967; Cazel and Girardie, 1968). The antidiuretic effect at this level is directed towards the conservation of water, and seems to be reasonably well established. Hormonal control of water absorption by the rectum is intimately associated with the state of hydration of the insect (Wall, 1965; 1966; 1967), which in turn may be reflected by changes in haemolymph volume and solute concentration (Wall, 1966; Pichon, 1970).

Thus, from an excretory point of view the rectum is responsible for accomplishing certain important tasks. It modifies the primary filtrate by removing ions, water and other useful small molecules such as amino acids and sugars. It completes the acidification of the excreta which is necessary to precipitate uric acid. In addition, it carries out osmoregulation in response to the physiological needs of the insect as mediated through the hormonal system.

Uric acid

It has been known for perhaps 150 years that uric acid is produced by insects (Wigglesworth, 1972), and it is now established as the most important single nitrogenous waste material. However, within the past 2 to 3 decades it has been recognized that several other closely related compounds are also excreted by one or another insect species. Among them are hypoxanthine, xanthine, guanine, allantoin and allantoic acid. As will be made clear in subsequent paragraphs, these compounds are all members of one or more pathway(s) leading to the formation or subsequent breakdown of uric acid. Insects which excrete them usually differ from uric acid excreters only by the absence or presence of one or sometimes two enzyme activities. Thus, it is logical to treat all of these excretory products together, and as variants of uric acid excretion. Bursell (1970) has already broadened the definition of uricotelism to include those species which excrete uric acid, allantoin, allantoic acid or any combination of the three. It seems reasonable to extend the concept to include those species, developmental stages, and mutant stocks which lack xanthine dehydrogenase (XDH), for example, and accordingly excrete hypoxanthine or xanthine. The origin of the hypoxanthine, xanthine or guanine is an important question, which will be treated later in this section, but is essentially irrelevant to the concept under consideration. Therefore, the production and excretion of uric acid will be treated here under the broadened concept. This general area has been extensively reviewed in the past (Gilmour, 1961, 1965; Kilby, 1963, 1965; Chefurka, 1965; Wigglesworth, 1972; Razet, 1966; Bursell, 1967, 1970; Corrigan, 1970; Schoffeniels and Gilles, 1970).

Because of the widespread occurrence of uricotelism (as defined above) among insects, it is important to know what compounds are excreted by the various species. Bursell (1967) has tabulated such information according to insect order and, where possible, developmental stage. No attempt will be made to repeat this effort. Rather, the matter will be discussed only to the extent of establishing the diversity which occurs.

Many species are truly uric acid excreters, or at least have been so reported. Among them are *Aëdes, Anopheles* (Irreverre and Terzian, 1959), *Locusta, Acheta* (Nation and Patton, 1961), *Lucilia* pupae and adults (Brown, 1938a; Birt and Christian, 1969), *Bombyx* (Fukuda and Nishitsutsuji, 1961; Tojo, 1971), *Mamestra* (Tojo and Hirano, 1966), *Glossina* (Bursell, 1965a) and *Rhodnius* (Wigglesworth, 1931). In addition, Razet (1961) has catalogued a large number of uric acid excreting species. In these species uric acid is presumably the principal excretory product, although in *Aëdes* only small amounts are voided (DeGuire and Fraenkel, 1973). In *Aeshna* nymphs uric acid is eliminated, but ammonia is the chief excretory product (Staddon, 1959). The published reports on *Periplaneta* as a uric acid excreter (Nation and Patton, 1961; Srivastava and Gupta, 1961; McEnroe, 1966a) appear to be in error. Wharton and Wharton (1961) stated that only small amounts of excreta nitrogen from this species could be uric acid. Subsequent reports have shown that under carefully controlled conditions none is voided (Mullins and Cochran, 1972, 1973b; Cochran, 1973). The implications of this finding will be explored in connection with internal uric acid storage.

In *Galleria* it has been shown that excreta from larvae, pupae and adults contain not only uric acid but also hypoxanthine and xanthine (Nation, 1963). The partitioning of purine nitrogen was essentially the same in haemolymph and excreta. For larvae it was 75 per cent uric acid, 7 per cent hypoxanthine, and 18 per cent xanthine. In pupal meconium and adult excreta uric acid constituted more than 90 per cent of the purines. Feeding high protein diets shifted the purine partition toward hypoxanthine and xanthine (Nation and Thomas, 1965). Similarly, *Melophagus* females excrete uric acid, hypoxanthine and xanthine, while males produce only uric acid (Nelson 1958). Because of the characteristic of viviparity in this species attention was paid to larval excretion which proved to be the source of hypoxanthine and xanthine. In *Pieris* Harmsen (1966a, b) has shown that a shift in metabolism of purine compounds occurs at pupation to favour the production of xanthine. In adults hypoxanthine, xanthine and uric acid are present, but only uric acid is actually excreted. In the same group Tartter (1940) found uric acid and hypoxanthine deposited as wing scale pigments. The available evidence from mutant stocks indicates that the absence of xanthine dehydrogenase results in the excretion and internal accumulation of hypoxanthine and sometimes xanthine (Morita, 1958; Glassman and Mitchell, 1959; Mitchell *et al.*, 1959; Kürsteiner, 1961; Zeigler, 1961a; Grell, 1962; Nation, 1963; Bonse, 1967). It is quite probable that certain other mutants as well as wild types of many species excrete varying amounts of hypoxanthine and xanthine in addition to uric acid (Mitlin and Vickers, 1964; Kokolis, 1968). It would be worthwhile to examine the extent of this phenomenon.

Guanine excretion in insects does not appear to be widespread. Mitlin and Vickers (1964) have reported low levels in the excreta of *Anthonomus*, but uric acid seemed to be the most prominent purine present. Descimon (1971) detected guanine and xanthine in the meconium of *Colias* following injection of labelled guanine. In this case the voiding of these two substances may have been due to overloading the system. Guanine is a major excretory product of several related groups such as spiders, mites, scorpions and centipedes (Schmidt *et al.*, 1955; McEnroe, 1961; Anderson, 1966; Horne, 1969).

The excretion of allantion and allantoic acid singly, together, or in combination with uric acid is very common in insects. It occurs in a large number of species, in immature and adult stages, and in most of the major orders (Razet, 1961). Much of the information available in this area comes from the extensive work of Razet which both he (Razet, 1966) and Bursell (1967) have summarized. Indeed, this excretory pattern appears to be so extensive as to challenge the supremacy of simple uric acid excretion. Much additional work needs to be accomplished to further our understanding of its scope within the Insecta. An evaluation of possible microbial degradation of uric acid in the gut would also be helpful.

The mechanism of excretion for the compounds involved in uricotelism is almost completely unknown except for uric acid. With the latter, a substantial theory was proposed many years ago and appears to have largely withstood the test of time. It will be briefly summarized to put the matter in historical perspective. Working with the blood-sucking insect *Rhodnius,* Wigglesworth (1931) proposed that a soluble salt of uric acid, probably monopotassium or monosodium urate, passes through the cells of the distal portion of the Malpighian tubules into the lumen. This clear watery urine which is slightly alkaline gradually passes into the proximal portion of the tubule. Here the tubule is reabsorptive in function and withdraws water and an alkaline substance, perhaps potassium bicarbonate, from the urine. As the solution becomes more acid and as ions are removed, uric acid and possibly some urates begin to crystallize out forming the granular urine characteristic of the proximal end of the tubules. Further removal of water may occur both in the tubule and in the rectum (Ramsay, 1958; Wigglesworth, 1972).

Evidence of several types was adduced in support of the theory. In *Rhodnius* a most striking histological distinction occurs between the proximal portion of the tubules, where the cells bear a brush border, and the distal portion containing cells with a honeycomb border. The change from one cell type to the other is abrupt, and corresponds with the change in type of urine. Experiments with ligatures and dyes corroborated these findings (Wigglesworth, 1931). In addition, Ramsay (1952, 1953) demonstrated the active secretion of potassium into the distal portion of the tubule, and its reabsorption in the proximal region. A corresponding active transport of sodium was not demonstrated. Of course, there now exists a substantial body of evidence supporting a potassium mediated active flow of water and urate into the distal part of the tubule, and in some species a flow of water and ions out of the lumen proximally. Where no histological dimorphism of the tubules is apparent, it is probable that the rectum is largely responsible for the function of reabsorption.

An interesting modification of Wigglesworth's theory has been proposed by Miles (1966). It was suggested that active transport of ions may result from a separation of charge by a selective barrier to protons in conjunction with an electron-transport chain. This modification could alleviate the necessity of introducing an anion, like bicarbonate, into the system. One could suggest a K^+ for H^+ exchange distally, and a H^+ for K^+ exchange proximally to explain the observed results. Berridge (1966a) has demonstrated the functioning of the Krebs cycle and its associated electron-transport system in the Malpighian tubules of *Calliphora.* Perhaps this system not only furnishes the energy for active transport, but also fits in with the proposed barrier. Alternatively, Miles (1966) suggested that the flavins which are present in the Malpighian tubules of many insect species (Drilhon and Busnel, 1939; Busnel and Drilhon, 1942; Wessing and Eichelberg, 1968) could play a role in the proposed electron-transport mediated movement of ions.

It is not clear how the currently established urate excretion mechanism can accommodate the excretion of such substances as hypoxanthine or allantoin. Perhaps they become ionized in the same way that uric acid does, thereby serving as an anion to balance the active transport of K^+. Alternatively, several of these products are somewhat more soluble than is uric acid (Bursell, 1967). This could be an aid to their passive movement into the lumen of the Malpighian tubule, but would probably lower the efficiency of the rectum. Berridge (1965) has presented evidence bearing on this point. He demonstrated that in *Dysdercus* the Malpighian tubules have a low permeability to allantoin, its principal excretory product. However, this species voids a watery urine perhaps in response to the greater solubility of allantoin. Further experimentation is required to elucidate the excretory mechanism involved with these products.

In addition to the production and voiding of uric acid, some insects have developed an elaborate alternative mechanism of dealing with this waste product. This is the process of internal deposition of uric acid and sometimes other purines. The phenomenon appears to be rather common as is illustrated in the following paragraphs. Harmsen (1966a; 1966b; 1966c) has shown that in *Pieris,* uric acid, xanthine and hypoxanthine are all retained internally. Uric acid and xanthine are found in substantial quantities primarily in the fat body. Likewise, fat body storage of uric acid has been reported for *Aphaereta* (Salkeld, 1967), *Chilo* (Tojo and Hirano, 1968), *Chrysopa* (Spiegler, 1962), *Habrobracon* (Clark and Smith, 1967), *Hyalophora* (Jungreis and Tojo, 1973), and *Nemeritis* (Corbet and Rotheram, 1965). Deposition of uric acid in or associated with the cuticle and its epithelial lining, where it may function in pigmentation, is also well documented (Brighenti, 1940; Yoshitake and Aruga, 1952; Tsujita and Sakurai, 1964; 1965; Friauf and Edney, 1969; Caveney, 1971). In addition, a rather general internal distribution of uric acid is sometimes found as in *Galleria* larvae where it occurs in blood, muscles, fat body and integument (Zielinska, 1957). Muscular involvement with internal uric acid has also been reported for *Locusta* (Kermack and Stein, 1959).

Cockroaches are a group in which the storage phenomenon has been extensively studied. Haydak (1953) examined *Blattella, Blatta* and *Periplaneta* fed on diets varying in protein content. On the higher protein diets (49 and 74 per cent protein) he found that *Periplaneta* females had abdomens distended by enlarged fat bodies, and that white deposits, presumed to be uric acid, were broadly distributed within the insect's body. Similar results were obtained on *Periplaneta* by McEnroe (1956) and by Mullins (1971), who measured the amount of uric acid present. He found that fat body is the principal organ of storage, but in agreement with Haydak (1953), reported uric acid to be almost indiscriminately deposited when the insect is confronted with prolonged, excessive dietary nitrogen. In a survey paper, Cochran (1973) reported that the fat bodies of 14 species of cockroaches on a 24 per cent protein diet all contained significant amounts of uric acid. Obviously, uric acid storage is wide-spread among cockroaches.

Numerous other less well-defined examples of internal urate storage are also known (Wigglesworth, 1942; Bernard and Fixler, 1963; Mitlin and Mauldin, 1966). Contrarily, in still other species retention of uric acid in the insect's body is, at most, trivial (Staddon, 1959). Nevertheless, the internal storage phenomenon seems to be of importance in insects. Its implications and ramifications are numerous, and must be explored. Before doing so, however, it is appropriate to discuss the biochemical aspects of uric acid metabolism.

When considering the biochemistry of uric acid in insects, it appears that three generalized pathways are involved. These are: (1) The *de novo* synthetic process utilizing protein nitrogen, and usually called the uricotelic pathway; (2) The degradative pathway in which nucleic acids or their components are the starting material. This pathway has been referred to as the uricolytic pathway (Bursell, 1967), but the term nucleicolytic seems more appropriate; (3) The pathway whereby uric acid is degraded in insect tissues, and for which the term uricolytic should be reserved. Each of these pathways will be discussed in some detail.

It has long been assumed that a *de novo* biosynthetic pathway for the production of uric acid is necessary to explain the large amounts produced by insects. Among the earliest attempts to demonstrate such a pathway were those of Leifert (1935b) who proposed a mechanism involving urea and malonate, based on experiments with *Antheraea*. This was in agreement with earlier proposals for birds (Weiner, 1902). Subsequent work did not substantiate the concept in insects (Anderson and Patton, 1955; Desai and Kilby, 1958b), nor in birds (Schuler and Reindel, 1935). With the use of labelled precursors, the Buchanan group and others were able to establish an

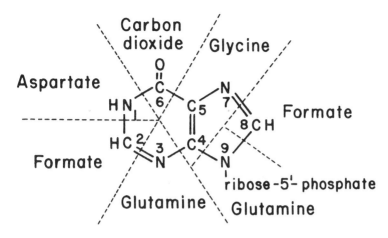

Fig. 3.1. Origin of the carbon and nitrogen atoms of inosine-5′-phosphate produced by the purine biosynthetic pathway (Adapted from Bursell, 1967.)

elaborate pathway for the production of uric acid in pigeon tissues (Buchanan and Sonne, 1946; Buchanan *et al.,* 1948; Sonne *et al.,* 1948; Buchanan, 1951). It is now recognized that the pathway starts with 5-phospho-D-ribosyl-1-pyrophosphate. To it are added substituents in the 1 position of the ribose molecule which replace the pyrophosphate moiety resulting ultimately in the production of inosine-5′-phosphate (IMP). The individual reactions along the pathway are known as well as the origin of each of the atoms in the rings (Fig. 3.1). The pathway is energetically demanding, since a total of approximately 8 high energy phosphate bonds are consumed in the production of one molecule of IMP (Mahler and Cordes, 1966). It is currently accepted as the *de novo* pathway for the production of the purine ring in birds (Barrett and Friend, 1970), and may be found illustrated in most of the recent biochemistry textbooks and metabolic charts. IMP may be considered a branchpoint along the pathway since numerous compounds can be produced from it (Magasanik and Karibian, 1960). For the uricotelic organism, however, the production of uric acid is quite straightforward involving the dephosphorylation of IMP to inosine, the hydrolytic or phosphorolytic removal of ribose for the production of hypoxanthine, and the action of xanthine oxidase on hypoxanthine and xanthine to form uric acid (Fig. 3.2).

Work on the *de novo* synthesis of uric acid in insects has lagged behind, and still relies heavily on the information derived from birds. However, it has been established that in insects there is a correlation

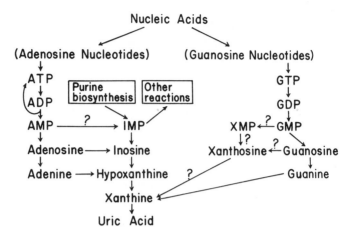

Fig. 3.2. Possible reactions in the nucleicolytic pathway for the production of uric acid in insects. The terminal steps from IMP to uric acid are also on the uricotelic pathway.

between the rate of uric acid production and the rate of degradation of proteins and amino acids (Terzian *et al.,* 1957; Tojo and Hirano, 1968; Birt and Christian, 1969; Mullins, 1971; Tojo, 1971). It has also been shown from *in vivo* studies with *Periplaneta* that carbon atoms 2 and 8 of the purine ring originate from labelled formate (McEnroe, 1956; McEnroe and Forgash, 1957, 1958). Similar results were obtained with *Drosophila* in which it was found that the C-6 of glucose and the C-3 of serine are primary sequential sources of the formate carbon (Brenner-Holzach and Leuthardt, 1965). In a more elaborate study with *Rhodnius,* employing appropriately labelled precursors, Barrett and Friend (1970) have demonstrated that the origin of all five carbon atoms of the uric acid molecule is the same as that proposed for birds (Buchanan, 1951). Other less direct evidence in support of this pathway comes from *in vitro* studies. Heller and Jezewska (1959) incubated homogenates of fat body and gut from *Antheraea* larvae and pupae with a mixture of precursors, patterned after that used by the Buchanan group in birds, and demonstrated increased uric acid production. Similarly, Desai and Kilby (1958b) found uric acid synthesis was stimulated by the addition of 4-amino-5-imidazole carboxamide to a homogenate of *Calliphora* larval fat body. While the free carboxamide is not on the prime biosynthetic pathway, its ribotide is, and the corresponding transformation was proposed. It is also clear from the *in vitro* work that fat body is implicated as an important tissue for the synthesis of uric acid. The gut may be involved, but little is known about the uricogenic capabilities of other tissues.

Thus, the evidence from recent work seems to support the hypothesis that uric acid synthesis by the *de novo* pathway in insects probably occurs in the fat body, and is similar to the mechanism established for birds. As pointed out by Bursell (1967), however, only the general outline of the pathway has been demonstrated in insects. Nothing is known of the individual steps prior to the formation of IMP. The stoichiometry of the system has not been worked out. Because of these and other facts, alternative pathways for the synthesis of uric acid cannot unequivocally be excluded. While the resolution of this problem may seem mundane, it remains one of the unanswered questions of insect biochemistry.

The second recognized mechanism for the production of uric acid in insects is the nucleicolytic pathway. This involves the degradation products of nucleic acids derived either from the process of digestion or the processes of tissue maintenance and repair (Bursell, 1967). Its significance in the overall production of uric acid is difficult to evaluate, but presumably its input is normally very small in contrast to the uricotelic pathway (Wigglesworth, 1972). In support of this contention, Terzian *et al.* (1957) reported that about 40 times more uric acid was excreted by *Aëdes* adults fed whole blood as compared with those fed sugar. The insects given sugar had not been fed with blood for one or two weeks, but it is possible that some of the uric acid was derived from these dietary proteins thus reducing even further the contribution of nucleicolytic sources. Similarly, Mullins (1971) found a difference of about the same magnitude between total body uric acid in *Periplaneta* adults fed high protein diets as opposed to those on a low protein diet. Neither of these reports addressed the question of the nucleic acid content in the high protein diets, but it was undoubtedly small in relation to protein content. If these comparisons are valid, then the normal contribution of nucleic acid-derived uric acid to the total production of uric acid is indeed small.

The pathway itself is a rather diffuse one, with a large number of enzymatic reactions possibly contributing products. This is especially true of the primary products of nucleic acid digestion. Many of the reactions involving these substances have other major functions, and their contribution to excretion may be more or less incidental. As the pathway proceeds, however, it funnels into the same reactions which constitute the terminal steps of the uricotelic pathway. These relationships are shown in Fig. 3.2.

More specifically, nucleic acid degradation has not been adequately investigated in insects. A DNAase has been reported from *Bombyx* pupae which liberates acid-soluble nucleotides from DNA (Koga and Akune, 1972). Also in *Bombyx* it has been shown that a decrease in both DNA and RNA occurs during the period from mature pupa to emerging adult (Chinzei and Tojo, 1972). This decrease was accompanied by a change in distribution of nucleic acids in the insect's body especially from the silk gland to the fat body. Nucleic acid synthesis in insects has also been demonstrated. Moriuchi *et al.* (1972) reported DNA synthesis in the silk gland of *Bombyx* larvae. Similarly, Miller and Collins (1973) found the incorporation of [^{14}C]adenine into DNA adenine and guanine in *Musca* ovaries. RNA synthesis in larval fat body cells of *Musca* followed the injection of [^{14}C]adenine and [^{14}C]orotic acid (Russo-Caia, 1963). In *Oncopeltus* eggs, RNA and DNA synthesis correlate well with the disappearance of inosine and guanosine (Forrest *et al.*, 1967). These results indicate that an ebb and flow of nucleic acid synthesis and degradation occurs in insects. What is happening at any given time may depend on factors such as life stage, age, tissue, nutritional status, sex and probably others. Nevertheless, it seems quite clear that nucleotides from nucleic acid degradation are available for conversion into uric acid, probably at a low but variable level.

The above conclusion leads logically into the question of what form the nucleic acid-derived adenosine and guanosine nucleotides will take. Presumably, they are liberated at the monophosphate level. If this is correct, it might make the conversion to uric acid easier and less expensive metabolically. However, the cells and tissues would not be expected to distinguish between these nucleotides and the same molecules synthesized by the *de novo* pathway. It is also recognized that cells of many tissues abound with enzymes which can catalyze the modification of nucleotides, especially those containing adenosine. A plethora of ATPases have been described from insects (Sacktor, 1953; Sacktor *et al.*, 1953; Maruyama, 1954; McShan *et al.*, 1955; Sacktor and Cochran, 1957; Avi-Dor and Gonda, 1959; Mills and Cochran, 1967; Berridge and Gupta, 1968; Akai, 1970). Many of these enzyme systems will also catalyze the dephosphorylation of all the common nucleoside triphosphates. In addition, the interconversion of ATP, ADP, and AMP has been amply documented (Ray and Heslop, 1963; Mills and Cochran, 1966; Hofmanova *et al.*, 1967; Carney, 1969). In *Periplaneta* specific ATPases and a myokinase have been identified and purified from thoracic muscle tissue (Mills and Cochran, 1966; 1967). Myokinases from several other insects are also known (Gilmour and Calaby, 1953). The removal of the second phosphate group from GTP has been demonstrated in *Musca* (Sacktor and Cochran, 1957).

From these results, and in reference to Fig. 3.2, it is apparent that interconversion and/or degradation of adenosine and guanosine nucleotides to the monophosphate level can readily be accomplished. Presumably some of these nucleoside monophosphates will be further degraded into uric acid. It must be remembered, however, that many of the reactions described above are usually thought of in connection with energy conversions within the cell. In this context these reactions are probably directed towards the conservation of nucleotide molecules rather than their destruction. Again, the indication is that only small amounts of nucleoside monophosphates from energy conversion-type reactions are further degraded. In combination with similar molecules from nucleic acid degradation, they presumably constitute the nucleicolytic source of uric acid precursors.

Before leaving the nucleotides, some mention should be made of those containing pyrimidines as the base moiety. The degradation of nucleic acids would liberate nucleotides containing uridine, thymidine and cytidine. Very little information is available on their fate in insects. Undoubtedly, they could be used in the synthesis of additional nucleic acids. It is known that *Periplaneta* has several enzymes which can act upon pyrimidine nucleotides (Bruno and Cochran, 1965). Among them are dephosphorylating enzymes, cytidine deaminase, and uridine phosphorylase. By the action of these enzymes CTP and UTP can be converted into uracil. Sacktor and Cochran (1957) reported the complete removal of phosphate from UTP and CTP by *Musca* flight muscle mitochondria. Unfortunately, neither of these reports provided evidence on the possible significance of pyrimidines as nitrogenous end products. Mitlin and Wiygul (1972) showed that [^{14}C]thymidine and [^{14}C]uridine injected into *Anthonomus* resulted in the appearence of labelled uric acid in the faeces. Faecal guanine was not labelled. While this evidence is suggestive of a direct incorporation of the pyrimidine ring into uric acid, an alternative explanation is possible. The faeces were collected over a 3-day period which is probably long enough to allow the extensive degradation of the pyrimidines, and re-incorporation of the label into uric acid. Apparently, the consideration of pyrimidines as end products of nitrogen metabolism must remain an open question.

The dephosphorylation of AMP, IMP and GMP has been reported with several enzyme preparations from *Periplaneta* (Cochran, 1961; Cochran and Bruno, 1963; Bruno and Cochran, 1965). These included a thoracic muscle mitochondrial source (Cochran 1961) that was in sharp contrast to a similar preparation from *Musca* which failed to dephosphorylate purine nucleoside monophosphates (Sacktor and Cochran, 1957). Similar results have been reported for other species (Gilmour and Calaby, 1952). While no explanation for these species differences is at hand, it seems likely that 5'-nucleotidases are not uncommon enzymes if they are sought in appropriate tissues. The dephosphorylation of IMP is, of course, on the uricotelic pathway.

The products of 5'-nucleotidases of importance to this discussion are adenosine, inosine and guanosine. Of these, adenosine and guanosine are subject to deamination. The enzyme adenosine deaminase has been identified in insects (Florkin and Frappez, 1940a, Lennox, 1941b; Wagner and Mitchell, 1948; Desai and Kilby, 1958b; Cochran, 1961; Cochran and Bruno, 1963; Cordero and Ludwig, 1963; Bruno and Cochran, 1965; Hodge and Glassman, 1967a, b). It appears to be widely distributed, and converts adenosine to inosine. The existence of its counterpart, guanosine deaminase, has not been established for insect tissues (Duchâteau *et al.*, 1940). Xanthosine is the product of such a reaction, and it too is rarely reported from insects. The related enzymes AMP deaminase (adenylic acid deaminase) and GMP deaminase are also of questionable significance. Cochran (1961) reported that AMP was dephosphorylated prior to deamination by *Periplaneta* muscle mitochondria. On the other hand, Hodge and Glassman (1967a) presented evidence indicating the presence of AMP deamination in *Drosophila* homogenates. The corresponding enzyme for GMP has not been reported, and again its product, XMP, is apparently a rare compound in insect tissues.

Nucleosides are also subject to the removal of ribose by the action of nucleosidases. Corrigan (1970) has pointed out that these reactions can either be of the hydrolytic or phosphorolytic type. Studies of such enzymes in insects are very limited. Inosine phosphorylase has been demonstrated in *Periplaneta* (Cochran and Bruno, 1963; Bruno and Cochran, 1965) and in *Drosophila* (Hodge and Glassman, 1967a). This enzyme is responsible for the conversion of inosine to hypoxanthine and is, therefore, one of the principal terminal reactions in the synthesis of uric acid. That it has not been studied in more detail is surprising. Similarly, the conversion of adenosine to adenine, and guanosine to guanine are also important reactions which have remained largely unstudied. However, the enzymes adenase and guanase, which deaminate adenine to hypoxanthine and guanine to xanthine, are reasonably well known (Duchâteau *et al.*, 1940; Anderson and Patton, 1955; Desai and Kilby, 1958b; Lisa and Ludwig, 1959; Hayashi, 1961a; Prota, 1961; Cline and Pearce, 1963; Cordero and Ludwig, 1963; Pierre, 1965; Hopkins and Lofgren, 1968). They occur in a wide diversity of species, but adenase appears to be absent from *Drosophila* homogenates (Hodge and Glassman, 1967a, b).

The final reactions considered here are the conversion of hypoxanthine to xanthine and xanthine to uric acid, both catalyzed by xanthine dehydrogenase (XDH). This is the most extensively studied reaction in either of the pathways discussed in this section. Irzykiewicz (1955) demonstrated that the enzyme is a dehydrogenase rather than an oxidase in several insect species. It cannot couple with molecular oxygen directly, but reduces methylene blue. Its activity was stimulated by nicotinamide-adenine dinucleotide (NAD), and NAD plus pyruvate in the absence of methylene blue. XDH and several other enzymes will be discussed next in connection with identifying the tissues where nucleicolytic and terminal uricotelic activities take place.

In surveying the enzymes discussed above, some of them, such as ATPases, are found virtually everywhere they are sought. This fact, together with their primary involvement in energy conversion reactions, make them unlikely subjects to shed much light on sites of uric acid production. Certain other enzymes are better sources of information and from these, adenosine deaminase, adenase, guanase, and XDH have been selected for more detailed study. Unfortunately, many reports dealing with these enzymes are based on the use of homogenates or extracts which are of no help in localizing sites of activity. The available information on these enzymes is presented in Table 3.1. Razet (1966) has also tabulated much of this information, but without reference to enzyme localization. From the table it appears that adenosine deaminase may be generally distributed over the insect's body. The ease with which this activity is found in thoracic and abdominal homogenates lends support to this concept (Bruno and Cochran, 1965). The information is scanty, however, and additional studies are needed. Adenosine deaminase is probably also involved in ammonia metabolism (p. 219), which may have a bearing on its distribution. At present however, it does not provide much insight into sites of uric acid production.

Table 3.1
Tissue distribution of selected enzymes involved in uric acid production in insects

Enzyme	Species	Tissue	Reference
Adenosine deaminase	*Dytiscus*	Haemolymph	Florkin and Frappez, 1940b
	Calliphora	Larval fat body (f.b.)	Desai and Kilby, 1958b
	Lucilia	Larval gut, Malpighian tubules (M.t.) f.b.	Lennox 1940, 1941b
	Periplaneta	Adult thoracic muscle	Cochran and Bruno, 1963
		Adult gut and M.t.	Cordero and Ludwig, 1963
Adenase	*Prodenia*	Larval f.b.	Anderson and Patton, 1955
	Tenebrio	Larval f.b.	Anderson and Patton, 1955
	Antheraea	Larval and pupal f.b.	Leifert, 1935b
	Bombyx	Larval f.b.	Hayashi, 1961a
Guanase	*Bombyx*	Larval f.b.	Hayashi, 1961a
	Calliphora	Larval f.b.	Desai and Kilby, 1958b
	Leucophaea	Larval and adult f.b.	Lisa and Ludwig, 1959 Pierre, 1965
	Periplaneta	Adult gut, M.t., leg muscle	Cordero and Ludwig, 1963
	Prodenia	Larval f.b.	Anderson and Patton, 1955
	Tenebrio	Larval f.b.	Anderson and Patton, 1955
Xanthine dehydrogenase	*Calliphora*	Larval f.b.	Desai and Kilby, 1958b
	Tenebrio	Larval f.b.	Irzykiewicz, 1955
		Larval f.b.	Anderson and Patton, 1955
		Larval and adult haemolymph	Prota, 1961
	Prodenia	Larval f.b.	Anderson and Patton, 1955
	Periplaneta	Adult f.b.	Anderson and Patton, 1954
		Adult gut, M.t., leg muscle	Cordero and Ludwig, 1963
		Adult f.b., M.t., gut	Hayashi, 1961c

Table 3.1
continued

Enzyme	Species	Tissue	Reference
	Leucophaea	Larval and adult f.b.	Lisa and Ludwig, 1959
	Antheraea	Larval and pupal f.b.	Leifert, 1935b
	Drosophila	Larval f.b., M.t.	Ursprung and Hadorn, 1961
		Adult haemolymph, M.t., gut	Munz,1964
	Bombyx	Larval f.b., M.t., gut	Ito and Mukaiyama, 1964
			Hayashi, 1960, 1961a,b
	Cephonodes	Larval f.b., M.t.	Hayashi, 1961c
	Barathra	Larval f.b., M.t.	Hayashi, 1961c

However, the other three enzymes tabulated are more informative (Table 3.1). It is abundantly clear that the fat body is a major organ in which these enzymes are found, although the Malpighian tubules and gut are sometimes involved. In the many studies (summarized in Table 3.1), other tissues were also examined and found to have little or no activity. Thus, larval fat body has been the overwhelming choice of tissue for study. This, of course, presents an unbalanced picture, but there can be little doubt of the importance of this tissue for the production of uric acid in larvae. The evidence presented in Table 3.1 on adults and pupae, together with the studies reported above on *de novo* uric acid synthesis, are consistent with the hypothesis that the insect fat body is the major site of uric acid synthesis.

Of the enzymes discussed in the preceding paragraphs, only XDH is sufficiently well known for further consideration. Much of the interest surrounding this enzyme arises because it is also involved with the metabolism of pteridines (p. 252), and has been implicated as the biochemical lesion in certain genetic mutants, particularly in *Drosophila*. Many stocks of this insect are either completely deficient in XDH or have an altered level of this enzyme (Morita, 1958; Glassman and Mitchell, 1959; Mitchell *et al.*, 1959; Glassman, 1962b; Grell, 1962; Keller, 1964; Keller and Glassman, 1964a, b; Glassman, 1965, Yen and Glassman, 1967; King, 1969; Seybold, 1973). Various combinations of these mutants result in different expressions of the traits and/or levels of XDH. Wild-type *Drosophila* fed on semi-defined diets showed an increase in the activity of XDH, probably resulting from supplying additional necessary co-factors (Collins *et al.*, 1970). Phenocopies of certain mutants were produced by growing larvae in media containing the XDH inhibitor 4-hydroxypyrazolo (3,4-d) pyrimidine (Keller and Glassman, 1965; Boni *et al.*, 1967). Electrophoretic variants and isoenzymes of *Drosophila* XDH have been reported (Keller *et al.*, 1963; Smith *et al.*, 1963; Yen and Glassman, 1965; Glassman *et al.*, 1968). One study showed the conversion of XDH from one form to another when incubated with the appropriate *Drosophila* extracts (Shinoda and Glassman, 1968).

Perhaps of more direct concern to this review are certain experiments with *Bombyx* XDH. Here too, some work has been done on mutants deficient in XDH (Eguchi, 1961; Hayashi, 1961d). In addition, it has been confirmed that XDH is a dehydrogenase which couples readily with methylene blue, NAD and FAD, but not oxygen (Hayashi, 1960; 1961b, e, f). Hayashi (1961c) obtained similar results for two other lepidopterous species. In the work on *Bombyx*, homogenates from larval fat body which contained XDH, oxidized adenine, guanine, hypoxanthine and xanthine (Hayashi, 1961a). While this is to be expected from crude homogenates, it is somewhat surprising to find that the substrate specificity of a purified XDH from the same source was equally broad (Hayashi, 1962b). In this case xanthine, hypoxanthine, guanine, adenine, aldehydes, 2-amino-4-hydroxypteridine, xanthopterin and NADH were all attacked. This work raises a question concerning the identity of the enzymes called adenase and guanase. It is at least conceivable that they are merely additional activities of XDH. Hayashi (1962a) compared the *Bombyx* XDH with similar enzymes from milk and mammalian liver. The insect enzyme contains Fe, Mo and FAD in a ratio of 9:1:1, thereby showing a close resemblance to the liver enzyme. A highly purified XDH has also been prepared from *Drosophila* adults (Parzen and Fox, 1964). The enzyme utilized NAD, one mole being reduced for each mole of xanthine converted to uric acid. A 2:1 ratio resulted when hypoxanthine was the substrate. Unfortunately, no other purine or pteridine substrates were studied.

The synthesis of uric acid is usually considered to be a metabolically expensive operation, as mentioned above. In view of the information just presented, this conclusion may not be justified for insects. If the action of XDH is coupled with the reduction of NAD under *in vivo* conditions, as seems to be the case, then the subsequent oxidation of this NADH by an electron transport chain could result in the recovery of 6 moles of ATP per mole of hypoxanthine oxidized to uric acid. Should this prove to be correct, the synthesis of uric acid in insects may be less expensive metabolically than is the production of urea in mammals. Perhaps this is why insects have developed XDH as opposed to xanthine oxidase. A correlation may exist between the type of xanthine enzyme present and uricotelism versus ureotelism. This possibility should certainly be investigated further.

The third pathway in insects involving uric acid is the uricolytic pathway, wherein uric acid is degraded to simpler compounds. Several steps on this pathway are well established. Certain other reactions are less well documented, and may be attributable either to the insects themselves or to their intracellular symbionts. Fig. 3.3 illustrates the main part of the pathway.

The first reaction on the pathway is the conversion of uric acid to allantoin by the enzyme uricase. The reaction is complex, and apparently occurs in two steps (Florkin and Duchâteau, 1943; Razet and Barraud, 1965). The intermediate is presumed to be unstable so allantoin normally accumulates. However, Tojo and Yushima (1972) found a labelled substance in the wings of *Papilio* injected with 2-[^{14}C] uric acid which they reported to be an intermediate between uric acid and allantoin, or a conjugate of allantoin.

Uricase has been reported from a large number of species (Leifert, 1935b; Robinson, 1935; Truszkowski and Chajkinowna, 1935; Brown, 1938a; Rocco, 1938; Razet, 1952, 1953, 1954, 1956, 1961, 1964; Poisson and Razet, 1953; Lisa and Ludwig, 1959; Cordero and Ludwig, 1963; Nelson, 1964; Pierre, 1964; Berridge, 1965). Sometimes its occurrence is limited as in *Popillia* where it exists only in embryos and larvae (Ross, 1959), or in *Lucilia* where it is lost during the pupal stage (Brown, 1938a). Many other species do not have uricase and cannot produce allantoin (Florkin and Duchâteau, 1943; Razet, 1961). The known tissue distribution of uricase presents a confusing picture. In *Periplaneta* it occurred in all tissues tested but was most active in gut and Malpighian tubules (Cordero and Ludwig, 1963). It is also concentrated in Malpighian tubules in *Dysdercus* (Berridge, 1965). Fat body homogenates have uricase in *Leucophaea* (Lisa and Ludwig, 1959). In *Calliphora* and *Musca* uricase activity was confined to the Malpighian tubules (Razet, 1953; Nelson, 1964). In view of the small amount of available data, all that can be said is that the Malpighian tubules and probably other tissues, such as fat body and gut, are sites of uricase activity.

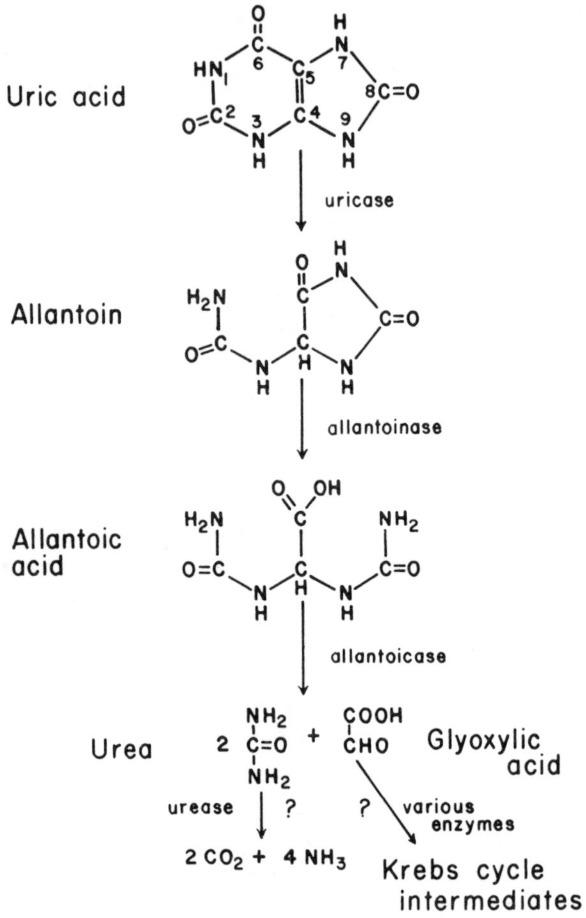

Fig. 3.3. The uricolytic pathway with possible additional steps. (Adapted from Florkin and Duchâteau, 1943.)

The conversion of allantoin to allantoic acid is catalyzed by the enzyme allantoinase. It has been described from several insect species (Rocco, 1936, 1938; Manunta, 1948, 1949; Razet, 1953, 1954, 1956, 1965, 1966). The general impression gained from this work is that allantoinase is much more restricted in its distribution than is uricase (Razet, 1961), with the possible exception of the Lepidoptera (Bursell, 1967). The tissue localization of allantoinase seems to be similar to that of uricase (Razet, 1961).

The third reaction in the sequence is the conversion of allantoic acid to urea and glyoxylic acid by the enzyme allantoicase. The information available on this enzyme in insects is meagre. Its occurrence in two species seems to be reasonably certain (Razet, 1952, 1961, 1965; Wang and Patton, 1969). As pointed out by Bursell (1967), the near absence of urea in insect excreta correlates well with the apparent scarcity of allantoicase. This reaction does not appear to be an important one in insect excretion.

The further breakdown of urea to ammonia and CO_2 by urease is of questionable occurrence in insects. The enzyme has been reported (Truszhowski and Chajkinowna, 1935; Baker, 1939; Robinson and Baker, 1939; Robinson and Wilson, 1939; Razet, 1961), but the results are usually discounted because of methodological short-comings (Gilmour, 1961; Wigglesworth, 1972; Bursell, 1967, 1970). Desai and Kilby (1958a) failed to find urease in *Calliphora* larvae, as did Lennox (1941b) in *Lucilia* larvae even though ammonia is produced by this latter insect.

Additional degradation reactions involving uric acid are also possible. Some of them have been attributed to intracellular symbionts (Donnellan and Kilby, 1967), and will be treated more fully in connection with further discussions of the urate storage phenomenon. Other reports suggest mechanisms of uric acid degradation outside the scope of the uricolytic pathway just described. For example, McEnroe (1966b) reported that [14C] uric acid was degraded to $^{14}CO_2$ by *Periplaneta*. The evidence indicated a preferential oxidation of 2-C over 8-C, but both carbons were oxidized. Similarly, Mitlin and Wiygul (1973) reported that 2-[14C] uric acid injected into *Anthonomus* resulted in the production of labelled RNA, DNA, amino acids and CO_2 within 2 hours. Both of these reports suggest that unknown pathways for the degradation of uric acid are operational in these species. Of course, microbial systems within the insect's body could also be the explanation here. However, proof of this suggestion must be provided before alternative insectan pathways can be excluded.

Before leaving the area of uric acid biochemistry, an additional word may be in order concerning excretory products. The reader was asked earlier to accept on faith the relatedness of the several uric acid-type materials found in excreta, and the ease with which species could evolve different excretory patterns. Hopefully, any lingering doubts will have been dispelled by discussion of the three bio-chemical pathways related to uric acid, and the mutant stocks. No fundamental alteration of the insect excretory system nor the biochemical pathways appears to be necessary to account for the observed excretory patterns. One or more single gene mutations, and their subsequent fixation in a population, would explain most of the diversity described in the introductory paragraphs of this section.

Returning now to the urate storage phenomenon, there are several areas still to be discussed, although Nolfi (1970) has summarized the storage concept for animals in general. He pointed out that stored purines and related substances are involved in storage excretion, pigmentation, a carbon and nitrogen reserve, a stable source of purines for nucleic acid synthesis, and probably other functions. With reference to insects, all of these possibilities exist. Evidence on the association of uric acid with pigmentation and nucleic acid synthesis is not extensive, but has been reviewed above. Storage excretion and the role of urates as carbon and nitrogen reserves needs further consideration.

Looking first at internal storage, one of the points to be considered is the form the stored material assumes. In previous paragraphs the terms uric acid and urates have been used somewhat interchangeably. This is imprecise usage, but the exact form taken is not known with certainty. Presumably free uric acid and the urates of sodium, potassium, ammonium, and possibly other cations are the most likely configurations. The tautomeric form assumed by uric acid may also be important (Inwang, 1971). Seegmiller (1969) has indicated that in biological fluids the urate of the principal cation present will predominate. This usually means a mono-substituted urate with the cation occurring at the C-9 position. Substitutions at C-3 also occur but much more sparingly (Bergmann and Dikstein, 1955). When forming a precipitate, however, the situation may be somewhat different because free uric acid and ammonium urate are the least soluble of the substances listed (Porter, 1963a). Also solubility is greatly influenced by pH, and NH_4^+ are known to take part in ion exchange reactions resulting in the precipitation of ammonium urate (Porter, 1963b). Thus, it might be expected that a mixed urate-uric acid complex is the storage form, and that it could vary in composition in different parts of the body depending on tissue pH and predominating cations.

Wigglesworth (1931) and Ramsay (1952; 1953) presented evidence that potassium urate is important in the excretory process. This finding also gives a probable indication of the internal form of urate, at least in *Rhodnius* haemolymph. In work with *Hyalophora*, Jungreis and Tojo (1973) demonstrated a simultaneous increase in uric acid and potassium in the fat body at the larval-pupal molt. The uric acid apparently arose by *de novo* synthesis, but the K^+ increase corresponded with a loss of K^+ from the integument. These authors concluded that fat body is a major site for the storage of both uric acid and potassium during the pupal stage, the implication being that potassium urate was involved. Similarly, Mullins (1971) found that in *Periplaneta* fed diets high in protein, uric acid accumulated in the body. The accumulation was mirrored by increases in potassium, and sometimes sodium. Also the increases in uric acid were accompanied by unexplained increases in nitrogen content. Based on these findings Mullins (personal communication) postulated urate storage as potassium and ammonium urates in the fat body, and as sodium urate in certain other parts of the insect's body. Pichon (1970) found unequal cation distribution in the body of this insect. Much more work remains to be done in this area, but one of the ramifications suggested by the above results in that the urate storage complex may also serve as an 'ion sink' during certain life stages or in response to physiological stress. This could be one of the unknown sources of haemolymph solutes mentioned by Wall (1970).

In addition to the storage forms already discussed, Hopkins and Lofgren (1968) presented evidence that labelled urates in the fat body of *Leucophaea* are strongly bound to proteins and possibly other substances, such as peptides, at physiological pH. Along similar lines, chromogranule proteins which combine with pteridines and uric acid have been isolated from cuticular epithelial cells of *Bombyx* larvae (Tsujita and Sakurai, 1964, 1965). Three specific proteins were involved which combined with sepiapterin, isoxanthopterin and uric acid. Amino acid analysis has revealed that the protein combining with uric acid is rich in glycine. From these results it is apparent that more complex storage forms exist than was previously recognized. Future workers in this area must take these complexities into account.

Assessing the partitioning of stored urates into their theoretically possible functions (Nolfi, 1970) is not an easy task, when based on the scant current knowledge. Almost certainly these substances are involved (to varying degrees) in the synthesis of nucleic acids, pigmentation, and permanent sequestration (storage excretion). However, the frequently stated concept that the principal function of stored urates is excretory in nature (Maddrell, 1971) is now subject to serious challenge. There is considerable evidence showing the dynamic state of stored urates, and the ability of the insect to mobilize these reserves in times of dietary stress. Gier (1947) was one of the first to recognize this possibility. He suggested that in *Periplaneta* the fat body crystals (uric acid) represent accumulations of protein waste which are either eliminated or utilized when dietary protein is limited. Subsequent work on the same species has shown in much more detail that urates accumulate extensively, particularly in the fat body, in insects on high protein diets, and are rapidly mobilized when dietary protein is restricted or absent (Haydak, 1953; McEnroe, 1956; Mullins, 1971). It has also been proven in *Periplaneta* that uric acid is not eliminated from the body under a range of dietary protein levels including severe restriction (Mullins, 1971; Mullins and Cochran, 1972, 1973b). In view of its rapid disappearance under the latter condition, utilization is the only logical explanation. Similarly, uric acid has been proposed as a nitrogen source for tissue building in *Popillia* (Ludwig, 1954). In another beetle, *Anthonomus*, it has been demonstrated with labelled uric acid that the label subsequently appears in amino acids, nucleic acids and CO_2, again suggesting uric acid utilization (Mitlin and Mauldin, 1966; Mitlin and Wiygul, 1973). It remains to be determined how commonly this phenomenon occurs in the insect world, but it is obvious that urate storage can no longer be written off solely as storage excretion.

The concept of urate mobilization and subsequent utilization as a metabolic reserve raises additional questions. The principal one is how can mobilization occur if insect tissues have only limited ability to catabolize uric acid as described above? An answer which has been suggested many times in the past is the involvement of intracellular symbionts (Kilby, 1965; Donnellan and Kilby, 1967; Lanham, 1968). These organisms were discovered many years ago (Blochmann, 1887), and have been described as having bacterial and rickettsial characteristics (Brooks and Richards, 1966; Hinde, 1971b). They are usually considered to be of bacterial origin (Bush and Chapman, 1961; Brooks, 1970). They have been reported to occur widely in cockroaches, and in termites, weevils, and certain homopterans (aphids) (Brues and Dunn, 1945; Brooks, 1963; Lanham, 1968). The fat body mycetocyte cells (Brooks and Richards, 1966; Brooks, 1970; Hinde, 1971a b) and the ovaries (Bush and Chapman, 1961; Brooks and Richards, 1966; Brooks, 1970) are the main sites where symbionts are found. Brooks (1970) stated that only one symbiont species has been definitely established for cockroaches, while Hinde (1971b) presented evidence for two types in aphids. Highly evolved relationships between host and symbiont have been developed in both the fat body (Malke and Schwartz, 1966) and ovaries (Bush and Chapman, 1961) of cockroaches.

In attempting to elucidate the metabolic contribution of

symbionts versus insect tissues, certain problems arise. Some authors have endeavoured to culture symbionts separately from the insect (Keller, 1950; Pierre, 1964, 1965; Donnellan and Kilby, 1967; Hinde, 1971c). Others have tried to produce insects free of symbionts by feeding or injecting antibiotics or microbial disrupting agents, such as lysozyme (Brooks and Richards, 1955; Malke, 1964; Malke and Schwartz 1966; Brooks 1970; Wharton and Lola 1969a, 1970). Following a critical review of the literature, Brooks (1970) contended that all previous reports on cultured symbionts were based on work with contaminants. Similarly, insects which were sterilized by one or another method suffered abnormalities in growth rate, colour, behaviour and probably metabolism (Brooks and Richards, 1955; Harshbarger and Forgash, 1964; Brooks and Richards, 1966; Brooks, 1970; Daniel and Brooks, 1972). Finally, Brooks and Richards (1966) discounted the evidence on re-infecting aposymbiotic cockroaches with cultured microbes, and instead attributed the re-appearance of such symbionts to incomplete sterilization. These are very severe criticisms of previous work in this field, and serve to point out how difficult it is to obtain meaningful results with symbionts or on the host-symbiont relationship.

Bearing these limitations in mind, many metabolic activities have been attributed to microbial symbionts. Among them are: cleavage of aromatic rings (Murdock *et al.*, 1970); possessing guanase, uricase, malic enzyme, and malic dehydrogenase (Pierre, 1964, 1965; Dubowsky and Pierre, 1966; Tarver and Pierre, 1967); contributing vitamins (Ludwig and Gallagher, 1966); incorporating inorganic sulphur into organic sulphur compounds (Henry and Block, 1960); and synthesizing certain amino acids (Henry and Block, 1962). Of direct pertinence to uric acid mobilization is the work of Donnellan and Kilby (1967), who reported uric acid degradation through allantoin, allantoic acid, ureidoglycolate, glyoxylic acid, tartronic semialdehyde to glycerate and into the Krebs cycle by cultivated fat body symbionts from *Periplaneta*. While this report is subject to the criticism levelled by Brooks (1970), it provides an interesting insight into what the *in vivo* symbionts might be doing. In addition, Milburn (1966) has reported that morphological changes in cockroach symbionts may be a key to their metabolic activity. She showed rather striking changes in their appearance at times when demands on lipid and glycogen stores in the fat body were high. These changes may be under the control of the corpora cardiaca.

With the lysozyme injection approach, Malke (1964) was able to show a great reduction in *Periplaneta* fat body symbiont numbers.

He claimed complete elimination was achieved. Using insects in this condition, Malke and Schwartz (1966) demonstrated a 20-fold increase in fat body uric acid in comparison with untreated cockroaches. They proposed an integrated functioning of the symbionts, urate cells and other fat body cells in the utilization of stored urates so that their concentration is regulated under normal conditions, whereas fat body cells in aposymbionts accumulate excess uric acid (Brooks and Richards, 1956; Harshbarger and Forgash, 1964; Pierre, 1964), and must utilize stored lipids, proteins and carbohydrates. This brings about a depletion of these reserves. Malke and Schwartz (1966) concluded that cockroach symbionts play a central role in fat body intermediary metabolism. This conclusion is probably valid for those insects maintaining symbionts, in spite of the methodological problems mentioned above. At least the theoretical, and probably the actual, basis for the mobilization and catabolism of stored urates has been established. This area of research promises to be quite exciting in the years ahead.

In certain parts of the preceding discussions there has been a clear indication that uric acid and/or urates are moved from one part of the insect's body to another. Indeed, such movements have been documented (Wigglesworth, 1942; Haydak, 1953). The question is how such insoluble substances are transported. Obviously, the haemolymph is involved (Florkin, 1937; Heller and Szarkowska, 1957; Barrett and Friend, 1966; Hilliard and Butz, 1969; Wharton and Lola, 1969b; Jungreis and Tojo, 1973). It is well recognized that insect haemolymph contains an abundance of amino acids and proteins (Wyatt, 1961; Wigglesworth, 1972). Seegmiller (1969) has pointed out that in the presence of serum proteins uric acid readily forms highly supersaturated solutions. These solutions can be quite stable, but tend to precipitate in the presence of urate crystals. In addition, complex solutions of this type often form lyophobic colloids which greatly enhances their stability (Porter, 1963b). Thus, the concept of haemolymph carrying the necessary amounts of urates in solution seems quite plausible. For these possibilities to be important in physiological transport, however, they must be subject to general or localized control. The flocculation of colloids can be accomplished by altering the pH or by the presence of certain cations, notably ammonium ions (Porter, 1963b). These and possibly other methods (Hilliard and Butz, 1969), together with crystallization around pre-existing urate crystals, offer adequate opportunity to bring about the necessary control in pertinent tissues. Hormonal control of the level of urate available for transport may also occur (Bodenstein, 1953; Milburn, 1966; Thomas and Nation, 1966). In addition, protein bound urates may be a means of transport (Hopkins and Lofgren, 1968; Whitmore and Gilbert, 1972). 3-Ribosyluric acid has also been isolated from insect tissues and suggested as a transport vehicle (Heller and Jezewska, 1960; Jezewska *et al.*, 1967; Tojo and Hirano, 1968). From this sizeable number of potential contributing factors, it seems clear that controlled transport of urates or uric acid by the insect circulatory system can easily be accomplished. The actual mechanisms used by insects remain largely undetermined.

Ammonia

One of the dogmas of biology holds that a correlation exists between the type of nitrogenous material excreted by animals and the habitat they occupy (Needham, 1938). Thus, aquatic forms usually excrete ammonia, while their terrestrial cousins may excrete a variety of materials less demanding of water elimination. Useful as dogmas may be, in research they often serve to inhibit the development of an area. This is probably the case for ammonia excretion in insects, and possibly for certain other animals. Coupled with this is the highly volatile nature of the chemical itself which virtually insures a major loss from excreta unless specific measures are taken to prevent it. Documentation for both of these points will be provided below.

From the older literature it is well established that certain aquatic larvae and the larvae of blowflies, which live in moist environments, are strongly ammonotelic. Weinland (1906) was perhaps the first to demonstrate this fact using larvae of *Calliphora*. Microbial production of ammonia was found to be small by growing insects under sterile conditions with little or no diminuation in ammonia production (Brown, 1938b; Lennox, 1941a). In these insects as much as 90 per cent of the excretory products may be in the form of ammonia (Gilmour, 1961). Similarly, Staddon (1955) has shown that up to 90 per cent of the nitrogen excreted by *Sialis* larvae is present as ammonia, under certain conditions. However, if excretion was prevented, tissue ammonia did not increase appreciably, indicating that other routes for its dispersal are available. Staddon (1959) also found that ammonia is the main nitrogenous component of excreta from dragonfly nymphs, and that it increases sharply when the insects feed on a high protein diet. In addition to these reports, it has been established by various authors that ammonia is present in small amounts in the faeces of diverse insect species (Payne, 1936; Brown,

1936, 1937; Powning, 1953; Irreverre and Terzian, 1959; Razet, 1961). This work has been well summarized (Gilmour, 1961; Wigglesworth, 1972; Razet, 1966; Bursell, 1967) and need not be repeated here. Suffice to say that ammonia is not an uncommon excretory product among insects, even though it is usually reported to be present in small amounts. In *Rhodnius* it is said to be absent from the urine (Wigglesworth, 1931).

In some of the more recent literature the question of ammonia excretion by terrestrial animals has again been raised. Speeg and Campbell (1968) have reported the volatilization of ammonia gas from terrestrial snails. The methods of production and volatilization are different from those expected in insects (see below), but these animals serve as a case in point. It is known that terrestrial isopods excrete ammonia with the faeces and through the general body wall as a gas (Hartenstein, 1968; Wieser and Schweizer, 1970; Wieser, 1972). Study of the enzymes of *Onicus* again indicates the origin of ammonia may be different from that expected in insects, and that these animals can tolerate levels of ammonia generally described as being toxic (Hartenstein, 1968). Several species of diplopods excrete ammonia in amounts ranging from 20–57 per cent of the total non-protein nitrogen eliminated (Hubert and Razet, 1965; Bennett, 1971). Finally, American cockroaches, held under various nutritional regimes, were shown to excrete ammonia in amounts ranging from 12–91 per cent of the total nitrogen voided (Mullins and Cochran, 1972). Blight (1969) also reported variable amounts of ammonia emanating from certain colonies of the desert locust.

While the ramifications of these findings are not entirely clear at present, they amply demonstrate that terrestrial animals, particularly arthropods, can employ ammonia as a major vehicle for the excretion of excess nitrogen. Future research may show that some special condition(s) must accompany ammonia voiding in order to minimize its toxicity. As a possible example, freshly voided faeces from the American cockroach are often very wet. Subsequent drying occurs and is accompanied by loss of ammonia. Mullins and Cochran (1973b) have shown that from 48–72 per cent of the ammonia present in wet excreta from this insect is lost upon drying. Not only could this be a mechanism for minimizing ammonia toxicity to tissues, it could also help explain why rather small amounts of ammonia are normally reported from faeces of terrestrial insects. This is not to imply that ammonia is always an important nitrogenous waste for such species (McNally *et al.*, 1965). Rather, it should serve to forewarn future researchers that adequate precautions must be taken to guard against loss of faecal ammonia in evaluating total nitrogen excretion. Examples from past literature where such precautions were apparently not taken, are numerous.

The biochemistry of ammonia production in insects is an important, but not completely resolved, question. Several possible mechanisms exist and may be involved under appropriate conditions or in some species. One route that does not seem to be of significance in insects is the breakdown of urea to ammonia and carbon dioxide by the enzyme urease. This is probably true for at least two reasons. The enzyme urease itself has only rarely been reported in insects (Truszkowski and Chajkinowna, 1935; Baker, 1939; Robinson and Baker, 1939; Robinson and Wilson, 1939), and its presence there is usually discounted (Lennox, 1941b; Gilmour, 1961; Wigglesworth, 1972; Bursell, 1970). In addition, the principal pathways by which urea is produced seem to be blocked or are of questionable significance, as is discussed more fully in other sections (pp. 208, 225).

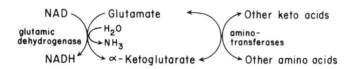

Fig. 3.4. An illustration of amino acid interconversions and ammonia production in insects. (Adapted from Bursell, 1967.)

Of those mechanisms that do occur in insects the one which appears to be of greatest importance is the complex of reactions involving transaminases (amino-transferases) and glutamate dehydrogenase (Fig. 3.4). With this scheme, glutamate is oxidatively deaminated by a coupled reaction involving NAD, resulting in the production of α-ketoglutarate, ammonia, and NADH. In consequence, α-ketoglutarate is available as a substrate for transaminase reactions, which in turn assure a continuing supply of glutamate (Bursell, 1970). McAllen and Chefurka (1961a) demonstrated that extensive transaminase activity is present in *Periplaneta* homogenates when α-ketoglutarate and a variety of amino acids are used as substrates. In most cases glutamate was detected without difficulty. They partially purified the glutamate-aspartate transaminase activity and detected the presence of glutamate dehydrogenase in cockroach fat body. Using this same insect Mills and Cochran (1963) identified and partially purified glutamate-aspartate transaminase, glutamate-alanine transaminase, and glutamate dehydrogenase from muscle mitochondria. From *Schistocerca*, homogenates of fat body contain transaminase activity for α-ketoglutarate and numerous amino acids (Kilby and Neville, 1957). Glutamate-aspartate and glutamate-alanine transaminases were most active, and were widely distributed within the insect's tissues. In a related study the occurrence of glutamate dehydrogenase was also shown in *Calliphora* (Desai and Kilby, 1958a). Chen and Bachmann-Diem (1964) have demonstrated the presence of active transaminases from *Drosophila* larval fat body. Again the glutamate-alanine and glutamate-aspartate reactions appeared to be most active. These authors briefly reviewed the earlier literature concerning insect transaminases, and clearly showed that this type of enzyme is widely distributed among insects. More recently Crabtree and Newsholme (1970) have surveyed muscle tissue from about a dozen insect species for amino-transferase activity and the dehydrogenases of glutamate and proline. The enzymes were usually present. The potential for proline utilization as an energy source was pointed out (also see p. 227). Bheemeswar (1959), Gilmour (1965), and Bursell (1970) have stressed the significance of this transaminase-dehydrogenase system, and its probable involvement in ammonia production *in vivo*.

Another mechanism which is potentially important in the production of ammonia is the action of amino acid oxidases on amino acids. This is a simpler mechanism since amino acids are directly deaminated. The difficulty here is proving its significance for the 'natural' L-amino acids in the presence of the transaminase-dehydrogenase system described above (Gilmour, 1965). Also complicating the situation is the existence of very active D-amino acid oxidases in a variety of species (Kilby and Neville, 1957; Desai and Kilby, 1958a; Auclair, 1959; Corrigan *et al.*, 1963). Indeed, Gilmour (1965) has questioned the existence of L-amino acid oxidases, while maintaining that the function of the D-amino acid oxidases is not known. Certain functions have now been suggested for the D-amino acids themselves in insects, and in other animals (Corrigan, 1969). One obvious suggestion is that their oxidases play a role in regulating the level of specific D-amino acids in the blood. Whether this is a physiologically significant source of ammonia remains to be determined. L-amino acid oxidases have been found in insects (Garcia *et al.*, 1957; Kilby and Neville, 1957; Desai and Kilby, 1958a), although some of the peculiarities reported in the older literature have not been satisfactorily explained (Bheemeswar, 1959). It appears doubtful that their action is a major source of ammonia.

Several other deaminating enzymes are known to occur in insects. Perhaps the best studied of these is adenosine deaminase. This enzyme has been known in vertebrates for many years (Conway and Cooke, 1939). Shortly thereafter it was reported from the blowfly, *Lucilia*, in which it was adjudged to be an important enzyme in the production of ammonia (Lennox, 1941b). Florkin and Frappez (1940b) reported it from the blood of the beetle *Dytiscus*. It is present in the larvae of *Drosophila* and increases with age (Wagner and Mitchell, 1948). Cochran (1961) demonstrated its occurrence in the muscle mitochondria of the American cockroach. In this same insect, the enzyme has been shown to be present in a variety of tissues including gut and Malpighian tubules (Cordero and Ludwig, 1963). In other work on *Drosophila*, Hodge and Glassman (1967a, b) have pointed out that guanine deaminase (guanase) occurs in addition to adenosine deaminase but not adenine deaminase (adenase). The enzymes adenase and guanase have been reported from other insects (Florkin and Frappez, 1940a; Duchâteau *et al.*, 1941). Monoamine oxidase is known from the Malpighian tubules of two species of cockroaches (Boadle and Blaschko, 1968), and Corrigan (1970) mentioned the non-oxidative deaminase, serine dehydrase. Sepiapterin deaminase has been studied in the silkworm mutant *lemon* (Tsusue, 1967, 1971). The action of all of the enzymes discussed in this paragraph result in the production of ammonia. For most of them, critical evaluation of their physiological significance in ammonia excretion has not been accomplished.

In considering ammonia as an excretory product, there are two areas that need further discussion. One is the tissue distribution of the enzymes involved; the other is evidence for the accumulation of ammonia within the insect's body, and its relation to sites of elimination. Because of the chemical nature and toxicity of ammonia, it is likely to be dispersed quickly or undergo chemical reaction. From this it follows that the mere occurrence of an enzyme capable of producing ammonia does not prove the importance of that tissue in ammonia excretion, but the accumulation of ammonia in a particular tissue or organ should be evaluated as a possible signal concerning its excretion.

A tabular review of the tissue distribution of ammonia-producing enzymes has been prepared (Table 3.2 – also see Table 3.1 for adenosine deaminase data). As is often the case, many studies involved the use of whole-body homogenates and are of no help here. From the tables it can be seen that transaminases and adenosine

Table 3.2
Tissue distribution of ammonia-producing enzymes in insects

Enzyme	Species	Tissue	Reference
Transaminases (amino transferases)	*Calliphora*	Larval fat body (f.b.)	Desai and Kilby, 1958a
		Flight muscles (F.m.)	Crabtree and Newsholme, 1970
	Schistocerca	Nymphal and adult f.b.	Kilby and Neville, 1957
	Hyalophora	Adult muscle; low activity in blood, f.b., Malpighian tubules (M.t.), gut	McAllen and Chefurka, 1961b
	Periplaneta	Muscle mitochondria	Mills and Cochran, 1963
		M.t., f.b., nerves, muscle, gut	McAllen and Chefurka, 1961b
		F.m.	Crabtree and Newsholme, 1970
	Melolontha	F.m.	Crabtree and Newsholme, 1970
	Glossina	F.m.	Crabtree and Newsholme, 1970
	Locusta	F.m.	Crabtree and Newsholme, 1970
	Lethocerus	F.m.	Crabtree and Newsholme, 1970
	Apis	F.m.	Crabtree and Newsholme, 1970
	Bombus	F.m.	Crabtree and Newsholme, 1970
	Laothoe	F.m.	Crabtree and Newsholme, 1970
	Sarcophaga	F.m.	Crabtree and Newsholme, 1970
Glutamate dehydrogenase	*Calliphora*	Larval f.b.	Desai and Kilby, 1958a
	Schistocerca	Nymphal and adult f.b.	Kilby and Neville, 1957
	Periplaneta	Muscle mitochondria	Mills and Cochran, 1963
		F.b.	McAllen and Chefurka, 1961a
		F.m.	Crabtree and Newsholme, 1970
	Melolontha	F.m.	Crabtree and Newsholme, 1970

Table 3.2
continued

Enzyme	Species	Tissue	Reference
	Locusta	F.m.	Crabtree and News-holme, 1970
	Lethocerus	F.m.	Crabtree and News-holme, 1970
	Bombus	F.m.	Crabtree and News-holme, 1970
	Laothoe	F.m.	Crabtree and News-holme, 1970
	Sarcophaga	F.m.	Crabtree and News-holme, 1970
L-Amino acid oxidase	*Calliphora*	Larval f.b.	Desai and Kilby, 1958a
	Schistocerca	Nymphal and adult f.b.	Kilby and Neville, 1957
D-Amino acid oxidase	*Calliphora*	Larval f.b.	Desai and Kilby, 1958a
	Schistocerca	Nymphal and adult f.b.	Kilby and Neville, 1957
	Galleria	F.b., M.t., midgut	Corrigan *et al.*, 1963
	Calliphora	F.b., M.t., midgut	Corrigan *et al.*, 1963
	Oncopeltus	F.b., M.t., midgut	Corrigan *et al.*, 1963
	Periplaneta	F.b., M.t., midgut	Corrigan *et al.*, 1963
	Byrsotria	F.b., M.t., midgut	Corrigan *et al.*, 1963
	Eurycotis	F.b., M.t., midgut	Corrigan *et al.*, 1963
	Periplaneta	F.b., M.t.	Auclair, 1959
	Blattella	F.b., M.t.	Auclair, 1959
Adenosine deaminase (See Table 3.1)			
Monoamine oxidase	*Periplaneta*	M.t.	Boadle and Blashko, 1968
	Blaberus	M.t.	Boadle and Blashko, 1968
Sepiapterin deaminase	*Bombyx* (lemon mutant)	Midgut, M.t., f.b., body fluid, silk gland, integument, gonads	Tsusue, 1971

deaminase have the widest tissue distribution according to available information. The other enzymes are less well studied and interpretations are somewhat difficult. However, the general impression gained from the tables is that ammonia-producing enzymes are widespread within the insect's body. This implies that the blood could be of importance in the excretion of ammonia by serving as a transport vehicle from the site of production. Insect blood has been reported to contain ammonia (Florkin and Frappez, 1940b; Lennox, 1941a; Levenbook, 1950; Staddon, 1955). The difficulty, as Gilmour (1961) pointed out, is that ammonia is usually present in rather low concentrations which tends to negate the idea of blood transport. Perhaps a more likely alternative is that ammonia produced by some of these enzymes is utilized in synthetic reactions.

Several authors have shown that the main site of ammonia accumulation within the insect is the gut, especially the hind-gut. In *Lucilia* (Lennox, 1941a) and *Sarcophaga* (Prusch, 1971) larvae the concentration of ammonia increases steadily in the hind-gut from the anterior to the posterior region. In *Sialis* larvae Staddon (1955) also found high concentrations of ammonia in the hind-gut and suggested that it got there by way of the Malpighian tubules. Ammonia may be present in other parts of the gut as well. In *Lucilia* it occurs in the middle section of the mid-gut (Lennox, 1941a). It was proposed that this ammonia is absorbed into the haemocoel and does not contribute, by direct passage within the gut, to that which is present in the hind-gut. Lennox (1941a) attributed ammonia excretion into the hind-gut to the Malpighian tubules. Prusch (1971) showed that ammonia is present in very high concentrations in the fore-gut of *Sarcophaga*, decreases to very low levels in the posterior part of the mid-gut and is present in high concentrations again in the hind-gut. He also provided direct evidence for a net ammonia efflux from isolated mid-gut and a net ammonia influx into isolated hind-gut. According to this author the Malpighian tubules are not involved. In a subsequent paper, Prusch (1972) described the process more precisely. He found that NH_4^+ are moved across the isolated hind-gut against a concentration gradient. Maximum movement required the presence of NH_4^+ and amino acids in the outside medium. NH_4^+ movement into the hind-gut lumen is accompanied by movement of Cl^- in the same direction, and Na^+ and K^+ out of the hind-gut into the blood. Thus, in addition to voiding excess nitrogen this process may be concerned with ion regulation and osmoregulation (see p. 187). Prusch (1972) also found evidence that the hind-gut can deaminate amino acids. Given the high level of amino acids in insect blood (Florkin, 1959; Wyatt, 1961) and the low level of blood ammonia in this and other species (Prusch, 1971), the implication is that deamination is accomplished by the tissue which does the excreting, i.e. the hind-gut. If correct, this would obviate much of the difficulty associated with ammonia excretion. The hind-gut, and especially its posterior parts, would be the only tissue required to deal with high concentrations of ammonia. For those species, including terrestrial forms, which void wet excreta the problem would not seem to be insurmountable. Whether these findings are generally applicable to insects excreting ammonia must await further experimentation. The type of deamination reaction(s) occurring in the hind-gut also needs clarification. In addition, the contribution of hind-gut microflora to ammonia excretion should be re-examined (Todd, 1944).

From the research reviewed, it appears that the gut is involved in a major way in ammonia physiology. However, in terrestrial snails it was shown that the respiratory system accomplishes the release of ammonia (Speeg and Campbell, 1968). That this could also be the case in insects is an obvious possibility. Lennox (1941a) performed a simple experiment in which he attempted to indicate sites of ammonia voiding. He secured blowfly larvae on Nessler-impregnated filter paper and observed the locations where a reaction occurred. The paper darkened significantly at the posterior end of the larvae and very slightly at their anterior end. General cuticular loss of ammonia was not detected. The experiment may have some value on the question of respiratory voiding, but most muscoid larvae have modified respiratory systems with only anterior and posterior dorsal spiracular openings (Snodgrass, 1935) which generally correspond in position to the gut openings. A modification of the technique might produce more definitive results. Mullins (personal communication) has studied the site of ammonia release in American cockroaches by sealing the mouth and anus with melted parafin. Under these conditions some ammonia is evolved but in greatly reduced amounts over the controls. Additional study will be required to prove the point, but the evidence available indicates the hind-gut as the principal organ of ammonia excretion in insects. Apparently, ammonia is subsequently released from the rectum along with the faecal pellets.

Urea

The most common nitrogenous excretory product among terrestrial vertebrates is urea. It is well suited for that purpose among animals that have reasonable access to drinking water, but can be excreted in fairly concentrated solutions without undue toxicity. The synthesis of urea in vertebrates is mediated primarily through the classical ornithine cycle, a description of which can be found in virtually any textbook of biochemistry. Perhaps because the case is so well known for vertebrates, insect biochemists have tended to take a rather restricted view of urea production and excretion. This appears to have delayed the development of a satisfactory rationale concerning urea production in insects. While the problem is still not completely understood, considerable progress has recently been made.

The occurrence of urea in insect excreta is well known. Only a few general comments will be made here because the area has been thoroughly reviewed (Gilmour, 1961; Wigglesworth, 1972; Razet, 1966; Bursell, 1967; Corrigan, 1970; Schoffeniels and Gilles, 1970). The usual finding is that urea comprises a small percentage of the total nitrogen in excreta, sometimes being present in only trace amounts. Wigglesworth (1972) stated that it is always present in the urine of insects. The results of certain other workers do not necessarily support that generalization (Brown, 1938b; Nation and Patton, 1961; McNally et al., 1965), but, of course, when one considers trace amounts and levels of detectability the question is more difficult to resolve. On the other hand, certain species excrete relatively large amounts of urea (Powning, 1953; Berridge, 1965). This finding is intriguing, and both authors concluded that the urea is of metabolic origin. For example, in nymphs of the plant-feeding bug, *Dysdercus*, the gut is discontinuous, thereby preventing direct voiding of ingested urea with the faeces (Berridge, 1965). In addition, there is ample evidence for the occurrence of urea in the blood and internal tissues of numerous insect species (Ludwig, 1954; Ludwig and Cullen, 1956; Ramsay, 1958; Hayashi, 1961g; Mitlin *et al.*, 1964a; Wang and Patton, 1969), which presumably also argues for the metabolic origin of urea. Contrarily, Wigglesworth (1931) stated that urea excreted by the blood-feeding bug *Rhodnius* is derived largely from ingested blood. In spite of this latter finding, Bursell (1967) concluded that food is not the source of urea in most species of insects. Rather, they apparently produce and excrete it themselves (Gilmour, 1965). In support of this concept, Ramsay (1958) found that urea and certain other low molecular weight materials are present in the primary filtrate of *Carausius* and diffuse into the Malpighian tubule lumen with this fluid.

Two biochemical pathways are often suggested as possible routes leading to the production of urea. They are: (1) degradation of purines via uric acid, and (2) involvement of the ornithine cycle. Both pathways have certain attendant problems. It has previously been pointed out (p. 207) that many insect species are capable of producing and excreting certain degradation products of uric acid. From the extensive studies of Razet (1953, 1954, 1956, 1964, 1965, 1966, 1970), and certain other workers, the occurrence of uricase and allantoinase in insect tissues is well documented. However, further degradation of the purine structure beyond allantoic acid by insect tissues has only rarely been reported (Razet, 1965; Gilmour, 1965; Kilby, 1965; Bursell, 1967; Wang and Patton, 1969). Thus, the available evidence indicates a metabolic block at this point preventing the production of urea from allantoic acid due to the apparent absence of the enzyme allantoicase. Further evidence could negate this conclusion, but in view of the extensive studies already reported it seems unlikely. Accordingly, for most species this pathway does not appear to be of importance in urea production. Of course, gut microbes could carry out the allantoicase reaction, possibly explaining the exceptional species mentioned above.

The second pathway to be discussed is related to the ornithine cycle. It should be made clear, however, that the complete ornithine cycle has not been reported for any insect species. Rather, certain of the reactions which form a part of the vertebrate ornithine cycle are present in insect tissues. The enzyme arginase, which catalyzes the formation of ornithine and urea from arginine, is the best known example. It has been reported from many insect species, especially in extracts derived from fat body and flight muscles (Garcia et al., 1956, 1958; Kilby and Neville, 1957; Szarkowska and Porembska, 1959; Porembska and Mochnacka, 1964; Kameyama and Miura, 1968; Reddy and Campbell, 1969a, b; Powles et al., 1972). It appears to be present throughout the life cycle of several species (Garcia et al., 1956), and increases in activity with adult emergence in Hyalophora (Reddy and Campbell, 1969a). Insect arginases vary in their properties, but generally appear to be similar to their vertebrate counterparts (Reddy and Campbell, 1969b). The intermediates citrulline, arginine and ornithine are often found in insect tissues (Garcia et al., 1956; Porembska and Mochnacka, 1964). In the apparent absence of a functioning ornithine cycle, the biochemical role and importance of these intermediates is not fully understood.

Razet (1966) stated that urea production alone is sufficient justification for the wide-spread presence of arginase in insects. Indeed, present information indicates the arginase reaction to be the major mechanism of urea production. In view of the typically small amounts of urea found in insects, however, an alternate hypothesis is that the production of urea by insects is almost incidental, and that arginase may have other more important functions. Support for this idea is now available from certain biochemical and nutritional studies.

Reddy and Campbell (1969a) have demonstrated the occurrence of arginase, ornithine transaminase and pyrroline-carboxylate reductase in fat body and flight muscles of *Hyalophora*. These enzymes catalyze the reactions mediating the conversion of arginine into proline (Fig. 3.5). The functional significance of the system was also demonstrated for intact fat body by its conversion of labelled arginine into labelled proline. Similar results have also been obtained by Inokuchi (1969) and Inokuchi *et al.* (1969) using *Bombyx*. In nutritional studies they were able to show a proline-sparing effect by the addition of ornithine or arginine to the diet, as well as an arginine-sparing effect by the addition of citrulline. A proline-sparing effect was also produced by added glutamate. In general, the sparing effects were more pronounced when low proline diets were employed. Confirmation of these results was obtained from parallel

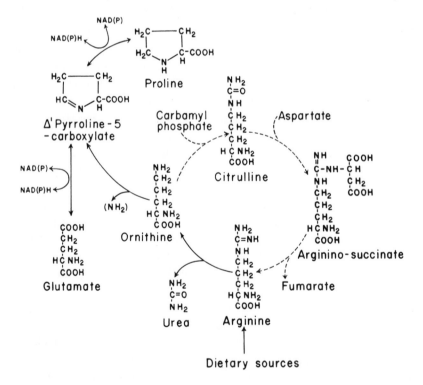

Fig. 3.5. Insect pathway for the conversion of arginine to proline. Broken arrows indicate reactions of questionable significance in insects.

studies using labelled intermediates. Inokuchi *et al.* (1969) concluded that a pathway exists in *Bombyx* from citrulline to proline via arginine and ornithine, which will result in the production of urea (Fig. 3.5). Porembska and Mochnacka (1964) have attempted to demonstrate the synthesis of citrulline from ornithine and carbamyl phosphate using a tissue extract from *Celerio*. The results were negative even though these authors easily demonstrated the reaction in ox liver extracts. In the latter system the addition of *Celerio* tissue extract to active preparations had no inhibitory effect. Nutritional studies on *Argyrotaenia* have allowed Rock (1969) to reach essentially the same conclusion. He demonstrated that the arginine requirement for this insect can be satisfied by the addition of citrulline to the diet, but not ornithine. This agrees with the findings of Inokuchi *et al.* (1969). Powles *et al.* (1972) have reported on enzyme studies of the ornithine cycle in the bug *Nezara*. They found an active ornithine transcarbamylase resulting in the incorporation of bicarbonate into citrulline. This is the clearest demonstration of this enzyme in an insect yet reported. On the other hand, these authors could not demonstrate the existence of the enzyme argininosuccinate synthetase. If these authors are correct, citrulline cannot be converted into arginine by this insect. Similarly, carbamyl phosphate synthetase could only be demonstrated by use of radio-active tracer techniques, indicating a very low activity. However, Abe and Miura (1972) reported a stable preparation of this enzyme in *Aldrichina*. Carbamyl phosphate is, of course, one of the reactants in the production of citrulline, and is the compound into which protein-derived ammonia is incorporated in urea excreting organisms. Thus, while *Nezara* has one very active ornithine cycle enzyme not usually found in insects, it still does not have the complete functioning cycle. Powles *et al.* (1972) concluded that urea produced by this insect is the result of arginase acting on excess arginine in the diet.

Dadd (1973) pointed out the irrationality of anticipating a functioning ornithine cycle in a group of animals which could most nearly be classified as uricotelic. This again brings us back to the idea that urea production may be little more than a by-product of metabolism. If this is true, then what is the significance of the arginine to proline pathway described above (Fig. 3.5)? While it is probably too early to give a definitive answer, one suggestion is that this is a mechanism to ensure a ready supply of proline for energy production or other metabolic pools. The central role of proline in flight metabolism has been documented (Bursell, 1963, 1966; Sacktor and Childress, 1967). It seems to be one of the few metabolites which can satisfactorily permeate flight muscle mito-chondria. Once inside, it is rapidly converted to glutamate and intermediates of the Krebs cycle leading to oxaloacetate (Brosemer and Veerabhadrappa, 1965; Sacktor and Childress, 1967). This in turn assures an adequate supply of precursors for the complete oxidation of pyruvate via the Krebs cycle. Given the very common occurrence of arginine in insect blood (Corrigan, 1970), this pathway could be of considerable importance as an adjunct to energy production. In view of these considerations, it appears that future research on urea production could better be directed towards understanding its relationship to energy metabolism and the dietary or metabolic availability of arginine, rather than towards continued efforts to demonstrate a complete ornithine cycle.

Amino acids

The existence of amino acids in insect excreta has been well documented (Gilmour, 1961; Wigglesworth, 1972; Razet, 1966; Bursell, 1967, 1970). However, the question of their origin is not so well established. It has often been stated that they represent food constituents ingested in excess of dietary requirements (Powning, 1953). This implies that they find their way into the excreta either by passing through the gut unabsorbed or by being excreted from the blood after absorption. While these possibilities obviously exist, this view is too circumscribed to account for all of the known facts. One must also consider as physiologically normal the excretion of small amounts of amino acids and specific amino acids excreted at a high level as major excretory products.

Amino acid excretion in insects is a difficult area in which to work because of the widespread potential presence of amino acids in foodstuffs, and their ready biochemical interconversion once inside the insect's body (Bursell, 1967; Corrigan, 1970; Schoffeniels and Gilles, 1970). Because of these interconversions, and the possible role of gut microbes in producing amino acids from gut contents, clear cut distinctions between origins are not easy to make. For this reason a variety of criteria should be employed in evaluating the significance of amino acids as excretory products. Among them, comparisons of the type and concentration of amino acids in excreta, haemolymph, and food sources are useful. Bearing in mind the previously mentioned vagaries, it is sometimes possible to determine, at least generally, the origin of voided amino acids. Morphological criteria can also be useful as will be pointed out below.

One of the most characteristic features of insect blood is its high titre and great diversity of free amino acids (Wyatt, 1961; Stevens, 1961; Gilmour, 1965; Chen, 1966; Wang and Patton, 1969; Abdel-Wahab, 1971). Because of this, it is not surprising that amino acids are present in the primary filtrate produced by the Malpighian tubules, sometimes in substantial amounts (Maddrell, 1971). In *Carausius* their transfer through cells of the Malpighian tubules was reported to be passive (Ramsay, 1958). Active re-absorption of some amino acids occurs in the rectum of *Periplaneta* and *Schistocerca* (Wall and Oschman, 1970; Balshin and Phillips, 1971). Under these circumstances, small to moderate amounts of amino acids would be expected to occur in the excreta, and may be taken as the so-called unavoidable loss due to failure of the rectum to completely re-absorb them (Maddrell, 1971).

While it is satisfying to view this pattern as 'normal' for insects on a low amino acid diet, there is some doubt concerning faecal contamination. The discontinuous gut of larval *Dysdercus* avoids this problem. Yet a large number of ninhydrin positive chromatographic spots were detected in their excreta, with amino nitrogen constituting the second largest amount of the total nitrogen eliminated (Berridge, 1965). Similarly, Singh (1964) was able to collect uncontaminated tubule fluid from *Gryllotalpa* and demonstrate the presence of 6 or 7 amino acids in it. These findings lend support to the significance of the pattern described above. They also make it clear that amino acids voided in this manner are excretory in nature, even though they may be present in small amounts and their loss attributable to failure of re-absorption by the rectum. In this sense the individual amino acids should be viewed as minor or even trace excretory products.

A number of insect species appear to fit this pattern of minor excretion of numerous amino acids. Mitlin *et al.* (1964b) found 20 amino acids in excreta from the boll weevil, but in total they accounted for less than 3 per cent of the excreta nitrogen. In this case the food source contained no free amino acids. Similarly, McNally *et al.* (1965) reported that honey-bees fed protein-free diets voided a little over 1 per cent of their excreta nitrogen as alpha-amino nitrogen. Finally, Irreverre and Terzian (1959) studied three species of mosquitoes and found amino acids constituting less than 5 per cent of the total excreta nitrogen, although they did not indicate which amino acids were involved. A few other reports have also appeared in which somewhat higher or undetermined levels of excreta amino acids were found (Brown, 1937; Yoshitake and Aruga, 1950;

Harington, 1961; Pant and Agrawal, 1963; Mullins and Cochran, 1973b).

Information on the voiding of amino acids ingested in excess of dietary requirements comes primarily from plant-sucking insects. Frequently, data on amino acid composition of sap and honeydew are provided, but only rarely for haemolymph. For example, Lamb (1959) reported 72–90 per cent of honeydew nitrogen to occur as free amino acids and amides in *Brevicoryne.* Glutamine, glutamate, aspartic acid and about 20 other amino acids were present in the honeydew in about the same proportion as found in the host plant. Mittler (1953) found 11 amino acids and amides in the honeydew of *Tuberolachnus.* They were the same as those present in the willow sap on which the aphid lives, but were excreted in considerably smaller quantities than found in the sap (Mittler, 1958). The pea aphid *Acyrthosiphon* yields a honeydew containing 19 amino acids and amides which were also present in the host plant and in aphid blood. Their total concentration in honeydew was similar to that in aphid blood but was 2–10 times higher than in pea plant juice (Auclair, 1958). Burns and Davidson (1966) found 8 amino acids and amides in the honeydew of the tuliptree scale *Toumeyella* and in the sap of its host. The relative proportions of the amino acids appeared to be different in the two liquids, and honeydew also contained 2 amino acids not found in sap. Tamaki (1964a) demonstrated 14 amino acids in the honeydew of *Ceroplastes*, but only 10 from its host the tea plant. Honeydew from *Kerria* had 17 amino acids and amides, as compared to 9 in its host plant sap (Srivastava and Varshney, 1966). Powles *et al.* (1972) reported 27 amino acids from the excreta of *Nezara,* while only 19 were present in its host plant. In this case 54 per cent of the total excreta nitrogen was in the form of amino nitrogen. A review pertinent to this area has been published (Auclair, 1963).

From these results it seems that excretion of excess dietary amino acids not only occurs, but probably must occur in order for these sap-feeding insects to maintain internal osmotic balance. However, sequestering amino acids in haemocytes is a possible alternative (Evans, 1972). The variations in excretory pattern reported above may also be attributable to species differences, seasonal changes, nutritional state of the insect, biochemical interconversions of amino acids, and probably other factors. In any case, it is clear that these insects can generally and selectively remove amino acids from sap, excrete amino acids not found in sap, pass sap through their bodies relatively unchanged, or they may concentrate or dilute it according to their physiological needs. Of course, this again does not resolve the question of faecal *v.* excretory origin of the amino acids. Harington (1961) has provided an example which appears to show a faecal origin. The blood-sucking bug *Rhodnius* excretes a copious urine low in amino acids except for a short period soon after feeding. A rather distinct faecal residue is also produced and contains small quantities of about a dozen amino acids and related compounds. These apparently represent the residue from the blood meal and are faecal in origin. Histidine and histamine are the most prominent compounds present (Harington, 1956). On the other hand, the work of Auclair (1958) almost certainly demonstrates at least a partial excretory origin for high levels of amino acids since their concentrations in honeydew and blood were much higher than in food.

The last pattern to be considered here is the high level excretion of specific amino acids. Those mentioned prominently in this connection are histidine and arginine, both of which have a high nitrogen content (Bursell, 1967). Kondo (1967) reported that excreta from older 5th instar *Bombyx* larvae contains 5–6 per cent histidine. This is a concentration 30–100 times greater than other amino acid components. Because he was unable to demonstrate histidine breakdown or synthesis, Kondo suggested that it is accumulated in larval blood from food and subsequently excreted. Levenbook *et al.* (1971) found histidine to be a major component of *Manduca* meconium. Ornithine was also present in substantial quantities. These authors stated that amino acid excretion is a selective process, and that meconial amino acids do not simply reflect concentrations present in the free amino acid pool(s). In *Glossina*, Bursell (1965a, b) demonstrated that histidine and arginine together comprise about 20 per cent of the total excretory products. Based on a comparison of the composition of human blood and tsetse fly excreta, he suggested that these two amino acids, together with haematin, are quantitatively eliminated by this insect. This again implies selective elimination by some excretory mechanism. Sidhu and Patton (1970) presented evidence for a significant accumulation of asparagine, aspartic acid, arginine and glutamine in the honeydew of *Lipaphis* as compared to the host plant leaf sap. Smaller accumulations of several other amino acids also occurred. Insects attacking wool encounter the problem of dealing with its high sulphur content. *Tineola* and *Attagenus* have apparently solved it by excreting cystine in amounts exceeding 10 per cent of total dried excreta (Powning, 1953).

Unfortunately, these results convey little information on selective excretory mechanisms for amino acids. It is possible that their accumulation in blood is sufficient to assure high level excretion. They may flow passively into the tubule lumen and become lost (Ramsay, 1958). Alternatively, some extremely interesting selective mechanisms might be involved. Future research will be required to decide which is correct.

Active transport of amino acids by the insect rectum was discussed and an apparent breakthrough in this area has been reported. Meister (1973) has described an enzymatic cycle (the gamma-glutamyl cycle) in vertebrate kidney, which offers a meaningful explanation for amino acid transport. A key enzyme in the cycle, gamma-glutamyl transpeptidase, is membrane bound. It is localized in the brush border of the proximal convoluted tubule, which is presumed to be a primary site of amino acid re-absorption. It catalyzes the reaction between glutathione inside the cell and any one of a number of protein amino acids outside the cell. The product is a gamma-glutamyl-amino acid located within the cell. By the action of intracellular gamma-glutamyl cyclotransferase the translocated amino acid is liberated, the other product of the reaction being 5-oxyproline. Subsequently, a series of ATP-requiring reactions occur intracellularly in which 5-oxyproline is converted stepwise to glutamic acid, gamma-glutamyl-cysteine, and glutathione, thus completing the cycle. While this work needs to be confirmed and extended to other groups, such as insects, it may lead to a rapid improvement in our understanding of active transport in general.

Tryptophan derivatives

Compounds derived from the amino acid tryptophan are of considerable importance in insect biochemistry. For example, it has been known for many years that the ommochrome pigments of insect eyes and other tissues are formed by the kynurenine, 3-hydroxykynurenine pathway (Butenandt, 1952; Kikkawa, 1953). Also, certain amine derivatives of tryptophan are reported to carry out hormonal functions in insects (Corrigan, 1970). The recognition of the involvement of tryptophan derived compounds in insect excretion has occurred much more slowly. Indeed, it is probably correct to state that the full significance of their excretion is not yet appreciated. Nevertheless, knowledge in this area has evolved far enough to warrant a fuller discusion than has been previously presented.

It is recognized that tryptophan may be involved in at least four different metabolic pathways (Henderson *et al.*, 1962): (1) oxidation to formylkynurenine, (2) hydroxylation to 5-hydroxytryptophan, (3) conversion into indole-3-acetic acid, and (4) fission to indole. In addition, from the first of these pathways through 3-hydroxy-anthranilic acid and glutaryl-CoA the ring structure of tryptophan can be completely oxidized to CO_2 (Gholson *et al.*, 1960). This latter pathway has been studied in cockroaches and grasshoppers by assaying for $^{14}CO_2$ following injection of 5-[^{14}C] DL-tryptophan and by measuring enzyme reaction products at three points along the pathway (Lan and Gholson, 1965). All of the evidence from insect and other arthropod tissues was negative, while the pathway was clearly demonstrated in a variety of vertebrate species. In addition, insects do not appear to convert 3-hydroxyanthranilic acid into nicotinic acid, thus accounting for the dietary requirement for the latter (Brunet, 1965). It has also been reported that pathways (3) and (4) above are not present in animals (Brunet, 1965). This position may have to be modified, however, in view of the finding that indole acetic acid is present in the salivary glands of certain plant-feeding Hemiptera, and that it is derived from tryptophan (Miles and Lloyd, 1967; Miles, 1968). Present evidence does not indicate the involvement of this pathway in excretory processes.

The reactions leading from tryptophan through 5-hydroxy-tryptophan to 5-hydroxytryptamine (serotonin) and thence to a variety of other compounds are of importance in insects. 5-Hydroxy-tryptamine is a hormone of major significance having been reported from gut, heart, Malpighian tubules, brain, corpora cardiaca and ventral nerve cord of *Periplaneta* (Colhoun, 1963, 1964). It plays a role in fluid secretion in salivary glands (Berridge and Patel, 1968; Berridge, 1970b; Prince and Berridge, 1973), movement (Pilcher, 1971) and secretion (Maddrell *et al.*, 1969, 1971) in Malpighian tubules, and apparently in the circadian rhythmicity of locomotory activity (Cymborowski, 1970, 1973). It is present in insect venoms (Beard, 1963). 5-Hydroxytryptamine and other indolealkylamines also appear to be involved in additional roles which are not yet understood (Brunet, 1965). Enzymatic studies have been conducted on the decarboxylation of 5-hydroxytryptophan. The reaction is known to occur in the American cockroach (Colhoun, 1963) and the blowfly (Marmaras *et al.*, 1966). The hydroxylation of tryptophan has not been examined in detail although it obviously occurs (Corrigan, 1970).

In keeping with the hormonal role ascribed to 5-hydroxy-tryptamine one would expect it to function in very dilute concentrations. This seems to be the case since Colhoun (1963) reported activity from concentrations as low as 10^{-8} to 10^{-9} M. He stated further that it is present in extremely small amounts in brain and nerve cord, in comparison with the neural-transmitter substance acetylcholine, and suggested a localized distribution in the nervous system. Substances which can be described in this manner are not likely to be detected as excretory products. This is the case here, although intermediates along this pathway have usually not been sought in studies of excretion.

It is in the remaining pathway of tryptophan metabolism that the primary interest for excretion lies. Brunet (1965) has stated the situation well when he surmized that ommochrome synthesis may have proved to be so important in the evolutionary history of insects that they channeled tryptophan metabolism into this pathway to the near exclusion of all others. What has not been adequately recognized so far is the variety and extent to which components of this pathway can be involved in excretory processes. The situation is further complicated because ommochromes and sometimes intermediates such as kynurenine, are deposited in various parts of the body where they function in cuticular pigmentation, visual acuity, and other roles (Kikkawa, 1953; Linzen, 1958; Butenandt *et al.*, 1960; Harmsen, 1966b; Langer and Hoffman, 1966; Umebachi and Katayama, 1966; Burnet *et al.*, 1968). This dual or multiple functioning of a single chemical substance has sometimes caused difficulty in understanding the true nature of the processes involved.

The ommochrome pathway has been described in detail by many authors. For convenience to the reader, it is presented here in Fig. 3.6. It is of historical significance because some of the earliest correlations between enzyme action and single gene mutations were made in mutant insects possessing blocks along this pathway (Butenandt, 1952, 1959; Kikkawa, 1953; Ziegler, 1961a; Brunet, 1965; Corrigan, 1970). This aspect of the problem will be referred to briefly in conjunction with review of the pertinent enzymes. Of more immediate importance is the establishment of tryptophan derivatives as excretory products.

From studies of a large number of insect species Butenandt (1959) and Butenandt *et al.* (1960) have described the occurrence of several ommochrome pigments in various parts of the insect body. They occur in eyes, cuticle, wings, internal organs and eggs. Other authors have reported them from testes, ovaries, and Malpighian tubules

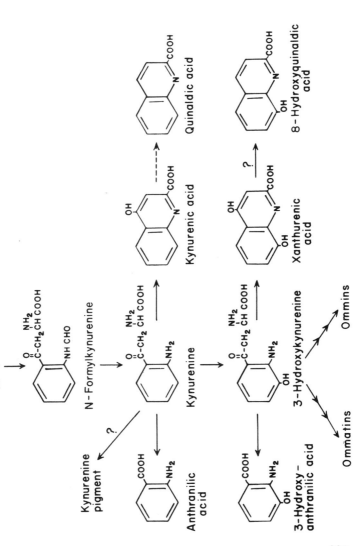

Fig. 3.6. The tryptophan-ommochrome pathway in insects (Adapted from Brunet, 1965 and Mullins and Cochran, 1973a). The broken arrow indicates a reaction not yet demonstrated in insects. Multiple arrowheads imply a sequence of reactions.

235

(Wolfram, 1949; Kikkawa, 1953; Wessing and Bonse, 1966; Umebachi and Uchida, 1970). According to the concept that ommochrome synthesis serves as a means of removal of excess tryptophan from insect tissues (Brunet, 1965; Buckmann *et al.*, 1966), it is not surprising that they have been reported in excreta from several species (Goodwin and Srisukh, 1950; Egelhaaf, 1956; Brunet, 1965; Butenandt *et al.*, 1960; Harmsen, 1966a, c). What is of more interest is that intermediates occurring much earlier in the pathway are either excreted directly, accumulated internally, deposited in wing scales, or converted to other compounds which are themselves accumulated or excreted. Thus, Kikkawa (1953) found that adult males of certain mutants of *Bombyx* excrete kynurenine or 3-hydroxykynurenine in the urine, while adult females deposit these materials in eggs. The deposition of 3-hydroxykynurenine in ovarian follicles has been related to diapause (Sonobe and Ohnishi, 1970). In the butterfly *Papilio*, tryptophan, kynurenine, and 3-hydroxy-kynurenine have been identified from the tissues with kynurenine being excreted in significant quantities, particularly at adult emergence (Umebachi and Yamada, 1964; Umebachi and Katayama, 1966). In the same group of butterflies kynurenine is deposited in the yellow-white wing scales (Umebachi and Takahashi, 1956), perhaps as 'kynurenine pigment' (Umebachi, 1962). Kynurenine has also been found in fat body, haemolymph and the meconium of *Pieris* (Harmsen, 1966c), as well as in the urine of the hemipteran *Dysdercus* (Berridge, 1965).

In *Drosophila*, Rizki (1961, 1963) reported the accumulation of kynurenine in fat body cells. Umebachi and Tsuchitani (1955) found kynurenic acid in wild type and several mutants, and Danneel and Zimmermann (1954) showed the accumulation of kynurenine which was partly degraded to kynurenic acid. Kynurenic acid is present in housefly pupae, but declines in adults (Grigolo, 1969b). In silkworm pupae xanthurenic acid and 4,8-dihydroxyquinoline have been found (Inagami, 1955). A new tryptophan metabolite, referred to as 3-hydroxykynurenine glucoside, has also been discovered in this species (Linzen and Ishiguro, 1966). Similarly, Brown (1965) reported L-dihydroxanthurenic acid as a new tryptophan metabolite in the wings of many species of butterflies. Leibenguth (1967) has demonstrated the excretion of kynurenic and xanthurenic acids from prepupae of *Habrobracon*.

The Malpighian tubule cells of *Drosophila* contain tryptophan, kynurenine, 3-hydroxykynurenine and kynurenic acid (Wessing and Eichelberg, 1968). Storage of large quantities of 3-hydroxykynurenine occurs in vesicles or dilations of the endoplasmic reticulum of tubule cells, and, at least in certain mutants, it is converted into ommatins and ommins (Wessing and Eichelberg, 1972). These vesicles contain pigments which are responsible for the colour of the tubules. Sometimes they rupture into the lumen of the tubule and the contents are lost in the urine (Wessing and Bonse, 1966). This finding fits in well with the concept of formed-body secretion discussed above (p. 187). Little other specific information is available on the mechanism of excretion of compounds from this pathway, but it seems likely that some more direct process may be involved.

From this evidence it is clear that many intermediates in the ommochrome biosynthetic pathway are implicated as excretory products. What is most remarkable is that in a large portion of the research reviewed no overt effort was reported to determine if tryptophan derivatives are present in the excreta. In a recent report in which this was a major objective, Mullins and Cochran (1973a) found that *Periplaneta* excretes kynurenic acid, xanthurenic acid, and 8-hydroxyquinaldic acid. The amounts of these materials excreted were small but increased significantly when the cockroaches were fed diets high in protein or tryptophan. Unpublished data from my laboratory indicates that an array of fluorescing compounds are present in excreta from twenty species of cockroaches. These compounds are presumed to be tryptophan metabolites, and some of them have been tentatively so identified. In two of these species, *Eurycotis* and *Supella*, the amounts present in faeces are quite significant even when the insects are fed diets low in tryptophan.

Caution must be exercised in interpreting these results, since it was shown that when American cockroaches were fed the intestinal antibiotic neomycin sulfate, 8-hydroxyquinaldic acid no longer appeared in the excreta (Mullins and Cochran, 1973a). The implication is that the insects produced the kynurenic and xanthurenic acids, but gut microbes carried out the dehydroxylation of xanthurenic acid to 8-hydroxyquinaldic acid. Similar results have been reported from rats and rabbits (Takahashi and Price, 1958; Kaihara and Price, 1963). *In vitro* studies with these two species have confirmed that the gut microflora, under anaerobic conditions, are responsible for the dehydroxylation reaction (Booth *et al.*, 1965; Tamada *et al.*, 1967). Investigations of this type are essential in insects in order to establish which excretory products they produce themselves.

The first reaction in the tryptophan-ommochrome pathway is the conversion of tryptophan to kynurenine by a two-step reaction with N-formyl-kynurenine as the intermediate. Some confusion exists over naming of the enzymes. As originally described from studies on mammalian liver, the term tryptophan pyrrolase was used in a general way to indicate the enzyme system catalyzing the entire two step conversion (Knox and Mehler, 1950; Mehler and Knox, 1950). The individual enzymes were called tryptophan peroxidase and kynurenine formamidase (formylase) (Knox, 1955). In more recent work the name tryptophan pyrrolase has been applied to the first enzyme (Prasad and French, 1971), while kynurenine formamidase seems to have been retained for the second enzyme (Grigolo, 1969a). The latter names will be used in the following discussion.

The tryptophan pyrrolase reaction is complex. Intermediate products have been proposed, but have eluded identification. The reaction results in the incorporation of two atoms of oxygen into the tryptophan molecule, and a cleavage of the indole ring. Peroxide may be involved and catalase is often an inhibitor of the reaction (Hayaishi *et al.*, 1957).

This enzyme has been studied extensively in insects. It is known to be the lesion in various mutant eye color traits (Baglioni, 1959, 1960; Egelhaaf, 1958; Kaufman, 1962; Grigolo and Cima, 1969), and was implicated in some of the earliest studies of biochemical genetics (Tatum, 1939; Becker, 1939; Kikkawa, 1941; Caspari, 1949). Because kynurenine formamidase is usually present in excess in the crude homogenate enzyme source, it is kynurenine which accumulates as the end product. The system has been studied in *Drosophila* (Marzluf, 1965; Phillips *et al.*, 1967; Ghosh and Forrest, 1967b; Rizki and Rizki, 1963, 1964; Kaufman, 1962; Baglioni, 1959, 1960), *Musca* (Lancaster and Sourkes, 1969; Grigolo and Cima, 1969), *Culex* (Prasad and French, 1971), *Habrobracon* (Leibenguth,1967), and *Schistocerca* (Pinomonti and Petris, 1966; Pinamonti *et al.*, 1966). Normally, ascorbate or some other activator, such as 2-mercaptoethanol, must be present for the system to show good activity. However, with *Culex* homogenates no exogenous activators nor haem prosthetic groups were necessary (Prasad and French, 1971). Since this is contrary to findings with other species, one might assume that a low native catalase activity or some other factor is involved. In an intriguing study, Jacobson (1971) found that, contrary to the generally accepted idea, the *Drosophila* mutant *vermillion* possesses tryptophan pyrrolase, but that it is inhibited by a specific iso-acceptor tyrosine t-RNA. It was presumed that the structure of the mutant enzyme was altered slightly so that association with the t-RNA could occur, thereby causing an inhibition, which did not occur with the wild-type enzyme. This work throws new light on how a mutation effect may be produced. Another interesting finding concerning this enzyme in *Drosophila* is its inhibition by certain naturally occurring pteridines (Ghosh and Forrest, 1967b). This result, together with other evidence, which will be cited later in connection with the further metabolism of kynurenine and 3-hydroxykynurenine, indicates a definite relationship between the tryptophan and pteridine pathways.

Kynurenine formamidase is the second enzyme in the tryptophan-ommochrome pathway. Its presence has more often been inferred than actually demonstrated. Glassman (1956) studied it in cell-free extracts from *Drosophila*. He showed that it is an active enzyme in wild-type as well as in a number of eye color mutants from which it might logically have been absent. Subsequently, Kaufman (1962) and Ghosh and Forrest (1967b) carried out studies on tryptophan pyrrolase by assaying for kynurenine, assuming that the formamidase enzyme is present in excess. From the housefly, Grigolo (1969a) has reported a mutant which may be deficient in kynurenine formamidase. It does not transform N-formylkynurenine into kynurenine. The same mutant also does not convert tryptophan to kynurenine, but whether N-formylkynurenine accumulates in this case was not reported (Grigolo and Cima, 1969). It would be of value to have this final piece of information in order to determine if this mutant is deficient in kynurenine formamidase.

The conversion of kynurenine to 3-hydroxykynurenine is carried out by the enzyme kynurenine-3-hydroxylase. In the *Drosophila* mutant *cinnabar*, this enzyme capability is lacking (Ghosh and Forrest, 1967a). With wild-type flies, activity was reported to be associated with the mitochondrial fraction. A pteridine co-factor requirement was suggested, and in the mutant *white* it was presumed not to be reduced, resulting in inhibition of the reaction. Accumulation of the oxidized form of the pteridine cofactor may inhibit tryptophan pyrrolase (Ghosh and Forrest, 1967b). Ziegler and Harmsen (1969) have also commented on the importance of the pteridine-ommochrome relationship. Studies on homogenates from the blowfly *Calliphora* have cast doubt on the association of the hydroxylase with mitochondria (Linzen and Hertel, 1967; Hendricks-Hertel and Linzen, 1969). These authors found the enzyme to be present in the 105 000 g supernatant with some activity in the 1200 g pellet. The enzyme required NADPH to react maximally and was stimulated by azide. The highest activity was in the Malpighian tubules, and it was suggested that the enzyme may be confined to these structures. However, this suggestion is difficult to accept in view of the widespread distribution of ommochromes in the insect body and the primary role of Malpighian tubules in excretion. Indeed, Linzen and Hendricks-Hertel (1970) have subsequently found that the enzyme is present in several tissues and various stages of the *Bombyx* mutant *rb*, although the activity was often low.

Other reactions of interest are the conversion of 3-hydroxy-kynurenine into ommochromes, and certain lateral reactions involving kynurenine and 3-hydroxykynurenine (Fig. 3.6). Butenandt *et al.* (1956) suggested that, in the production of ommatins, 3-hydroxykynurenine and dopa undergo a coupled reaction in which xanthommatin and a dopa-melanin are produced. Tyrosinase is probably involved. Glassman (1957) has suggested a similar 'non-enzymatic' reaction for both kynurenine and 3-hydroxykynurenine. According to the former authors, dopa is first transformed into an ortho-quinone which then oxidizes 3-hydroxykynurenine to a phenoxazinone molecule. Phillips and Forrest (1970), using a more highly refined system from *Drosophila*, have demonstrated the conversion of 3-hydroxykynurenine and 3-hydroxyanthranilic acid into a phenoxazinone molecule. The methods used separated the condensation reaction from phenol oxidase activity, and demonstrated that the formation of xanthommatin is at least a two step reaction. Subsequent modifications of the initially produced ommatins and ommins can readily occur. Butenandt *et al.* (1958) have also discussed the ommins, and Butenandt (1959) has presented a thorough review of the entire area.

Of the lateral reactions mentioned above, one is catalyzed by kynurenine transaminase. Both kynurenine and 3-hydroxy-kynurenine serve as substrates producing kynurenic and xanthurenic acids, respectively. In *Schistocerca*, the enzyme requires pyridoxal phosphate and an NH_2 acceptor (Pinamonti *et al.*, 1970). Elimination of the NH_2 is coupled with closing of the ring, resulting in the production of a quinoline molecule. Similar reactions have also been reported from immature stages of *Habrobracon* (Leibenguth, 1965, 1967). The enzyme is apparently present throughout most of the life cycle, but its activity seems to be reduced as adulthood is approached. It was suggested that inhibition may result from a lack of co-factors or a shift to the production of anthranilic or 3-hydroxyanthranilic acids. Since nicotinic acid synthesis does not seem to occur in insects (Brunet, 1965), a relevant question is what function do the anthranilate compounds have. Ishiguro and Linzen (1966) demonstrated a blue fluorescent substance in fat body from locusts and *Bombyx* after the injection of 3-hydroxyanthranilic acid. They showed that it is the O-β-glucoside of 3-hydroxyanthranilate. Compounds of this sort could play a role in pigmentation, although conjugation reactions are often involved in the excretion of foreign compounds (p. 257).

A summary of the tissue distribution of enzymes involved in this pathway is presented in Table 3.3. Many of the studies reviewed were conducted with enzymes obtained from whole body homogenates. In spite of this limitation, the table shows very clearly that the fat body is involved in all the reactions described with the possible exception

Table 3.3
Tissue distribution of tryptophan-metabolizing enzymes in insects

Enzyme	Species	Tissue	Reference
Tryptophan pyrrolase	*Schistocerca*	Fat body (f.b.) and haemolyph of gravid females.	Pinamonti *et al.*, 1966
		Central f.b. only	Pinamonti and Petris, 1966
	Drosophila	f.b. only	Rizki and Rizki, 1963
Kynurenine formamidase	*Drosophila*	f.b. only*	Rizki and Rizki, 1963
	Periplaneta	f.b.	Cochran (unpublished)
Kynurenine-3-hydroxylase	*Calliphora*	f.b. (low activity) Malpighian tubules (M.t.)	Hendrichs-Hertel and Linzen, 1969
	Bombyx	M.t., larval f.b., ovaries, spinning gland, pupal heart (last 3 have low activity)	Linzen and Hendrichs-Hertel, 1970
Kynurenine transaminase	*Schistocerca*	f.b. only	Pinamonti *et al.*, 1970
3-Hydroxyanth-ranilate	*Bombyx*	f.b.	Ishiguro and Linzen, 1966
conjugating enzyme	*Locusta*	f.b.	

*By inference because kynurenine production was used to determine the activity of tryptophan pyrrolase.

of kynurenine-3-hydroxylase. Some of the studies quoted were extensive enough for the authors to conclude that fat body alone contained the enzyme (e.g. Pinamonti and Petris, 1966), although in other instances only the fat body was examined. In addition to the importance of this tissue for the pathway, Hyde (1972) demonstrated quite clearly the capability of individual epidermal and eye cells of *Periplaneta* to synthesize pigments, which are presumed to be of the ommochrome type. Here the pigments are unquestionably for use within the ommatidial cells.

In attempting to generalize the information presented in this section to the problem of excretion, several points arise. It will almost certainly be shown in the future that tryptophan derivatives are of more importance as excretory products than has been previously recognized. They probably represent the means used by the insect to void excess tryptophan. Whether one or more of these derivatives will emerge as the main excretory product of the pathway remains to be determined. The fat body appears to play a central role in the reactions discussed and is probably the principal tissue which is capable of carrying out the reactions rapidly enough to produce sufficient quantities of product for detection in the faeces. Perhaps cells of other tissues can carry out these reactions at specific times in keeping with their own needs. The actual transport and excretion of tryptophan derivatives have not been studied in sufficient detail. Presumably the haemolymph and the Malpighian tubule-rectum complex are involved.

The last point to be discussed here is the mutagenicity and carcinogenicity of certain tryptophan derivatives. 3-Hydroxy-kynurenine and 3-hydroxyanthranilic acid have been shown to be mutagenic to mice and to cultured human cells (Kuznezova (1969). Furthermore, these two compounds plus xanthurenic acid, 8-hydroxyquinaldic acid, and the 8-methyl ether of xanthurenic acid have all been implicated as mouse bladder carcinogens (Bryan *et al.*, 1964). In view of these findings it appears that normal (Boyland and Watson, 1956) or abnormal (Claudatus and Ginori, 1957) metabolism results in the production of compounds which may be detrimental to the animal. For insects, possessing an apparent exaggeration of this pathway (Brunet, 1965), similar questions must arise. It has been shown in cockroaches that recurrent nerve severance and anal blockage, both of which prevent the voiding of wastes, produce tumor-like lesions along the alimentary canal (Taylor, 1969; Taylor and Freckleton, 1969; Asokan, 1972). Feeding diets high in tryptophan may also produce tumors (Burnet and Sang,

1968; Mullins and Cochran, 1973a). These findings imply that carcinogen level, or length of tissue exposure, or both may be important. One would presume that normal amounts of tryptophan derivatives are innocuous, but that elevated concentrations could be detrimental.

Pteridines

The pteridines are another group of compounds which must be considered in the context of excretion. Present information indicates that they are minor excretory products, but as with the tryptophan derivatives and ommochromes their role in excretion has not been adequately explored. In general, their function in insects seems to relate primarily to pigmentation, thereby extending the parallel between pteridines and ommochromes. Especially in the compound eyes and integument of insects, the two types of pigments frequently occur together (Gilmour, 1965). Pteridines of a yellow, white, or red colour are frequently deposited in integumentary areas with low metabolic activity (Wigglesworth, 1972). In *Bombyx* they are associated with specific proteins in chromogranules (Tsujita and Sakurai, 1965). In addition, they assume an exaggerated importance in certain lepidopteran insects in their role as wing pigments and possible end products of metabolism. Indeed, it is in this connection that the pteridines were first studied in insects, although they were identified as uric acid and/or urates (Hopkins, 1895). From that early beginning, the increase in knowledge about this group of compounds has been extremely slow. Eventually the names pterines (Wieland and Schöpf, 1925) and pteridines (Schöpf and Becker, 1936) were proposed. The structure of leucopterin became known somewhat later (Purrmann, 1940; Schöpf and Reichert, 1941). Gradually, the fund of knowledge about these compounds has grown, finally reaching a substantial level (Pfteiderer and Taylor, 1964; Ziegler, 1965). The role of pteridines in insect biochemistry and physiology has been extensively discussed (Butenandt, 1959; Ziegler, 1961a; Gilmour, 1961, 1965; Kilby, 1963; Chefurka, 1965; Razet, 1966; Wigglesworth, 1972; Bursell, 1967; Ziegler and Harmsen, 1969; Corrigan, 1970; Chen, 1971). The reader is referred particularly to the review of Ziegler and Harmsen (1969) for a coverage more detailed than can be undertaken here.

The compounds of interest to this discussion will be considered briefly at this point. Most of them are derivatives of 2 amino-4-hydroxypteridine, with substituents occurring at the 6 and/or 7 positions according to the following general formula:

This specialized group of pteridines are usually referred to as pterines. Some of the more common examples are shown in Fig. 3.7. As can be seen, most of them have rather simple substituent or R groups. Other pteridines, not shown in the figure, are more complex. Folic acid is an example of a conjugated pterine which is known to be a very important vitamin and co-factor. Insects are unable to synthesize this molecule and must, therefore, rely on dietary sources. Folic acid is of interest here because one of its degradation products contributes to the general pterine pool. Riboflavin is a derivative of 2,4-dihydroxypteridine or lumazine, and is often found in Malpighian tubules (Drilhon and Busnel, 1939; Busnel and Drilhon, 1942; Wessing and Eichelberg, 1968). Sepiapterin and isosepiapterin are examples of dihydrogenated pterines. Again the reader is referred to the extensive review of Ziegler and Harmsen (1969) for a fuller consideration of pteridine structural relationships and nomenclature.

Information on the role of pterines in insect excretion is quite limited. This may be the result of two concepts which developed quite early in the literature. The first came from the original paper of Hopkins (1895) in which he clearly indicated the presence of excretory substances that function in ornamentation. Out of this work there seems to have developed the idea that in some insects, particularly the Pieridae, 'excess' pteridines are disposed of as products stored in the wing scales, while simultaneously serving in pigmentation. Only trace amounts of excreted pteridines had been found up to that time (Wigglesworth, 1972). The second influential report was that of Becker (1937) in which he stated that pterines do not pass the Malpighian tubule barrier and, therefore, cannot be excreted. Thus, the stage was set for a rather complete acceptance of the concept of storage excretion of pteridines wherein highly oxidized molecules containing nitrogen are permanently immobilized, especially in wing scales (Harmsen, 1966a).

More recently it has been emphasized that the faecal material of

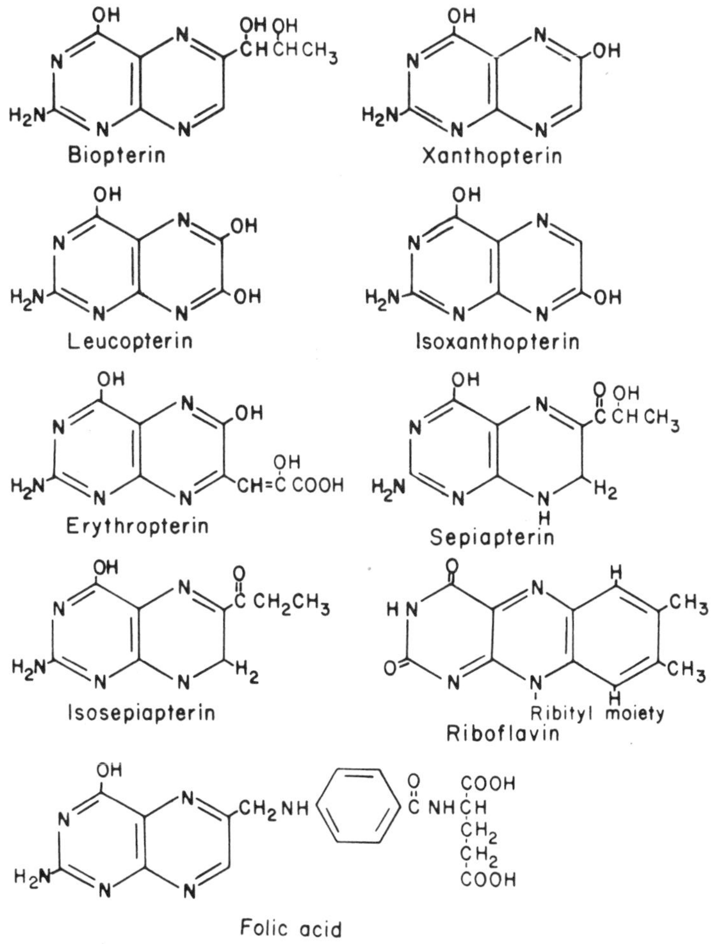

Fig. 3.7. Structural formulae for certain biologically important pteridines. (Adapted from Forrest and Nawa, 1962; Gilmour, 1965, and Ziegler and Harmsen, 1969.)

insects does contain certain pterines. Xanthopterin and iso-xanthopterin are present in the faeces of *Oncopeltus* where they account for about 0.2 per cent of the total faecal material and over 5.0 per cent of faecal nitrogen (Bartel *et al.*, 1958; Hudson *et al.*, 1959; Craig, 1960). In the meconium of wild-type *Drosophila*, isoxanthopterin and 2-amino-4-hydroxypteridine are present, whereas the pterine content of the meconium from certain mutants changes significantly (Kursteiner, 1961). Also, in wild-type *Drosophila* Wessing and Eichelberg (1968) found tetrahydrobiopterin, xanthopterin, isoxanthopterin, drosopterin, neodrosopterin and riboflavin to be present in the Malpighian tubules, thereby accounting in part for the fluorescence of these organs. The occurrence of riboflavin in the Malpighian tubules of insects was mentioned above.

In the genus *Pieris*, pterines accumulate to quite a significant level particularly in the wings and certain other body parts (Harmsen, 1966a). Leucopterin is the principal compound involved, but isoxanthopterin and certain others are also present (Harmsen, 1966b). Only about 2.0 per cent of the total pterine content is excreted with the meconium (Harmsen, 1966,a, c). Because of the large increases in pterine content which occur during pupation, *in vivo* synthesis is the only satisfactory explanation for their origin. Removal of some of the storage sites (one wing, e.g.) increased the excretion and fat-body storage of pterines only slightly, but significantly reduced total pteridine synthesis. Similar results were obtained with a mutant stock characterized by reduced wing scales (Harmsen, 1964, 1966a). A different pattern occurred with *Vanessa* in which about one tenth as much pteridines were produced, but nearly half were excreted. However, the total amounts excreted by *Pieris* and *Vanessa* were not greatly different.

From this work, and that of others, Harmsen (1966a) proposed that a strong but undefined excretion barrier for pteridines exists, especially at the time of meconium formation. Presumably, this barrier occurs at the level of the Malpighian tubules (Becker, 1937), and would favour a retention of pterines within the body to the extent that in *Pieris* large accumulations occur. In this sense, Harmsen (1966a) considered pterine production and storage in this genus (particularly in the wing scales), as an example of pterines being important end products of metabolism, while also recognizing their role as pigments. Corrigan (1970) criticized this approach for not adequately recognizing the adaptive significance of such pigmentation as documented by Brower (1969). Nolfi (1970) considered their primary role to be pigmentation. Perhaps the correct interpretation is that this represents another example in insects where certain compounds serve dual roles, one of which has excretory significance. In *Vanessa* large accumulations do not occur, but this is mainly because total synthesis is reduced.

The only known exceptions to the excretion barrier concept are in *Mylothsis* which does not have such a barrier for 2-amino-4-hydroxypteridine (pterin) (Harmsen, 1969), and in certain insects, such as *Phonoctonus*, which ingest large quantities of pteridines (Ziegler and Harmsen, 1969). In the latter case, the essential process may be faecal elimination rather than excretion via the Malpighian tubules. Rembold and Hanser (1960) have shown that dietary biopterin is not utilized by honey-bee larvae, and is completely eliminated from the body. Additional studies on diverse insect groups are needed to gain a better understanding of the importance of pteridines in insect excretion, including that portion referred to as storage excretion.

Among the functions carried out by pteridines in insects are their presence as pigments in the compound eyes, wing scales, integument, and certain internal organs, and their importance as co-factors or vitamins. Mutant forms of several species have been instrumental in elaborating certain of these roles (Ziegler, 1961a; Egelhaaf, 1962). Because this area has been recently reviewed (Ziegler and Harmsen, 1969), only a few comments will be made here.

It is clear that in the compound eyes of insects light perception is not dependent in a significant way on either pteridines or ommochromes (Goldsmith, 1958; Goldsmith and Warner, 1964). Rather, these pigments are present in eye pigment cells and are often distributed in a very specific manner within the cells (Zeutzshel, 1958; Ziegler, 1961a, b). Their function is that of screening pigments and as light-filtering substances (Langer and Hoffmann, 1966). They also play a role in contrast perception as demonstrated from a series of *Drosophila* eye colour mutants (Hengstenberg and Gotz, 1967; Wehner *et al.*, 1969).

It has been shown that the wasp *Vespa* contains xanthopterin, isoxanthopterin, biopterin, violapterin and an unknown pterin as general body and organ pigments (Ikan and Ishay, 1967). These pigments were broadly but specifically distributed in the wasp's body. Isoxanthopterin, as well as certain ommochromes, are involved in body pigmentation in *Locusta* (Bouthier, 1966). In a study of eleven species of Hemipteroid insects, Merlini and Nasini (1966) reported the presence of isoxanthopterin, violapterin, and 7-methylxanthopterin in some species. Erythropterin was present in all species examined and these authors emphasized its importance in the pigmentation of Hemiptera. A yellow mutant strain of *Oncopeltus* lacks this pigment (Smissman and Orme, 1969). Descimon (1965) identified sepiapterin in the wings of *Colias* as well as in 33 of 48 other species of pierid butteflies examined. Along similar lines, Hubby and Throckmorton (1960) have conducted an elaborate study of the systematic distribution of pteridines in the genus *Drosophila*. The eggs of *Oncopeltus* contain isoxanthopterin, erythropterin and xanthopterin as a reduced derivative (Forrest *et al.*, 1966). Similarly, the eggs of *Pyrrhocoris* contain erythropterin, isoxanthopterin, violapterin, chrysopterin and a derivative of violapterin. The content of these pterines changed with development, but no simple precursor-product relationships were discovered (Smith and Forrest, 1969).

Using three genotypes of *Ephestia*, Egelhaaf (1956) studied the pterine content of several organs. He found that all organs examined contained isoxanthopterin, and all except the mid-gut contained xanthopterin, 2-amino-6-hydroxypterine and an unidentified pterine. All pterines were integrated into the cellular structures. The highest concentrations were found in the eyes followed by the Malpighian tubules, testis, and ovary. One of the mutants differed from wild-type in having a higher isoxanthopterin content and an altered xanthopterin/isoxanthopterin ratio in nearly all organs.

Pterine production and distribution may change during the development of an insect. Descimon (1966) reported on the concentration of four pterines during the larval development of *Colias*. Sepiapterin seemed to predominate early in development, whereas leucopterin, isoxanthopterin, and to a lesser degree xanthopterin were more prominent toward the end of the larval period. Similarly, Harmsen (1966c) studied the distribution of leucopterin and isoxanthopterin in pupae and pharate adults of *Pieris*. He found that both substances accumulate in the wings just prior to adult emergence. Immediately preceding this the haemolymph concentration of both substances increases dramatically and then falls rather sharply. The presence of both substances, but especially leucopterin, in the intestine and meconium also increases as emergence nears. At the same time, the fat body content falls steadily again especially for leucopterin. The author concluded that this is strong evidence for the fat body being the site of synthesis for simple pteridines with the haemolymph serving to transport them to their sites of excretion or deposition.

Pteridines are also important in certain other respects. That folic acid and related compounds are conjugated pterines has already been noted. The role of pteridines as co-factors, especially in the conversion of phenylalanine to tyrosine, has recently been reviewed (Kaufman 1967). The prominence and accumulation of riboflavin in insects, while not well understood, is very evident, and is often influenced by genetic traits (Weber and Roberts, 1967; Nickla, 1972). Certain pteridines have been shown to interfere with the synthesis of ribosomal RNA and DNA in *Oncopeltus* eggs (Harris and Forrest, 1967). As mentioned in the previous section (p. 239), some pterines of the biopterin type inhibit the activity of tryptophan pyrrolase (Ghosh and Forrest, 1967b), possibly accounting for the often noted relationship between pterines and ommochromes (Ziegler, 1961a).

Discussion of the biochemical origins and metabolism of pteridines in insects is rendered a much easier task because of the work of others (Ziegler and Harmsen, 1969; Corrigan, 1970). Only a rather generalized account will be given here to make the broader outlines discernible. For more comprehensive coverage, the reader should consult the above references.

It has been quite well established that pterine biosynthesis occurs in insects from a purine precursor. The compounds usually mentioned are guanosine or GMP, with the conversion having been demonstrated using labelled precursors (Weygand *et al.*, 1961; Simon *et al.*, 1963, Liaci, 1965; Ziegler, 1965; Sugiura and Goto, 1967; Watt, 1967). The conversion is apparently initiated by the release of C-8 from the purine ring (Levenberg and Kaczmarek, 1966). The product is a reactive hydrogenated and perhaps phosphorylated pyrimidine intermediate (Ziegler and Harmsen, 1969). Ring closure occurs and what Watt (1967) called the 'initial pteridine' is produced (Fig. 3.8). From this point, a number of metabolic reactions appear

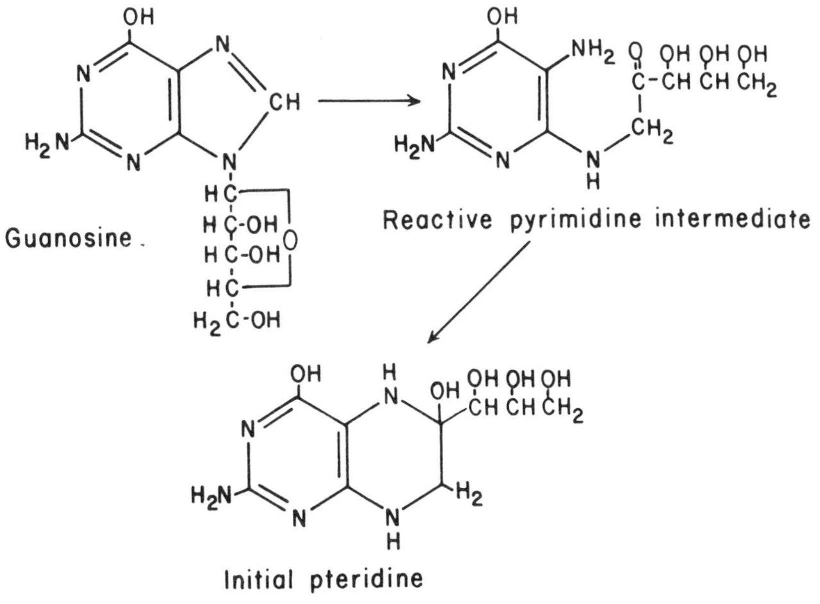

Fig. 3.8. Pathway for the conversion of purine into pteridine (Adapted from Watt, 1967 and Ziegler and Harmsen, 1969.)

to be possible. However, Watt (1967) demonstrated that in *Colias* larvae the C-6 side chain of sepiapterin originates from the guanosine ribose and is not secondarily added. This C-6 side chain can obviously be modified or eliminated, perhaps from the 'initial pteridine', to produce simpler pterine molecules (Watt, 1967; Rembold *et al.*, 1969; Ziegler and Harmsen, 1969). Some of the specific changes at the C-6 and C-7 positions have been demonstrated (Rembold and Gutensohn, 1968; Harmsen, 1969).

The C-7 substituted pterines appear to have a different origin. For example, erythropterin is formed by the addition of a 3 carbon side chain to xanthopterin (Forrest *et al.*, 1966; Watt, 1967). A number of molecules, such as pyruvate, malate, lactate (Watt, 1967) and oxaloacetate (Forrest *et al.*, 1966), are appropriate sources of the 3 carbon chain. There is some question as to the form of the molecule to which the side chain is added. Ziegler and Harmsen (1969) proposed 7,8-dihydroxanthopterin based on the findings of several authors (Forrest *et al.*, 1966; Descimon, 1967). There is also evidence that the 7,8-dihydro form may be important in the action of XDH (Rembold & Gutensohn, 1968). In any case, the C-6 and C-7 substituted pterines are quite different molecules both in origin (Fig. 3.9) and biological functioning (Harmsen, 1964; 1966b).

Lumazine-type compounds can be produced by insects from 4-hydroxypteridine (Forrest *et al.*, 1961). The oxidative deamination of sepiapterin is another method (Tsusue, 1967, 1971). Whether insects can synthesize riboflavin does not appear to be definitely established, but it is usually a dietary requirement (Friend, 1958). Riboflavin can be produced from purine precursors (McNutt, 1956). The final step is the production of one molecule of riboflavin from 2 molecules of 6,7-dimethyl-8-ribityllumazine catalysed by the enzyme riboflavin synthetase (Wacker *et al.*, 1964). Similarly, the conjugated pterines, such as folic acid, are not produced by insects and must be ingested (Ziegler and Harmsen, 1969). In *Galleria* the enzyme formyltetrahydrofolate synthetase has been identified (Zielinska and Grzelakowska-Sztabert, 1968). It carries out the reversible conversion of tetrahydrofolate to formyltetrahydrofolate. A folate reductase enzyme, separate from sepiapterin reductase, has been found in *Bombyx* (Matsubara *et al.*, 1963). The corresponding enzyme activity is also known from *Drosophila* (Taira, 1961).

A generalized scheme or pathway for the metabolism of pterines in insects has been proposed (Fig. 3.9). Some of the specific reactions have been studied to a reasonable extent while others have not. Of these reactions, the ones catalysed by XDH are the best

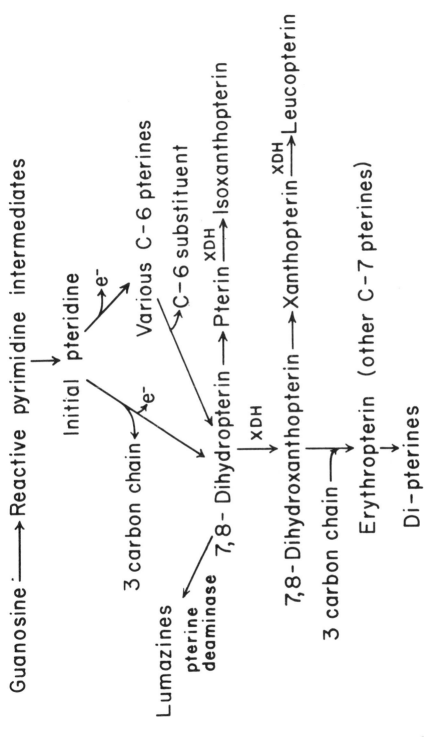

Fig. 3.9. Generalized pterine biosynthetic pathway in insects (Adapted from Watt, 1967 and Ziegler and Harmsen, 1969.)

known. A number of authors have demonstrated the conversion of 2-amino-4-hydroxypteridine, xanthopterin and xanthine to isoxanthopterin, leucopterin and uric acid, respectively, using cell free extracts or purified enzymes from a variety of *Drosophila* wild-type strains and eye-colour mutants (Forrest *et al.*, 1956; Glassman and Mitchell, 1959; Glassman, 1962a, b, 1965; Keller, 1964; Keller and Glassman, 1964a, b, 1965). Certain of the mutants have been shown to be deficient in the enzyme which results in the accumulation of precursors like 2-amino-4-hydroxypteridine (Glassman and Mitchell, 1959). The complexity of the genetic control of XDH has also been demonstrated (Glassman, 1965; Chen, 1971).

Munz (1964) examined the tissue distribution of XDH in adult *Drosophila*. He found no activity in ovaries, eggs, or testes. Some occurred in the gut and Malpighian tubules, but the highest activity was in the haemolymph. Considerable activity has been reported in the larval fat body (Ursprung and Hadorn, 1961). Over the course of development, XDH activity was high in larvae, reached a minimum in 4-day old pupae, and increased again in 3-day old adults (Munz, 1964). The reader is also referred to the section on uric acid for a fuller discussion of XDH in insects (p. 205).

Of considerable interest is the recent finding that xanthine oxidase from vertebrate sources can attack the 6 position of the pteridine molecule (Rembold and Gutensohn, 1968). Previously, it was considered that only the 2, 4 and 7 positions are vulnerable. In insects a similar situation may occur, at least under certain conditions (Harmsen, 1969). This reaction is of importance in the proposed scheme of pterine metabolism in insects for the conversion of 7,8-dihydropterin to 7,8-dehydroxanthopterin (Fig. 3.9).

Additional information on pterine metabolism and on other reactions in insects is available, but is limited (Ziegler and Harmsen, 1969; Corrigan, 1970). It appears that the most pressing need in this area is for detailed biochemical studies of specific reactions along the proposed pathway. From such work, a better understanding of the pathway itself would almost certainly emerge.

Other nitrogenous substances

Several other nitrogenous substances have been reported from the excreta of insects. Among them are creatine, creatinine, haematin, proteins and peptides. Usually they are minor constituents, and their origin may be in doubt. For example, Gilmour (1961) concluded that creatine is a true excretory product resulting from protein catabolism, while Bursell (1967) assumed it to be a product of digestion. Neither position was supported by evidence. Indeed, information on these substances in insects is quite limited.

Creatine has been reported from the excreta of *Lucilia* (Brown, 1938b) and *Rhodnius* (Wigglesworth, 1931). Creatine and creatinine were detected in the excreta of *Tineola* (Powning, 1953), *Anthonomus* (Mitlin *et al.*, 1964a) and *Apis* (McNally *et al.*, 1965). On the other hand, neither substance was found in the excreta of *Melanoplus* (Brown, 1937), *Bombyx, Apis* (Roche *et al.*, 1957) nor in the meconium of *Antheraea* (Leifert, 1935a). In all instances where one or both of these products were found, they were present in small to trace quantities. Brown (1938b) produced evidence that creatinine is converted into creatine by *Lucilia* larvae. As a result, only traces of creatinine could be detected in their excreta. The presence of small to trace amounts of creatine and creatinine in excreta perhaps also explains the discrepancy reported above for the honey-bee. Razet (1966) questioned whether the existence of these substances in insects is not due merely to an exogenous origin. This suggestion is not supported by the work of McNally *et al.* (1965) who found both creatine and creatinine in the excreta of honey-bees maintained on a diet of sugar, cellulose and water. Thus, there seems to be little reason at present to retreat from the position that creatine and creatinine are excretory products resulting from protein catabolism (Gilmour, 1961). A closer study of this area needs to be undertaken using modern biochemical techniques.

Information on the presence of haematin in insect excreta quite naturally comes from blood-feeding insects, but is very limited. Wigglesworth (1943) studied the fate of haemoglobin in *Rhodnius*. He found most of it to be digested to a haematin compound in the gut, and to be excreted without further change. A small amount of haemoglobin was absorbed and dealt with by the pericardial cells or by epithelial cells of the gut. At these sites it was converted to biliverdin, some of which appeared in the faeces. The gut cells apparently voided it into the lumen. Bursell (1965a,b) reported a substantial amount of haematin in the excreta of the tsetse-fly. Its concentration varied in relation to time of feeding and, as it was not detected in the haemolymph, the author concluded that it should be considered a faecal material.

The question of amino acid excretion has already been discussed (p. 228). The related topics of peptide and protein excretion must also be considered. Harington (1961) reported two peptides from the dry excreta of *Rhodnius*. His concept of dry excreta production would seem to indicate a faecal origin. Peptides and/or proteins have also been reported from the excreta of mosquitoes (Irreverre and Terzian, 1959; Thayer and Terzian, 1971) and the American cockroach (Mullins and Cochran, 1973b). In both cases free amino acids increased appreciably following hydrolysis of excreta extracts. The large amounts of food eaten by American cockroaches fed a diet containing 24 per cent crude protein fibre indicates that at least some of it passed through the gut undigested (Mullins and Cochran, 1973b). However, in adult female *Aëdes* significant levels of bound amino acids occurred in their excreta even though the insects were maintained on a sucrose diet (Thayer and Terzian, 1971). One possible explanation is the voiding of digestive enzymes which is known to occur in *Aëdes* larvae (Yang and Davies, 1971), although holding the larvae without food brought about a 10—15 fold reduction in the level of chymotrypsin in their excreta which may partially negate that idea. Proteins derived from gut microbial cells is another likely source. Approaching the problem from a different angle, Kessel (1970) reported on the permeability of dragonfly Malpighian tubule cells to horseradish peroxidase used as an exogenous protein source. The results showed that protein was taken up by the Malpighian tubule cells and accumulated internally. While this is contrary to findings on other insects (Berridge and Oschman, 1969; Locke and Collins, 1967, 1968), it suggests the possibility of excretion of peptides or small protein molecules by the Malpighian tubules. Work from a number of other laboratories has shown that Malpighian tubules can produce globules, called brochosomes, which consist of lipoprotein and other substances (Day and Briggs, 1958; Smith and Littau, 1960; Gouranton and Maillet, 1966, 1967; Mello and Bonzo, 1969). The process of formed-body excretion was discussed earlier (p. 187). Whether this is part of the same phenomenon remains to be proven. On the basis of these findings it appears that both a faecal and an excretory origin for excreta peptides and proteins must be held open until additional studies clarify the issue.

Carbohydrates

The presence of carbohydrates, particularly sugars, in insect honeydews is well known (Auclair, 1963; Strong, 1965). Usually they are considered to be digestive products (Gray and Fraenkel, 1954), which presumably means that they have passed through the digestive tract unabsorbed. In the process they may or may not have been modified from the plant sugars. Of course, another possibility is that excess absorbed sugars are excreted by the Malpighian tubules. Tamaki (1964b) suggested the excretion of ribitol by *Ceroplastes*, but it was not specified which of the above alternatives was involved. In addition, there is considerable evidence that honeydews excreted by insects differ significantly in carbohydrate content from that of host plant sap (Gray and Fraenkel, 1954; Wolf and Ewart, 1955; Ewart and Metcalf, 1956; Bacon and Dickinson, 1957; Mittler, 1958; Sidhu and Patton, 1970). In one instance rare sugars, very different from the sugars of the plant sap, were reported (Basden, 1970). Results of this nature suggest the possibility that certain insects can excrete carbohydrates, perhaps even as end products of metabolism. It is not clear how important this phenomenon may be, but it obviously needs further exploration. An examination of carbo-hydrate-metabolizing enzymes existing in the gut from the stand-point of excretion might also be useful.

Miscellaneous excretory products

Wigglesworth (1972) has pointed out the likelihood of insect urine containing large numbers of waste products in quantities too small to be readily detected. While this is undoubtedly correct, it does not seem worthwhile to pursue in the near absence of specific infor-mation. Nevertheless, attention should be drawn to the presence of calcium carbonate, calcium oxalate, and oxalic acid in the urine of certain plant-feeding insects either in solution or as crystals (Wiggles-worth, 1972; Takahashi *et al.*, 1969; Teigler and Arnott, 1972). Little is known about their significance, but they probably are normal minor excretory products in these insects. Wigglesworth (1972) has reviewed this, and certain other information on unusual urinary constituents, and it will not be repeated here. In addition, a report has appeared on the presence of lipids in the honeydew of an aphid (Strong, 1965). Free fatty acids, sterols, and triglycerides were identified. The author indicated that at least some of these substances came directly from the host plant. Similarly, Earle *et al.* (1965) found three types of steroids in the faeces of *Anthonomus* fed labelled cholesterol. Two of them were not free cholesterol, and apparently represent excretory products. Certainly, there is a strong need for research in such areas. In fact, it is virtually impossible to evaluate the importance of these compounds as excretory products using our current knowledge.

Xenobiotics

Insects normally encounter in their diet substances that are not useful as nutrients, and which may be toxic. Such materials probably form a part of the ancient insect-host relationship, which may have undergone many changes during evolutionary history (Whittaker and Feeny, 1971). Certain of these chemicals are complex molecules which must be eliminated from the insect's body. Some of them probably move through the gut unabsorbed, while others pass into the haemocoel. It might be expected that these latter substances are metabolically altered and then excreted, but little specific information concerning them is available (Whittaker and Feeny, 1971). However, more attention has been given to the excretion of other chemicals, such as synthetic dyes and insecticides, which insects do not normally encounter in nature. The biochemical and excretory mechanisms used to deal with these foreign compounds (Xenobiotics) may be related to those evolved to dispose of naturally occurring toxic or noxious chemicals. In this sense, knowledge concerning disposal of dyes and insecticides may be helpful in establishing the detoxication and excretory operations employed by insects. This particular topic would lend itself to a lengthy review, but the following discussion will be confined to outlining the principles involved in the disposal of these substances.

Information with reference to dyes is very limited and has been considered by Maddrell (1971). It appears that Malpighian tubules, gut epithelial cells, pericardial cells, haemocytes and haemolymph are involved with their disposal. From the work of Lison (1937, 1938) and Palm (1952) it is apparent that acid dyes are readily excreted by the Malpighian tubules, whereas basic dyes are excreted much more slowly. The rate at which dyes leave the system is greatly influenced by their concentration in the haemolymph. Since they usually arrive there by injection, concentration can be closely controlled. Also, the degree of involvement of the pericardial cells, the gut epithelial cells, and the haemocytes is correlated with the difficulty of dye excretion by the Malpighian tubules. Beyond this, excess dye simply remains in the circulating haemolymph. The pH of the haemolymph may be important, since Maddrell (1971) reported that making the artificial medium surrounding excised Malpighian tubules more alkaline increased their ability to excrete certain dyes. Very little other specific information about the biochemistry or physiology of dye excretion is available.

The situation with respect to insecticides is much more highly developed as it relates to biochemical reactions, but not to excretory mechanisms. Most modern insecticides are contact poisons, and presumably reach the internal organs by cuticular penetration. Generally speaking, they are highly water insoluble chemicals, but are soluble in non-polar solvents. This latter property may be related to their ability to penetrate cuticle. The fact that substantial amounts of various insecticides and/or their metabolites are present in the excreta of treated insects is well documented (Kikal and Smith, 1959; Gessner and Smith, 1960; Dorough and Casida, 1964; Chamberlain and Hamilton, 1964; Gatterdam *et al.*, 1964; Perry *et al.*, 1964; Self *et al.*, 1964; Rolofson, 1968; Watanabe and Kobara, 1968; Zayed *et al.*, 1968).

Knowledge of detoxication mechanisms in insects is quite extensive. Smith (1962, 1968) and Hook *et al.* (1968) have discussed the pertinent biochemical reactions. In general, two types of events seem to be of importance. There is the introduction of a centre of ionization or polarization into the highly non-polar insecticide molecule. This is often followed by a conjugation reaction with the resulting conjugate being readily excreted. Considering these events in more detail, Smith (1968) has shown that the initial event is frequently an oxidation-type reaction, but the specific reaction involved is subject to considerable variation depending upon the molecule being metabolized. Alkyl group oxidation, hydroxylation, N-demethylation, and epoxidation are a few examples. These reactions are usually carried out by the mixed-function oxidases characteristically found in microsomes. Reduction and dehydro-chlorination are other types of initial reactions known to occur in insects. Sometimes these reactions alone alter the molecule enough to allow excretion to occur. However, more often the product of the initial reaction undergoes a conjugation with simple sugars, amino acids or peptides, glutathione, sulfate or phosphate (Dutton and Duncan, 1960; Trivellani, 1960; Cohen and Smith, 1964; Cohen *et al.*, 1964; Smith and Turbert, 1964; Binning *et al.*, 1967; Smith, 1968). The products of these reactions are usually more water soluble than the original molecule or its primary metabolite.

Very little is known of excretory mechanisms aside from the fact that insecticides and their metabolites are found in the excreta. In the absence of specific information, one might presume that conjugated insecticide molecules are soluble enough to get into the Malpighian tubule lumen by passive diffusion. Certain of the primary insecticide metabolites may be ionizable. If they are present as anions in the haemolymph, they could pass into the tubule lumen in a manner similar to the mechanism suggested for the urate molecule. This type of mechanism would require the metabolite to have a net negative charge to balance the active transport of K^+. As insecticide molecules are insoluble and chemically inert, they could be eliminated unchanged. Unlikely as this event seems, it has been demonstrated by several research workers. (Gessner and Smith, 1960; Chamberlain and Hamilton, 1964; Rolofson, 1968; Watanabe and Kobara, 1968; Zayed *et al.*, 1968). Formed-body excretion is a possible explanation of the presence of insecticides in excreta but, it must be remembered that the topic of formed-body involvement in excretion is itself subject to question (p. 187). It appears likely that some unknown phenomenon, perhaps related to normal lipid transport and excretion, may be active in the excretion of unchanged insecticide molecules.

Once inside the tubule lumen or hindgut, insecticide and/or metabolite excretion seems reasonably well assured. These chemicals would be expected to mix with the digestive wastes, and pass from the body with the faecal pellets. The intima molecular seive could aid in this process. In addition, some unchanged insecticide might penetrate this cuticular intima as it does the general body cuticle. Physical or chemical binding to faecal wastes might counteract this tendency. Many questions remain to be answered before a full understanding of insecticide and/or metabolite excretion is achieved.

MINOR EXCRETORY SYSTEMS

Up to this point the discussion has been confined to the major excretory organs and the excretory substances which feed into them. It is also well recognized that minor excretory organs exist in insects (Wigglesworth, 1972). They are of various types and occur in different parts of the insect's body. Some of them form a part of or are derived from the gut. In that sense they could just as well have been discussed above along with the Malpighian tubules and rectum. In fact, the role of the hindgut in ammonia excretion was considered in that section. Others originate from certain ectodermal or mesodermal cells and are entirely distinct from the gut. The role of the pericardial cells in dealing with foreign proteins was mentioned previously. Thus, the grouping of these miscellaneous organs and cells is an arbitrary one, based primarily on their actual or presumed ancillary role in excretion.

The functioning of these minor excretory organs has been reviewed previously (Maddrell, 1971, Riegel, 1972). Much of the available information deals with their physiology and ultrastructure. For example, it appears that some of them may be involved in osmoregulation although perhaps only in a minor way (Harvey and Nedergaard, 1964; Kafatos, 1968). Their contribution to excretion, especially its biochemical aspects, has not been adequately explored. Most of the information that is available has to be gleaned from the physiological studies.

Labial glands

In certain insect groups, particularly the apterygotes, there are renal organs which are variously called antennal glands, labial glands, cephalic glands or by other names (Fain-Maurel and Cassier, 1971; Reigel, 1972). They usually lie in the head or forward part of the thorax, are paired, but may have a single external opening located somewhere on the lower part of the head (Maddrell, 1971). Studies of these glands are quite restricted, but they are presumed to play a role in excretion particularly among apterygotes lacking Malpighian tubules (Wigglesworth, 1972). In the silk moth *Hyalophora,* the glands are the same as those used by larvae to produce silk (Edwards, 1964). Both here and in moths of the genus *Antheraea* a principal function seems to be that of fluid (water) secretion perhaps as a mechanism for volume reduction (Edwards, 1964; Kafatos, 1968; Maddrell, 1971). In *Antheraea* the secreted fluid also contains K^+, HCO_3^- and Cl^- with the first two being present in considerable excess over their concentrations in blood (Kafatos, 1968). Edwards (1964) has ascertained that the Malpighian tubules of *Hyalophora* are functional, indicating that here labial gland secretion is not merely one organ replacing the function of another. In addition, labial gland secretion seems to play a role in ecdysis. It serves as a medium for the enzyme cocoonase, maintaining the correct pH for its activity (Kafatos and Williams, 1964). Fluid secretion can be induced by a number of stimuli including physical confinement which is presumed to simulate natural emergence conditions (Kafatos, 1968). Ziegler and Harmsen (1969) have suggested that the pharyngeal gland of *Apis* may function in pterine excretion.

In the thysanuran *Petrobius* the fine structure of the labial kidney cells is similar to that of the vertebrate nephron (Fain-Maurel and Cassier, 1971). Of particular importance are the findings that molecules dealt with by pinocytosis are degraded by enzymes, such as acid phosphatases, apparently located on the Golgi apparatus and on the cytosomes of labial kidney cells. Also the cells have apical and basal structures which are elaborate and contain numerous mito-chondria, perhaps suggesting their involvement in active transport. Kafatos (1968) attempted to demonstrate active transport in the labial glands of *Antheraea*. Acetazoleamid, an inhibitor of carbonic anhydrase, and ouabain, an inhibitor of $Na^+ - K^+$ exchange, were both ineffective at the concentrations used. In contrast, the respiratory inhibitor DNP markedly reduced secretion. This finding lends support to the process of secretion being active and respiration mediated. It does not clarify whether fluid or ion secretion is the primary process. Nevertheless, labial glands and other related structures, when they are present, seem to have an important function in those insects possessing them. They appear to be excellent examples of structures and/or processes which serve multiple functions. Further studies of these structures, including biochemical studies, are much needed in order to better understand their significance.

Pericardial and other cells

There exists in insects a group of cells of mesodermal origin which are usually presumed to play a role in excretion. These are the pericardial cells and certain others which are often referred to as nephrocytes (Maddrell, 1971). The pericardial cells, as their name implies, surround the heart and aorta, while the nephrocytes tend to be scattered in other parts of the body. Wigglesworth (1970) has published a review of these cells in which he compared the pericardial cells with the vertebrate reticulo-endothelial system. From this review it is clear that they possess certain properties which may be considered as excretory. Included here is their role in dealing with absorbed haemoglobin in the blood-sucking insect *Rhodnius* as was discussed earlier (p. 253). Perhaps this role should be considered more broadly as that of protein turnover, wherein cells of this type pick up useless or unwanted proteinaceous, colloidal particles, digest them, and return their constituent amino acids or peptides to the blood (Wigglesworth, 1970). They are also concerned with removal of certain types of injected dyes from the blood. While these functions can be considered a form of excretion, they may have broader implications. In addition, Maddrell (1971) has pointed out that other cells, such as haemocytes, apparently carry out essentially similar tasks. Furthermore, Wigglesworth (1970) has suggested that pericardial cells may have major undiscovered roles.

Little can be said about the biochemistry of these cells. It is clear that they possess granules, vacuoles and lysosomes, and that they absorb colloidal proteins (Wigglesworth, 1970). Presumably they digest these proteins through the action of lysosomal enzymes, but no precise studies have been reported. It seems that it should be possible to study this process more closely either by harvesting pericardial cells for *in vitro* work or by using intact insects and labelled proteins and dyes.

Utriculi majores

Certain species of cockroaches are known to have uric acid in the utriculi majores, which are part of the male accessory glands (Roth and Dateo, 1964, 1965; Roth, 1967, 1971; Ballan-Dufrancais, 1968). Upon mating the uric acid is poured out over the spermatophore and forms a white deposit. The entire mass is subsequently dropped by the female. This phenomenon has been described as being excretory in nature (Roth and Dateo, 1964, 1965; Maddrell, 1971). However, it appears that a fuller discussion of this and alternate possibilities might be useful.

The problem of uric acid excretion in cockroaches is complex, as has been mentioned earlier (p. 195). Many species do not void this material to the exterior, but rather store it internally primarily in the fat body (Mullins and Cochran, 1972, 1973b; Cochran, 1973). Those species which void uric acid do so in the form of discrete pellets produced by the rectum (Cochran, 1973). One of the species studied was *Shawella* in which both males and females void uric acid-containing pellets, but also store it internally. Thus, the patterns for dealing with excess uric acid in this group of insects seem to be established. The question is how or whether uric acid in the utriculi majores may be considered as an excretory mechanism.

The phenomenon has been studied best in *Blattella,* a species which does not void uric acid-containing pellets (Cochran, 1973). In this insect when adult males emerge they have no uric acid in the utriculi majores (Roth and Dateo, 1964), but within 1–2 days these glands are full. In males not allowed to mate, the glands continue to enlarge, while subsequent mating almost completely empties them. From this it seems clear that uric acid in the utriculi majores relates primarily to the mating process, and that it probably plays some yet undiscovered function in reproduction (Roth, 1967). One possibility is that it makes available to reproducing females an excellent source of dietary carbon and nitrogen at a critical time and still avoids coprophagy (Mullins, personal communication). It has been shown that dietary uric acid is absorbed by *Periplaneta* (Mullins and Cochran, 1972), and that these insects can utilize uric acid reserves in times of dietary stress (Gier, 1947; Haydak, 1953; McEnroe, 1956; Mullins, 1971). A multi-functional role for male accessory gland secretion in supporting female reproduction has been demonstrated in *Musca* (Terranova *et al.*, 1972). If this should be the case here, then accessary gland participation in excretion becomes debatable. Uric acid elimination via this route would be a mechanism to help ensure survival of the species, not the individual. The ability of unmated males to retain large amounts of glandular uric acid without apparent harm also argues against this being an important excretory function. Merely because the glandular material is the same as a major insectan excretory product does not prove that its role in this context is excretory. Ballan-Dufrancais (1970) considered the phenomenon to be both secretory and excretory.

Some information is available on the origin of the glandular uric acid. Labelled uric acid injected into the haemolymph of *Blattella* males began to appear in the soluble phase of the utriculi majores material within 15 minutes after injection (Ballan-Dufrancais 1970). In three hours the urate crystals were labelled. No radioactivity was found in the Malpighian tubules, but the post-injection excreta was strongly radioactive. These results indicate that uric acid is transported to the utriculi majores by the haemolymph. At the glandular site it is apparently rapidly transferred into the lumen of the glands. Ballan-Dufrancais (1970) suggested that the process is essentially similar to urate excretion by Malpighian tubule cells, and that active transport might be involved. The latter contention is supported by the finding that adult male *Blattella* void a very high percentage of the total body uric acid by this route (Roth and Dateo, 1964). Active transport may be necessary to accomplish this feat. Detection of radioactivity in the post-injection excreta, but not in Malpighian tubules, can probably be explained on the basis of faecal contamination by voided utriculi majores contents (Cochran, 1973). One of the most challenging problems in this area lies in the further development of a satisfactory rationale for the hierarchical distribution of uricose glands in cockroaches (Roth, 1967).

The gut

In developing this section, only those portions of the gut, exclusive of the Malpighian tubules and rectum will be considered. Primarily this means the midgut and the anterior portion of the hindgut.

Wigglesworth (1972) has discussed the role of the midgut in dealing with injected fluorescein in *Periplaneta,* and its involvement with absorbed haemoglobin in *Rhodnius.* Both of these examples involve excretion by cells of the midgut wall, but are of restricted interest since they concern a foreign compound and a highly specialized situation. Perhaps of more general importance is the ability of the midgut of certain plant-feeding insects to pump potassium from the blood into the gut lumen even against a steep electrochemical, potential gradient (Harvey and Nedergaard, 1964). Maddrell (1971) considered this to be a form of excretion. The pump is sodium independent (Harvey and Nedergaard, 1964), and is sensitive to several metabolic inhibitors such as DNP (Haskell and Clemons, 1963; Haskell *et al.,* 1965). Active transport of potassium appears to be involved. The goblet cells of the gut epithelium seem to be the most probable site of active transport based on their fine structure and certain kinetic studies (Anderson and Harvey, 1966; Harvey and Zerahn, 1969; Wood *et al.,* 1969). At present there is very little biochemical information concerning this process, except that energy-dependent active transport is apparently the vehicle for potassium transfer.

Potassium transfer is only one part of the regulatory capabilities of the midgut. For example, this structure is also capable of controlling water-flow from the haemolymph into the lumen (Mills *et al.,* 1970). In *Periplaneta,* this process appears to be active, is DNP sensitive, and is subject to regulation by hormonal substances from the terminal abdominal ganglion (Goldbard *et al.,* 1970; Sauer and Mills, 1971). Thus, the midgut in this species appears to be important for osmoregulation. Potassium transfer is a part of that process, but also fulfills the definition of excretion.

The role of the hindgut in ammonia excretion was described earlier (p. 222). Detailed biochemical analysis have not yet been reported, but the 'median portion' of the hindgut has been implicated (Prusch, 1972). In another study the ileum of *Blattella* has been shown to be important in the accumulation of minerals (Ballan-Dufrancais, 1972). These materials appear as concretions in the cytoplasm of the ileum epithelial cells. They contain P, Cl, Ca, Mg, K, and Fe in a glycoprotein matrix. It is not clear if the concretions are voided into the lumen, but the cells are described as having the fine structural characteristics typical of transit organs. Irrespective of whether the concretions are retained within the cells or are voided, the phenomenon appears to have excretory implications. In addition, the possible role of the hindgut anterior to the rectum, in modifying the Malpighian tubule fluid was discussed earlier (p. 187).

The anal papillae of certain mosquito larvae were considered by Maddrell (1971) to be possible ancillary excretory organs based on the work of Phillips and Meredith (1969). The suggestion is that under certain conditions the anal papillae may be able to pump chloride ions from the insect into its environment (Meredith and Phillips, 1973).

CONCLUDING REMARKS

There are several points which perhaps should be emphasized in concluding this review. One of them is the diversity of excretory materials produced by insects. Current knowledge is most highly developed with reference to nitrogen-containing compounds, but some information is beginning to appear on other types of excretory products. Clearly, in this largest group of animals a wide variety of chemicals are employed as end products of metabolism, and the firmly entrenched generalization, that insects are primarily uric acid excreting organisms, is no longer tenable. Even the broadened concept of uricotelism, used herein, does not adequately describe the situation as it is now understood. One of the most exciting findings is that certain terrestrial insects employ the highly toxic compound, ammonia, as a major excretory product. The biochemical and physiological mechanisms involved in this phenomenon are not completely understood. Therefore, this is an area that should be more extensively studied, since it could precipitate a major modification of the concepts surrounding ammonia excretion.

The re-cycling of water and ions by the Malpighian tubule-rectum complex should also be emphasized in relation to excretion. It has been demonstrated that this process, which is intimately associated with osmoregulation, is responsible for removing the numerous waste products from the haemolymph. The process is mediated by active ion transport in the Malpighian tubules, and also results in the removal of useful metabolites. Most of the useful metabolites are reclaimed by selective reabsorption, principally in the rectum. The resulting waste-containing, rectal fluid is typically hypertonic to the haemolymph, and is voided. Regulation of these complex processes occurs at several levels, many of which are not completely understood. Additional studies are needed to clarify how certain of the less well known excretory products fit into this concept of excretion.

Another point which deserves emphasis is the changing concept of urate storage within the insect's body. Previously, this phenomenon has been viewed as an example of storage excretion or permanent sequestration of a major waste product. However, evidence is beginning to accumulate, particularly from cockroaches, that stored urates are in dynamic equilibrium with active metabolic pools. Under certain conditions, stored urates are rapidly mobilized, and in some insect species these products are definitely not excreted, intimating that they are being utilized. It appears likely that intracellular symbionts are involved in metabolizing the urate molecule, particularly in the fat body. Presumably, the insect benefits from this relationship by having carbon and nitrogen reserves made available to it.

The last point concerns the considerable evidence now available to indicate the fat body as a major organ in excretory biochemistry. Many of the reactions necessary to the production of uric acid and other excretory products are known to occur in this tissue. It is also one of the two principal sites where the intracellular symbionts are harboured. However, much remains to be learned about enzyme localization, storage mechanisms, the role of symbionts in fat body metabolism, and the impact of stored reserves on fat body functioning. Of course other tissues, such as the gut and Malpighian tubules, are also involved in the biochemistry of certain excretory products.

ACKNOWLEDGEMENTS

I wish to express my appreciation to Dr. D. E. Mullins for many helpful discussions during the course of the preparation of this review. Both he and Dr. J. L. Eaton are extended recognition and thanks for critically reading the final manuscript. Mrs. June Mullins prepared the figures which appear in the review. Mrs. Jill James assisted in checking the references. Mrs. Jo Proco typed the manuscript. I express my grateful appreciation to each of these people. Some of my unpublished data, cited here, were obtained from research supported by NSF grant No. GB 34244.

REFERENCES

Abe, A and **Miura**, K. (1972) *Agr. biol. Chem.,* 36, 446–450.
Abdel-Wahab, A. M. (1971) *Bull. Soc. Entomol.* Egypte, 54, 81–85.

Akai, H. (1970) *Appl. Entomol. Zool.*, **5**, 112–117.
Anderson, A. D. and **Patton**, R. L. (1954) *Science*, **120**, 956.
Anderson, A. D. and **Patton**, R. L. (1955) *J. exp. Zool.*, **128**, 443–451.
Anderson, E. A. and **Harvey**, W. R. (1966) *J. Cell Biol.*, **31**, 107–134.
Anderson, J. F. (1966) *Comp. Biochem. Physiol.*, **17**, 973–982.
Asokan, H. (1972) *J. Insect Pathol.*, **19**, 171–178.
Auclair, J. L. (1958) *J. Insect Physiol.*, **2**, 330–337.
Auclair, J. L. (1959) *J. Insect Physiol.*, **3**, 57–62.
Auclair, J. L. (1963) *Ann. Rev. Entomol.*, **8**, 439–490.
Avi-Dor, Y. and **Gonda**, O. (1959) *Biochem. J.*, **72**, 8–14.
Bacon, J. S. D. and **Dickinson**, B. (1957) *Biochem. J.*, **66**, 289–297.
Baglioni, C. (1959) *Nature, Lond.*, **184**, 1084–1085.
Baglioni, C. (1960) *Heredity* **15**, 87–96.
Baker, F. C. (1939) *J. Parasitol.*, **25**, 280.
Ballan-Dufrancais, Ch. (1968) *Bull. Soc. Zool.* Fr., **93**, 401–421.
Ballan-Dufrancais, Ch. (1970) *Z. Zellforsch. Mikrosk. Anat.*, **109**, 336–355.
Ballan-Dufrancais, Ch. (1972) *Z. Zellforsch. Mikrosk. Anat.*, **133**, 163–179.
Balshin, M. and **Phillips**, J. E. (1971) *Nature, New Biol.*, **233**, 53–55.
Barrett, F. M. and **Friend**, W. G. (1966) *J. Insect Physiol.*, **12**, 1–7.
Barrett, F. M. and **Friend**, W. G. (1970) *J. Insect Physiol.*, **16**, 121–129.
Bartel, A. H., **Hudson**, B. W. and **Craig**, R. (1958) *J. Insect Physiol.*, **2**, 348–354.
Basden, R. (1970) *Proc. Linn. Soc. N.S.W.*, **95**, 9–10.
Beams, H. W., **Tahmisian**, T. N. and **Devine**, R. L. (1955) *J. biophys. biochem. Cytol.*, **1**, 197–202.
Beard, R. L. (1963) *Ann. Rev. Entomol.*, **8**, 1–18.
Becker, E. (1937) *Hoppe-Seyler's Z. physiol. Chem.*, **246**, 177–180.
Becker, E. (1939) *Insekten. Biol. Zbl.*, **59**, 597–627.
Bennett, D. S. (1971) *Comp. Biochem. Physiol.*, **39A**, 611–624.
Bergman, F. and **Dikstein**, S. (1955) *J. Amer. Chem. Soc.*, **77**, 691–696.
Bernard, G. R. and **Fixler**, D. E. (1963) *Physiol. Zool.*, **36**, 244–249.
Berridge, M. J. (1965) *J. exp. Biol.*, **43**, 535–552.
Berridge, M. J. (1966a) *J. Insect Physiol.*, **12**, 1523–1538.
Berridge, M. J. (1966b) *J. exp. Biol.*, **44**, 553–566.
Berridge, M. J. (1967) In: *Insects and Physiology* (J. W. L. Beament, J. E. Treherne, eds) pp. 329–347. Oliver & Boyd, Edinburgh and London.
Berridge, M. J. (1968) *J. exp. Biol.*, **48**, 159–174.
Berridge, M. J. (1969) *J. exp. Biol.*, **50**, 15–28.
Berridge, M. J. (1970a) *Chem. Zool.*, **5A**, 287–319.
Berridge, M. J. (1970b) *J. exp. Biol.*, **53**, 171–186.
Berridge, M. J. and **Gupta**, B. L. (1967) *J. Cell Sci.*, **2**, 89–112.
Berridge, M. J. and **Gupta**, B. L. (1968) *J. Cell Sci.*, **3**, 17–32.
Berridge, M. J. and **Patel**, N. G. (1968) *Science*, **162**, 462–463.
Berridge, M. J. and **Oschman**, J. L. (1969) *Tissue and Cell*, **1**, 247–272.
Bheemeswar, B. (1959) In: *Biochemistry of Insects* (L. Levenbook, ed.) Vol. XII, pp. 78–89. Pergamon Press, New York.
Binning, A., **Darby**, F. J., **Heenan**, M. P., and **Smith**, J. N. (1967) *Biochem. J.*, **103**, 42–48.
Birt, L. M. and **Christian**, B. (1969) *J. Insect Physiol.*, **15**, 711–719.
Blight, M. M. (1969) *J. Insect Physiol.*, **15**, 259–272.

Blochmann, F. (1887) *Insekten. Biol.* Zbl., 7, 606–608.
Boadle, M. C. and Blaschko, H. (1968) *Comp. Biochem. Physiol.,* 25, 129–138.
Bodenstein, D. (1953) *J. exp. Zool.,* 124, 105–115.
Boni, P., de Lerma, B. and Parisi, G. (1967) *Experientia,* 23, 186–187.
Bonse, A. (1967) *Z. Naturforsch.,* 22b, 1027–1029.
Booth, A. N., Robbins, D. J., Jones, F. T., Emerson, O. H. and Masri, M. S. (1965) *Proc. Soc. exp. Biol. Med.,* 120, 546–548.
Bouthier, A. (1966) *C. r. hebd. Séane. Acad. Sci.,* 262, 1480–1483.
Boyland, E. and Watson, G. (1956) *Nature, Lond.,* 177, 837–838.
Brenner-Holzach, O. and Leuthardt, F. (1965) *Helv. Chim. Acta.,* 48, 1147–1151.
Brighenti, A. (1940) *Boll. Soc. ital. Biol. sper.,* 15, 196–197.
Brooks, M. A. (1963) *Symp. Soc. Gen. Microbiol.,* 13, 200–231.
Brooks, M. A. (1970) *J. Invert. Pathol.,* 16, 249–258.
Brooks, M. A. and Richards, A. G. (1955) *Biol. Bull.,* 109, 22–39.
Brooks, M. A. and Richards, A. G. (1956) *J. exp. Zool.,* 132, 447–465.
Brooks, M. A. and Richards, K. (1966) *J. Invert. Pathol.,* 8, 150–157.
Brosemer, R. W. and Veerabhadrappa, P. S. (1965) *Biochim. biophys. Acta.,* 110, 102–112.
Brower, L. P. (1969) *Sci. Amer.,* 220, (2), 22–29.
Brown, A. W. A. (1936) *J. exp. Biol.,* 13, 131–139.
Brown, A. W. A. (1937) *J. exp. Biol.,* 14, 87–94.
Brown, A. W. A. (1938a) *Biochem. J.,* 32, 895–902.
Brown, A. W. A. (1938b) *Biochem. J.,* 32, 903–912.
Brown, K. S. (1965) *J. Amer. Chem. Soc.,* 87, 4202–4203.
Brues, C. T. and Dunn, R. C. (1945) *Science.,* 101, 336–337.
Brunet, P. C. J. (1965) In: *Aspects of Insect Biochemistry* (T. W. Goodwin, ed.), pp. 49–77. Academic Press, New York.
Bruno, C. F. and Cochran, D. G. (1965) *Comp. Biochem. Physiol.,* 15, 113–124.
Bryan, G. T., Brown, R. R. and Price, J. M. (1964) Cancer Res., 24, 596–602.
Buchanan, J. M. (1951) *J. Cell. comp. Physiol.,* 38, 143–171.
Buchanan, J. M. and Sonne, J. C. (1946) *J. biol. Chem.,* 166, 781–792.
Buchanan, J. M., Sonne, J. C. and Delluva, A. M. (1948) *J. biol. Chem.,* 173, 81–98.
Buckmann, D., Willig, A. and Linzen, B. (1966) *Z. Naturforsch.,* 21b, 1184–1195.
Burnet, B. and Sang, J. H. (1968) *Genetics,* 59, 211–235.
Burnet, B., Connolly, K. and Beck, J. (1968) *J. Insect Physiol.,* 14, 855–860.
Burns, D. P. and Davidson, R. H. (1966) *Ann. ent. Soc. Amer.,* 59, 1071–1073.
Bursell, E. (1963) *J. Insect Physiol.,* 9, 439–452.
Bursell, E. (1965a) *J. Insect Physiol.,* 11, 993–1001.
Bursell, E. (1965b) Proc. XII Internat. *Congr. Entomol.,* London, p. 797.
Bursell, E. (1966) *Comp. Biochem. Physiol.,* 19, 809–818.
Bursell, E. (1967) *Adv. Insect Physiol.,* 4, 33–67.
Bursell, E. (1970) *An Introduction to Insect Physiology.* Academic Press, London and New York.
Bush, G. L. and Chapman, G. B. (1961) *J. Bacteriol.,* 81, 267–276.
Busnel, R. G. and Drilhon, A. (1942) *Arch. Zool. exp. Gen.,* 82, 321–323.
Butenandt, A. (1952) *Endeavor,* 11, 188–192.

Butenandt, A. (1959) *Naturwissenschaften,* 46, 461–471.

Butenandt, A., Biekert, E. and Linzen, B. (1956) *Hoppe-Seyler's Z. physiol. Chem.,* 305, 284–289.

Butenandt, A., Biekert, E. and Linzen, B. (1958) *Hoppe-Seyler's Z. physiol. Chem.,* 313, 251–258.

Butenandt, A., Bickert, E., Kubler, H., and Linzen, B. (1960) *Hoppe-Seyler's Z. physiol. Chem.,* 319, 238–256.

Carney, G. C. (1969) *Life Sci.,* 8, 453–464.

Caspari, E. (1949) *Q. Rev. Biol.,* 24, 185–199.

Caveney, S. (1971) *Proc. Roy. Soc. Lond. B.,* 178, 205–225.

Cazal, M. and Girardie, A. (1968) *J. Insect Physiol.,* 14, 655–668.

Chamberlain, W. F. and Hamilton, E. W. (1964) *J. Econ. Entomol.,* 57, 800–803.

Chefurka, W. (1965) In: *The Physiology of Insecta* (M. Rockstein, ed.). Vol. II, pp. 669–768. Academic Press, New York and London.

Chen, P. S. (1966) *Adv. Insect Physiol.,* 3, 53–132.

Chen, P. S. (1971) *Biochemical Aspects of Insect Development.* S. Karger, Basel.

Chen, P. S. and Bachmann-Diem, C. (1964) *J. Insect Physiol.,* 10, 819–829.

Chinzei, Y. and Tojo, S. (1972) *J. Insect Physiol.,* 18, 1683–1698.

Clark, A. M. and Smith, R. E. (1967) *Exp. Gerontol.,* 2, 217–226.

Claudatus, J. and Ginori, S. (1957) *Science,* 125, 394–395.

Cline, R. E. and Pearce, G. W. (1963) *Biochemistry,* 2, 657–662.

Coast, G. M. (1969) *J. Physiol.,* 202, 102P–103P.

Cochran, D. G. (1961) *Biochim. biophys. Acta.,* 52, 218–220.

Cochran, D. G. (1973) *Comp. Biochem. Physiol.,* 46A, 409–419.

Cochran, D. G. and Bruno, C. F. (1963) *Proc. XVI Internat. Congr. Zool.,* Washington, 2, 92.

Cohen, A. J. and Smith, J. N. (1964) *Biochem. J.,* 90, 449–456.

Cohen, A. J., Smith, J. N. and Turbert, H. B. (1964) *Biochem. J.,* 90, 457–464.

Colhoun, E. H. (1963) *Experientia,* 19, 9–10.

Colhoun, E. H. (1964) In: *Comparative Neurochemistry* (D. Richter, ed.), pp. 333–339. Pergamon Press, New York.

Collins, J. F., Duke, E. J. and Glassman, E. (1970) *Biochim. biophys. Acta,* 208, 294–303.

Conway, E. J. and Cooke, R. (1939) *Biochem. J.,* 33, 479–492.

Corbet, S. A. and Rotheram, S. (1965) *Proc. Roy. ent. Soc. Lond.,* A40, 67–72.

Cordero, S. M. and Ludwig, D. (1963) *J. N. Y. ent. Soc.,* 71, 66–73.

Corrigan, J. J. (1969) *Science,* 164, 142–149.

Corrigan, J. J. (1970) In: *Comparative Biochemistry of Nitrogen Metabolism* (J. W. Campbell, ed.) Vol I, pp. 387–488. Academic Press, London and New York.

Corrigan, J. J., Wellner, D. and Meister, A. (1963) *Biochim. biophys. Acta,* 73, 50–56.

Crabtree, B. and Newsholme, E. A. (1970) *Biochem. J.,* 117, 1019–1021.

Craig, R. (1960) *Ann. Rev. Entomol.,* 5, 53–68.

Crowder, L. A. and Shankland, D. L. (1972) *Ann. ent. Soc. Amer.,* 65, 614–619.

Cymborowski, B. (1970) *Comp. Gen. Pharmacol.,* 1, 316–322.
Cymborowski, B. (1973) *J. Insect Physiol.,* 19, 1423–1440.
Dadd, R. H. (1973) *Ann. Rev. Entomol.,* 18, 381–420.
Daniel, R. S. and Brooks, M. A. (1972) *Exp. Parasitol.,* 31, 232–246.
Danneel, R. and Zimmermann, B. (1954) *Z. Naturforsch.,* 9b, 788–792.
Day, M. F. and Briggs, M. (1958) *J. Ultrastruct. Res.,* 2, 239–244.
DeGuire, D. M. and Fraenkel, G. (1973) *Ann. ent. Soc. Amer.,* 66, 475–476.
Desai, R. M. and Kilby, B. A. (1958a) *Arch. int. Physiol. Biochem.,* 66, 248–259.
Desai, R. M. and Kilby, B. A. (1958b) *Arch. int. Physiol. Biochem.,* 66, 282–286.
Descimon, H. (1965) *Bull. Soc. Chim. Biol.,* 47, 1095–1100.
Descimon, H. (1966) *C. r. Séanc. Soc. Biol.,* 160, 928–932.
Descimon, H. (1967) *Bull. Soc. Chim. Biol.,* 49, 1164–1166.
Descimon, H. (1971) *J. Insect Physiol.,* 17, 1517–1531.
Donnellan, J. F. and Kilby, B. A. (1967) *Comp. Biochem. Physiol.,* 22, 235–252.
Dorough, H. W. and Casida, J. E. (1964) *J. Agr. Food Chem.,* 12, 294–304.
Dressler, M. (1968) *Z. Wiss. Zool.,* 178, 40–71.
Drilhon, A. and Busnel, R. G. (1939) *C. r. hebd. Séanc. Acad. Sci.,* 208, 839–841.
Dubowsky, N. and Pierre, L. L. (1966) *Nature, Lond.,* 210, 1294.
Duchâteau, G., Florkin, M. and Frappez, G. (1940) *C. r. Séanc. Soc. Biol.,* 133, 436–437.
Duchâteau, G., Florkin, M. and Frappez, G. (1941) *Bull. Cl. Sci., Acad. Roy. Belg.,* 27, 169–173.
Dutton, G. J. and Duncan, A. M. (1960) *Biochem. J.,* 77, 18P.
Earle, N. W., Walker, A. B. and Burks, M. L. (1965) *Comp. Biochem. Physiol.,* 16, 277–288.
Edwards, J. S. (1964) *Nature, Lond.,* 203, 668–669.
Egelhaaf, A. (1956) *Z. Vererbungslehre,* 87, 769–783.
Egelhaaf, A. (1958) *Z. Naturforsch.,* 13b, 275–279.
Egelhaaf, A. (1962) *Fortschr. Zool.,* 15, 378–423.
Eguchi, M. (1961) *Jap. J. appl. Ent. Zool.,* 5, 163–166.
Evans, P. D. (1972) *J. exp. Biol.,* 56, 501–507.
Ewart, W. H. and Metcalf, R. L. (1956) *Ann. ent. Soc. Amer.,* 49, 441–447.
Fain-Maurel, M. A. and Cassier, P. (1971) *J. Microscopie,* 10, 163–178.
Ferreira, A. and de Cruz-Landin, C. (1969) *Ana. Acad. Brasil Cienc.,* 41, 591–600.
Florkin, M. (1937) *Arch. int. Physiol.,* 45, 241–246.
Florkin, M. (1959) In: *Biochemistry of Insects* (L. Levenbook, ed.), Vol. XII, pp. 63–77. Pergamon Press, New York.
Florkin, M. and Frappez, G. (1940a) *C. r. Séanc. Soc. Biol.,* 134, 117–118.
Florkin, M. and Frappez, G. (1940b) *Arch. int. Physiol.,* 50, 197–202.
Florkin, M. and Duchâteau, G. (1943) *Arch. int. Physiol. Biochem.,* 53, 267–307.
Forrest, H. S. (1962) *Comp. Biochem.,* 4, 615–641.
Forrest, H. S. and Nawa, S. (1962) *Nature, Lond.,* 196, 372–373.

Forrest, H. S., Glassman, E. and Mitchell, H. K. (1956) *Science,* **124**, 725–726.
Forrest, H. S., Hanly, E. W. and Lagowski, J. M. (1961) *Biochim. biophys. Acta,* **50**, 596–598.
Forrest, H. S., Menaker, M. and Alexander, J. (1966) *J. Insect Physiol.,* **12**, 1411–1421.
Forrest, H. S., Harris, S. E. and Morton, L. J. (1967) *J. Insect Physiol.,* **13**, 359–367.
Friauf, J. J. and Edney, E. B. (1969) *Proc. ent. Soc. Wash.,* **71**, 1–7.
Friend, W. G. (1958) *Ann. Rev. Entomol.,* **8**, 57–74.
Fukuda, T. and Nishitsutsuji, S. (1961) *Nippon Sanshigaku Zasshi,* **30**, 420–422.
Garcia, I., Tixier, M. and Roche, J. (1956) *C. r. Séanc. Soc. Biol.,* **150**, 632–634.
Garcia, I., Couerbe, J. and Roche, J. (1957) *C. r. Séanc. Soc. Biol.,* **151**, 1844–1847.
Garcia, I., Couerbe, J. and Roche, J. (1958) *C. r. Séanc. Soc. Biol.,* **152**, 1646–1649.
Gatterdam, P. E., De, R. K., Guthrie, F. E. and Bowery, T. G. (1964) *J. Econ. Entomol.,* **57**, 258–264.
Gersch, M. (1967) *Gen. comp. Endocrinol.,* **9**, 453.
Gessner, T., and Smith, J. N. (1960) *Biochem. J.,* **75**, 165–172.
Gholson, R. K., Hankes, L. V. and Henderson, L. M. (1960) *J. biol. Chem.,* **235**, 132–135.
Ghosh, D. and Forrest, H. S. (1967a) *Genetics.,* **55**, 423–431.
Ghosh, D. and Forrest, H. S. (1967b) *Arch. Biochem. Biophys.,* **120**, 578–582.
Gier, H. T. (1947) *Ann. ent. Soc. Amer.,* **40**, 303–317.
Gilmour, D. (1961) *The Biochemistry of Insects.* Academic Press, New York.
Gilmour, D. (1965) *The Metabolism of Insects.* Freeman, San Francisco.
Gilmour, D. and Calaby, J. H. (1952) *Arch. Biochem. Biophys.,* **41**, 83–103.
Gilmour, D. and Calaby, J. H. (1953) *Enzymologia,* **16**, 34–40.
Glassman, E. (1956) *Genetics,* **41**, 566–574.
Glassman, E. (1957) *Arch. Biochem. Biophys.,* **67**, 74–89.
Glassman, E. (1962a) *Science,* **137**, 990–991.
Glassman, E. (1962b) *Proc. Nat. Acad. Sci.* USA, **48**, 1491–1497.
Glassman, E. (1965) *Fed. Proc.,* **24**, 1243–1251.
Glassman, E. and Mitchell, H. K. (1959) *Genetics,* **44**, 153–162.
Glassman, E., Shinoda, T., Duke, E. J. and Collins, J. (1968) *Ann. N.Y. Acad. Sci.,* **151**, 263–273.
Goldbard, G. A., Sauer, J. R. and Mills, R. R. (1970) *Comp. Gen. Pharmacol.,* **1**, 82–86.
Goldsmith, T. H. (1958) *Proc. Nat. Acad. Sci.* USA, **44**, 123–126.
Goldsmith, T. H. and Warner, L. T. (1964) *J. gen. Physiol.,* **47**, 433–441.
Goodwin, T. W. and Srisukh, S. (1950) *Biochem. J.,* **47**, 549–554.
Gouranton, J. (1968) *C. r. hebd. Séanc. Acad. Sci.,* **266**, 1403–1406.,
Gouranton, J. and Maillet, P. L. (1966) *C. r. Séanc. Soc. Biol.,* **160**, 1724–1726.
Gouranton, J. and Maillet, P. L. (1967) *J. Miscroscopie,* **6**, 53–64.
Gray, H. E. and Fraenkel, G. (1954) *Physiol. Zool.,* **27**, 56–65.
Grell, E. H. (1962) *Z. Vererbungslehre,* **93**, 371–377.
Grigolo, A. (1969a) *Redia,* **51**, 169–178.
Grigolo, A. (1969b) *Redia,* **51**, 179–185.

Grigolo, A. and Cima, L. (1969) *Redia,* **51**, 211–218.

Gupta, B. L. and Berridge, M. J. (1966a) *J. Cell Biol.,* **29**, 376–382.

Gupta, B. L. and Berridge, M. J. (1966b) *J. Morphol.,* **120**, 23–81.

Harington, J. S. (1956) *Nature, Lond.,* **178**, 268.

Harington, J. S. (1961) *Parasitol.,* **51**, 319–326.

Harmsen, R. (1964) *Nature, Lond.,* **204**, 1111.

Harmsen, R. (1966a) *J. exp. Biol.,* **45**, 1–13.

Harmsen, R. (1966b) *J. Insect. Physiol.,* **12**, 23–30.

Harmsen, R. (1966c) *J. Insect. Physiol.,* **12**, 9–22.

Harmsen, R. (1969) *J. Insect. Physiol.,* **15**, 2239–2244.

Harris, S. E. and Forrest, H. S. (1967) *Proc. Nat. Acad. Sci.* USA, **58**, 89–94.

Harshbarger, J. C. and Forgash A. J. (1964) *J. Econ. Entomol.,* **57**, 994–995.

Hartenstein, R. (1968) *Amer. Zool.,* **8**, 507–519.

Harvey, W. R. and Nedergaard, S. (1964) *Proc. Nat. Acad. Sci.* USA, **51**, 757–765.

Harvey, W. R. and Zerahn, K. (1969) *J. exp. Biol.,* **50**, 297–306.

Haskell, J. A. and Clemons, R. D. (1963) *J. biol. Chem.,* **19**, 23A.

Haskell, J. A., Clemons, R. D. and Harvey, W. R. (1965) *J. Cell. comp. Physiol.,* **65**, 45–56.

Hayaishi, O., Rothberg, S., Mehler, A. H. and Saito, Y. (1957) *J. biol. Chem.,* **229**, 889–896.

Hayashi, Y. (1960) *Nature, Lond.,* **186**, 1053–1054.

Hayashi, Y. (1961a) *Nippon Sanshigaku Zasshi,* **30**, 305–312.

Hayashi, Y. (1961b) *Nippon Sanshigaku Zasshi,* **30**, 359–367.

Hayashi, Y. (1961c) *Jap. J. appl. Ent. Zool.,* **5**, 207–210.

Hayashi, Y. (1961d) *Nippon Sanshigaku Zasshii.,* **30**, 89–94.

Hayashi, Y. (1961e) *Nature, Lond.,* **192**, 756–757.

Hayashi, Y. (1961f) *Jap. J. appl. Ent. Zool.,* **5**, 282–283.

Hayashi, Y. (1961g) *Nippon Sanshigaku Zasshi.,* **30**, 13–16.

Hayashi, Y. (1962a) *Nippon Sanshigaku Zasshi.,* **31**, 25–31.

Hayashi, Y. (1962b) *Nippon Sanshigaku Zasshi.,* **31**, 32–36.

Haydak, M. H. (1953) *Ann. ent. Soc. Amer.,* **46**, 547–560.

Heller, J. and Szarkowska, L. (1957) *Bull. Acad. Pol. Sci., Ser. Sci. Biol.,* **5**, 111–113.

Heller, J. and Jezewska, M. M. (1959) *Bull. Acad. Pol. Sci., Ser. Sci. Biol.,* **7**, 1–4.

Heller, J. and Jezewska, M. M. (1960) *Acta Biochim. Pol.,* **7**, 469–473.

Henderson, L. M., Gholson, R. K. and Dalgliesh, C. E. (1962) *Comp. Biochem.,* **4**, 245–342.

Hendrichs-Hertel, U., and Linzen, B. (1969) *Z. Vergl. Physiol.,* **64**, 411–431.

Hengstenberg, R. and Gotz, K. G. (1967) *Kybernetik,* **3**, 276–285.

Henry, S. M. and Block, R. J. (1960) *Contrib. Boyce Thompson Inst. Plant Res.,* **20**, 317–330.

Henry, S. M. and Block, R. J. (1962) *Fed. Proc.,* **21**, 9.

Hilliard, S. D. and Butz, A. (1969) *Ann. ent. Soc. Amer.,* **62**, 71–74.

Hinde, R. (1971a) *J. Insect. Physiol.,* **17**, 1791–1800.

Hinde, R. (1971b) *J. Insect. Physiol.,* **17**, 2035–2050.

Hinde, R. (1971c) *J. Invert. Pathol.,* **17**, 333–338.

Hodge, L. D. and Glassman, E. (1967a) *Biochim. biophys. Acta,* **149**, 335–343.

Hodge, L. D. and Glassman, E. (1967b) *Genetics,* **57**, 571–577.

Hofmanova, O., Manowska, F., Pelouch, V. and Kubista, V. (1967) *Physiol. Bohemoslo.*, 16, 97–103.

Hook, G. E. R., Jordan, T. W. and Smith, J. N. (1968) In: *Enzymatic Oxidations of Toxicants* (E. Hodgson, ed.). pp. 27–41. N. C. State Univ. Press, Raleigh.

Hopkins, C. R. (1967) *J. Roy. Microscop. Soc.*, 86, 235–252.

Hopkins, F. G. (1895) *Phil. Trans. Roy. Soc.* (B), 186, 661–682.

Hopkins, T. L. and Lofgren, P. A. (1968) *J. Insect. Physiol.*, 14, 1803–1814.

Hopkins, T. L. and Srivastava, B. B. L. (1972) *J. Insect. Physiol.*, 18, 2293–2298.

Hopkins, T. L., Srivastava, B. B. L. and Bahadur, J. (1971) *J. Insect. Physiol.*, 17, 1857–1864.

Horne, F. R. (1969) *Biol. Bull.*, 137, 155–160.

Hubby, J. L. and Throckmorton, L. H. (1960) *Proc. Nat. Acad. Sci. USA*, 46, 65–78.

Hubert, M. and Razet, P. (1965) *C. r. hebd. Séanc. Acad. Sci.*, 261, 797–800.

Hudson, B. W., Bartel, A. H. and Craig, R. (1959) *J. Insect. Physiol.*, 3, 63–73.

Hyde, C. A. T. (1972) *J. Embryol. exp. Morphol.*, 27, 367–379.

Ikan, R. and Ishay, J. (1967) *J. Insect. Physiol.*, 13, 159–162.

Inagami, K. (1955) *Nippon Sanshigaku Zasshi*, 24, 295–299.

Inokuchi, T. (1969) *Bull. Seri. Expt. Sta.*, 23, 389–410.

Inokuchi, T., Horie, Y. and Ito, T. (1969) *Biochem. biophys. Res. Comm.*, 35, 783–787.

Inwang, E. E. (1971) *Comp. Biochem. Physiol.*, 39B, 569–577.

Irreverre, F. and Terzian, L. A. (1959) *Science*, 129, 1358–1359.

Irvine, H. B. (1969) *Amer. J. Physiol.*, 217, 1520–1527.

Irzykiewicz, H. (1955) *Aust. J. biol. Sci.*, 8, 369–377.

Ishiguro, I. and Linzen, B. (1966) *J. Insect. Physiol.*, 12, 267–373.

Ito, T. and Mukaiyama, F. (1964) *J. Insect. Physiol.*, 10, 789–796.

Jacobson, K. B. (1971) *Nature, New Biol.*, 231, 17–19.

Jezewska, M. M., Gorzkowski, B. and Sawicka, T. (1967) *Acta. Biochim. Pol.*, 14, 71–75.

Jungreis, A. M. and Tojo, S. (1973) *Amer. J. Physiol.*, 224, 21–26.

Kafatos, F. C. (1968) *J. exp. Biol.*, 48, 435–453.

Kafatos, F. C. and Williams, C. M. (1964) *Science*, 146, 538–540.

Kaihara, M. and Price, J. M. (1963) *J. biol. Chem.*, 238, 4082–4084.

Kameyama, A. and Miura, K. (1968) *Arch. int. Physiol. Biochem.*, 76, 615–623.

Kaufman, S. (1962) *Genetics*, 47, 807–817.

Kaufman, S. (1967) *Ann. Rev. Biochem.*, 36, 171–184.

Keller, E. C., Jr. (1964) *Z. Vererbungslehre*, 95, 326–332.

Keller, E. C., Jr. and Glassman, E. (1964a) *J. Elisha Mitchell Sci. Soc.*, 80, 130–133.

Keller, E. C., Jr. and Glassman, E. (1964b) *Science*, 143, 40–41.

Keller, E. C., Jr. and Glassman, E. (1965) *Nature, Lond.*, 208, 202–203.

Keller, E. C., Jr., Saverance, P. and Glassman, E. (1963) *Nature, Lond.*, 198, 286–287.

Keller, H. (1950) *Z. Naturforsch.*, 5b, 269–273.

Kermack, W. O. and Stein, J. M. (1959) *Biochem. J.*, 71, 648–654.

Kessel, R. G. (1970) *J. Cell Biol.*, 47, 299–303.

Kikal, T. and Smith, J. N. (1959) *Biochem. J.*, 71, 48–54.

Kikkawa, H. (1941) *Genetics*, **26**, 587–607.

Kikkawa, H. (1953) *Adv. Gen.*, **5**, 107–140.

Kilby, B. A. (1963) *Adv. Insect Physiol.*, **1**, 111–174.

Kilby, B. A. (1965) In: *Aspects of Insect Biochemistry* (T. W. Goodwin, ed.), pp. 39–48. Academic Press, New York.

Kilby, B. A. and Neville, E. (1957) *J. exp. Biol.*, **34**, 276–289.

King, J. C. (1969) *Proc. Nat. Acad. Sci. USA*, **64**, 891–896.

Knox, W. E. (1955) *Methods Enzymol.*, **2**, 242–253.

Knox, W. E. and Mehler, A. H. (1950) *J. biol. Chem.*, **187**, 419–430.

Koefoed, B. M. (1971) *Z. Zellforsch. Mikrosk. Anat.*, **116**, 487–501.

Koga, K. and Akune, S. (1972) *J. Fac. Agr., Kyushu Univ.*, **17**, 31–36.

Kokolis, N. (1968) *Comp. Biochem. Physiol.*, **25**, 683–691.

Kondo, Y. (1967) *Nippon Nogei Kagaku Kaishi*, **41**, 324–328.

Kürsteiner, R. (1961) *J. Insect. Physiol.*, **7**, 5–31.

Kuznezova, L. E. (1969) *Nature, Lond.*, **222**, 484–485.

Lamb, K. P. (1959) *J. Insect. Physiol.*, **3**, 1–13.

Lan, S. J. and Gholson, R. K. (1965) *J. biol. Chem.*, **240**, 3934–3937.

Lancaster, G. A. and Sourkes, T. L. (1969) *Comp. Biochem. Physiol.*, **28**, 1435–1441.

Langer, H. and Hoffmann, C. (1966) *J. Insect. Physiol.*, **12**, 357–387.

Lanham, U. N. (1968) *Biol. Rev.*, **43**, 269–286.

Leibenguth, F. (1965) *Z. Naturforsch.*, **20b**, 315–317.

Leibenguth, F. (1967) *Experientia*, **23**, 1069–1071.

Leifert, H. (1935a) *Zool. Jahrb. Physiol.*, **55**, 131–170.

Leifert, H. (1935b) *Zool. Jahrb. Physiol.*, **55**, 171–190.

Lennox, F. G. (1940) *Nature, Lond.*, **146**, 268.

Lennox, F. G. (1941a) *Aust. Council Sci. Ind. Res. Pamp.*, **109**, 9–35.

Lennox, F. G. (1941b) *Aust. Council Sci. Ind. Res. Pamp.*, **109**, 37–64.

Levenberg, B. and Kaczmarek, D. K. (1966) *Biochim. biophys. Acta*, **117**, 272–275.

Levenbook, L. (1950) *Biochem. J.*, **47**, 336–346.

Levenbook, L., Hutchins, R. F. N. and Bauer, A. C. (1971) *J. Insect Physiol.*, **17**, 1321–1331.

Liaci, L. (1965) *Riv. Biol.*, **58**, 49–55.

Linzen, B. (1958) *Deut. Zool. Gesell. Verh.*, **11**, 154–161.

Linzen, B. and Ishiguro, I. (1966) *Z. Naturforsch.*, **21b**, 132–137.

Linzen, B. and Hertel, U. (1967) *Naturwissenschaften*, **54**, 21.

Linzen, B. and Hendrichs-Hertel, U. (1970) *Wilhelm Roux Arch. EntwMech. Org.*, **165**, 26–34.

Lisa, J. D. and Ludwig, D. (1959) *Ann. ent. Soc. Amer.*, **52**, 548–551.

Lison, L. (1937) *Arch. Biol., Paris*, **48**, 321–360.

Lison, L. (1938) *Z. Zellforsch. Mikrosk. Anat.*, **28**, 179–209.

Locke, M. and Collins, J. V. (1967) *Science*, **155**, 467–469.

Locke, M. and Collins, J. V. (1968) *J. Cell. Biol.*, **36**, 453–483.

Ludwig, D. (1954) *Physiol. Zool.*, **27**, 325–334.

Ludwig, D. and Cullen, W. P. (1956) *Physiol. Zool.*, **29**, 153–157.

Ludwig, D. and Gallagher, M. R. (1966) *J. N. Y. ent. Soc.*, **74**, 134–139.

Maddrell, S. H. P. (1962) *Nature, Lond.*, **194**, 605–606.

Maddrell, S. H. P. (1963) *J. exp. Biol.*, **40**, 247–256.

Maddrell, S. H. P. (1964a) *J. exp. Biol.*, **41**, 163–176.

Maddrell, S. H. P. (1964b) *J. exp. Biol.*, **41**, 459–472.

Maddrell, S,. H. P. (1966) *J. exp. Biol.*, **45**, 499–508.

Maddrell, S. H. P. (1969) *J. exp. Biol.*, **51**, 71–97.

Maddrell, S. H. P. (1971) *Adv. Insect Physiol.*, **8**, 200–331.

Maddrell, S. H. P. and Klunsuwar, S. (1973) *J. Insect Physiol.*, **19**, 1369–1376.

Maddrell, S. H. P., Pilcher, D. E. M. and Gardiner, B. O. C. (1969) *Nature, Lond.*, **222**, 784–785.

Maddrell, S. H. P., Pilcher, D. E. M. and Gardiner, B. O. C. (1971) *J. exp. Biol.*, **54**, 779–804.

Magasanik, B. and Karibian, D. (1960) *J. biol. Chem.*, **235**, 2672–2681.

Mahler, H. R. and Cordes, E. H. (1966) *Biological Chemistry*, Harper and Row, New York and London.

Malke, H. (1964) *Nature, Lond.*, **204**, 1223–1224.

Malke, H. and Schwartz, W. (1966) *Z. Allg. Mikrobiol.*, **6**, 34–68.

Manunta, C. (1948) *Acad. Naz. Lincei Rend. Cl. Sci. Fis. Mat. Natur.*, **4**, 117–121.

Manunta, C. (1949) *Hereditas, Suppl.*, **624**, 624–625.

Marmaras, V. J., Sekeris, C. E. and Karlson, P. (1966) *Acta Biochim. Pol.*, **13**, 305–309.

Maruyama, K. (1954) *J. Fac. Sci., Tokyo Univ.* Sec IV, **7**, 231–271.

Marzluf, G. A. (1965) *Z. Vererbungslehre*, **97**, 10–17.

Matsubara, M., Tsusue, M. and Akino, M. (1963) *Nature, Lond.*, **199**, 908–909.

McAllan, J. W. and Chefurka, W. (1961a) *Comp. Biochem. Physiol.*, **3**, 1–19.

McAllan, J. W. and Chefurka, W. (1961b) *Comp. Biochem. Physiol.*, **2**, 290–299.

McEnroe, W. D. (1956) *Uric Acid Metabolism in the American Cockroach, Periplaneta americana* (L.). Doctoral Dissertation, Rutgers Univ., New Brunswick, N.J.

McEnroe, W. D. (1961) *Ann. ent. Soc. Amer.*, **54**, 925–926.

McEnroe, W. D. (1966a) *Ann. ent. Soc. Amer.*, **59**, 1012–1013.

McEnroe, W. D. (1966b) *Ann. ent. Soc. Amer.*, **59**, 1011.

McEnroe, W. D. and Forgash, A. J. (1957) *Ann. ent. Soc. Amer.*, **50**, 429–431.

McEnroe, W. D. and Forgash, A. J. (1958) *Ann. ent. Soc. Amer.*, **51**, 126–129.

McNally, J. B., McCaughey, W. F., Standifer, L. N. and Todd, F. E. (1965) *J. Nutr.*, **85**, 113–116.

McNutt, W. S. (1956) *J. biol. Chem.*, **219**, 365–373.

McShan, W. H., Kramer, S. and Olson, N. F. (1955) *Biol. Bull.*, **108**, 45–53.

Mehler, A. H. and Knox, W. E. (1950) *J. biol. Chem.*, **187**, 431–438.

Meister, A. (1973) *Science*, **180**, 33–39.

Mello, M. L. S. and Bozzo, L. (1969) *Protoplasma*, **68**, 241–251.

Meredith, J. and Phillips, J. E. (1973) *J. Insect Physiol.*, **19**, 1157–1172.

Merlini, L. and Nasini, G. (1966) *J. Insect Physiol.*, **12**, 123–127.

Milburn, N. S. (1966) *J. Insect Physiol.*, **12**, 1245–1254.

Miles, P. W. (1966) *J. theor. Biol.*, **12**, 130–132.

Miles, P. W. (1968) *J. Insect Physiol.*, **14**, 97–106.

Miles, P. W. and Lloyd, J. (1967) *Nature, Lond.*, **213**, 801–802.

Miller, S. and Collins, J. M. (1973) *Comp. Biochem. Physiol.*, **44B**, 1153–1163.

Mills, R. R. (1967) *J. exp. Biol.*, **46**, 35–41.

Mills, R. R. and **Cochran**, D. G. (1963) *Biochim. biophys. Acta,* 73, 213–221.

Mills, R. R. and **Cochran**, D. G. (1966) *Comp. Biochem. Physiol.,* 18, 37–45.

Mills, R. R. and **Cochran**, D. G. (1967) *Comp. Biochem. Physiol.,* 20, 919–923.

Mills, R. R. and **Nielson**, D. J. (1967) *Gen. comp. Endocrinol.,* 9, 380–382.

Mills, R. R., **Wright**, R. D. and **Sauer**, J. R. (1970) *J. Insect Physiol.,* 16, 417–427.

Mitchell, H. K., **Glassman**, E. and **Hadorn**, E. (1959) *Science,* 129, 268–269.

Mitlin, N. and **Vickers**, D. H. (1964) *Nature, Lond.,* 203, 1403–1404.

Mitlin, N. and **Mauldin**, J. K. (1966) *Ann. ent. Soc. Amer.,* 59, 651–653.

Mitlin, N. and **Wiygul**, G. (1972) *Ann. ent. Soc. Amer.,* 65, 612–613.

Mitlin, N. and **Wiygul**, G. (1973) *J. Insect Physiol.,* 19, 1569–1574.

Mitlin, N., **Vickers**, D. H. and **Gast**, R. T. (1964a) *Ann. ent. Soc. Amer.,* 57, 757–759.

Mitlin, N., **Vickers**, D. H. and **Hedin**, P. A. (1964b) *J. Insect Physiol.,* 10, 393–397.

Mittler, T. E. (1953) *Nature, Lond.,* 172, 207.

Mittler, T. E. (1958) *J. exp. Biol.,* 35, 74–84.

Mordue, W. (1969) *J. Insect Physiol.,* 15, 273–285.

Morita, T. (1958) *Science,* 128, 1135.

Moriuchi, A., **Koga**, K., **Yamada**, J. and **Akune**, S. (1972) *J. Insect Physiol.,* 18, 1463–1476.

Mullins, D. E. (1971) *An investigation into the Nitrogen Balance of an Insect, Periplaneta americana* (L.), *with Special Reference to Urate Storage and Mobilization, the Urate Storage Complex, and Nitrogenous Excretory Products.* Doctoral Dissertation, VPISU, Blacksburg, Virginia.

Mullins, D. E. and **Cochran**, D. G. (1972) *Science,* 177, 699–701.

Mullins, D. E. and **Cochran**, D. G. (1973a) *Comp. Biochem. Physiol.,* 44B, 549–555.

Mullins, D. E. and **Cochran**, D. G. (1973b) *J. Insect Physiol.,* 19, 1007–1018.

Munz, R. (1964) *Z. Vererbungslehre,* 95, 195–210.

Murdock, L. L., **Hopkins**, T. L. and **Wirtz**, R. A. (1970) *Comp. Biochem. Physiol.,* 34, 143–146.

Nation, J. L. (1963) *J. Insect Physiol.,* 9, 195–200.

Nation, J. L. and **Patton**, R. L. (1961) *J. Insect Physiol.,* 6, 299–308.

Nation, J. L and **Thomas**, K. K. (1965) *Ann. ent. Soc. Amer.,* 58, 883–885.

Needham, J. (1938) *Biol. Rev. Cambridge Phil. Soc.,* 13, 225–251.

Nelson, M. (1964) *Comp. Biochem. Physiol.,* 12, 37–42.

Nelson, W. A. (1958) *Nature, Lond.,* 182, 115.

Nickla, H. (1972) *Canad. J. Gen. Cytol.,* 14, 105–111.

Nolfi, J. R. (1970) *Comp. Biochem. Physiol.,* 35, 827–842.

Oschman, J. L. and **Wall**, B. J. (1969) *J. Morphol.,* 127, 475–510.

Palm, N. -B. (1952) *Ark. Zool.,* 3, 195–272.

Pant, R. and **Agrawal**, H. C. (1963) *Arch. int. Physiol. Biochem.,* 71, 605–613.

Parzen, S. D. and **Fox**, A. S. (1964) *Biochim. biophys. Acta,* 92, 465–471.

Payne, N. M. (1936) *Anat. Rec.,* 67, 37.

Perry, A. S., **Pearce**, G. W. and **Buckner**, A. (1964) *J. Econ. Entomol.,* 57, 867–872.

Pfteiderer, W. and **Taylor**, E. C. (1964) *Pteridine Chemistry,* Pergamon Press, London.

Phillips, J. E. (1964a) *J. exp. Biol.,* **41**, 15–38.
Phillips, J. E. (1964b) *J. exp. Biol.,* **41**, 39–67.
Phillips, J. E. (1964c) *J. exp. Biol.,* **41**, 68–80.
Phillips, J. E. (1965a) *Trans. Roy. Soc. Canada,* **3**, 237–254.
Phillips, J. E. (1965b) *Amer. Zool.,* **5**, 662.
Phillips, J. E. (1969) *Canad. J. Zool.,* **47**, 851–863.
Phillips, J. E. (1970) *Amer. Zool.,* **10**, 413–436.
Phillips, J. E. and Dockrill, A. A. (1968) *J. exp. Biol.,* **48**, 521–532.
Phillips, J. E. and Meredith, J. (1969) *Nature, Lond.,* **222**, 168–169.
Phillips, J. E. and Beaumont, C. (1971) *J. exp. Biol.,* **54**, 317–328.
Phillips, J. P. and Forrest, H. S. (1970) *Biochem. Genet.,* **4**, 489–498.
Phillips, J. P., Simmons, J. R. and Bowman, J. T. (1967) *Biochem. biophys. Res. Comm.,* **29**, 253–257.
Pichon, Y. (1970) *J. exp. Biol.,* **53**, 195–210.
Pierre, L. L. (1964) *Nature, Lond.,* **201**, 54–55.
Pierre, L. L. (1965) *Nature, Lond.,* **208**, 666–667.
Pilcher, D. E. M. (1970a) *J. exp. Biol.,* **52**, 653–665.
Pilcher, D. E. M. (1970b) *J. exp. Biol.,* **53**, 465–484.
Pilcher, D. E. M. (1971) *J. Insect Physiol.,* **17**, 463–470.
Pinamonti, S. and Petris, A. (1966) *Comp. Biochem. Physiol.,* **17**, 1079–1087.
Pinamonti, S., Petris, A. and Colombo, G. (1966) *J. Insect Physiol.,* **12**, 1403–1410.
Pinamonti, S., Petris, A. and Miliani, M. (1970) *Comp. Biochem. Physiol.,* **37**, 311–320.
Poisson, R. and Razet, P. (1953) *C. r. hebd. Séanc. Acad. Sci.,* **237**, 1362–1363.
Porembska, Z. and Mochnacka, I. (1964) *Acta Biochim. Pol.,* **11**, 109–117.
Porter, P. (1963a) *Res. Vet. Sci.,* **4**, 580–591.
Porter, P. (1963b) *Res. Vet. Sci.,* **4**, 592–602.
Powles, M. A., Janssens, P. A. and Gilmour, D. (1972) *J. Insect Physiol.,* **18**, 2343–2358.
Powning, R. F. (1953) *Aust. J. biol. Sci.,* **6**, 109–117.
Prasad, C. and French, W. L. (1971) *Comp. Biochem. Physiol.,* **38B**, 627–629.
Prince, W. T. and Berridge, M. J. (1973) *J. exp. Biol.,* **58**, 367–384.
Prota, C. D. (1961) *J. N.Y. ent. Soc.,* **69**, 59–67.
Prusch, R. D. (1971) *Comp. Biochem. Physiol.,* **39A**, 761–767.
Prusch, R. D. (1972) *Comp. Biochem. Physiol.,* **41A**, 215–223.
Purrmann, R. (1940) *Justus Liebigs Annln Chem.,* **544**, 182–190.
Ramsay, J. A. (1952) *J. exp. Biol.,* **29**, 110–126.
Ramsay, J. A. (1953) *J. exp. Biol.,* **30**, 358–369.
Ramsay, J. A. (1954) *J. exp. Biol.,* **31**, 104–113.
Ramsay, J. A. (1955a) *J. exp. Biol.,* **32**, 183–199.
Ramsay, J. A. (1955b) *J. exp. Biol.,* **32**, 200–216.
Ramsay, J. A. (1956) *J. exp. Biol.,* **33**, 697–708.
Ramsay, J. A. (1958) *J. exp. Biol.,* **35**, 871–891.
Ray, J. W. and Heslop, J. P. (1963) *Biochem. J.,* **87**, 39–42.
Razet, P. (1952) *C. r. hebd. Séanc. Acad. Sci.,* **234**, 2566–2568.
Razet, P. (1953) *C. r. hebd. Séanc. Acad. Sci.,* **236**, 1304–1306.
Razet, P. (1954) *C. r. hebd. Séanc. Acad. Sci.,* **239**, 905–907.

Razet, P. (1956) C. r. hebd. Séanc. Acad. Sci., 243, 185–187.
Razet, P. (1961) Bull. Soc. sci. Bretagne, 36, 1–206.
Razet, P. (1964) Proc. XII Int. Congr. Entomol., 220–221.
Razet, P. (1965) Bull. Soc. sci. Bretange, 40, 63–68.
Razet, P. (1966) Annee Biol., 5, 43–73.
Razet, P. (1970) C. r. Séanc. Soc. Biol., 164, 2627–2630.
Razet, P. and Barraud, J. (1965) C. r. Séanc. Soc. Biol., 159, 2492–2494.
Reddy, S. R. R. and Campbell, J. W. (1969a) Biochem. J., 115, 495–503.
Reddy, S. R. R. and Campbell, J. W. (1969b) Comp. Biochem. Physiol., 28, 515–534.
Rembold, H. and Hanser, G. (1960) Hoppe-Seyler's Z. physiol. Chem., 319, 213–219.
Rembold, H. and Gutensohn, W. (1968) Biochem. biophys. Res. Comm., 31, 837–841.
Rembold, H., Metzger, H., Sudershan, P. and Gutensohn, W. (1969) Biochim. biophys. Acta, 184, 386–396.
Riegel, J. A. (1966) J. exp. Biol., 44, 379–385.
Riegel, J. A. (1971) Chem. Zool., 6B, 249–277.
Riegel, J. A. (1972) Comparative Physiology of Renal Excretion. Oliver and Boyd, Edinburgh.
Rizki, T. M. (1961) J. biophys. biochem. Cytol., 9, 567–572.
Rizki, T. M. (1963) J. Cell Biol., 16, 513–520.
Rizki, T. M. and Rizki, R. M. (1963) J. Cell Biol., 17, 87–92.
Rizki, T. M. and Rizki, R. M. (1964) J. Cell Biol., 21, 27–33.
Robinson, W. (1935) J. Parasitol., 21, 354–358.
Robinson, W. and Baker, F. C. (1939) J. Parasitol., 25, 149–155.
Robinson, W. and Wilson, G. S. (1939) J. Parasitol., 25, 455–459.
Rocco, M. L. (1936) C. r. hebd. Séanc. Acad. Sci., 202, 1947–1948.
Rocco, M. L. (1938) C. r. hebd. Séanc. Acad. Sci., 207, 1006–1008.
Roche, J., Thoai, N. V. and Robin, Y. (1957) Biochim. biophys. Acta, 24, 515–519.
Rock, G. C. (1969) J. Nutr., 98, 153–158.
Rolofson, G. L. (1968) In vivo Studies of Suspected Mechanisms of DDT Resistance in Blattella germanica (L.). Doctoral Dissertation, VPISU, Blacksburg, Virginia.
Ross, D. J. (1959) Physiol. Zool., 32, 239–245.
Roth, L. M. (1967) Ann. ent. Soc. Amer., 60, 1203–1211.
Roth, L. M. (1971) Ann. ent. Soc. Amer., 64, 127–141.
Roth, L. M. and Dateo, G. P., Jr. (1964) Science, 146, 782–784.
Roth, L. M. and Dateo, G. P., Jr. (1965) J. Insect Physiol., 11, 1023–1029.
Russo-Caia, S. (1963) Rend. Ist. Sci. Univ. Camerina, 4, 216–228.
Sacktor, B. (1953) J. gen. Physiol., 36, 371–387.
Sacktor, B. and Cochran, D. G. (1957) J. biol. Chem., 226, 241–253.
Sacktor, B. and Childress, C. C. (1967) Arch. Biochem. Biophys., 120, 583–588.
Sacktor, B., Thomas, G. M., Moser, J. C. and Bloch, D. I. (1953) Biol. Bull., 105, 166–173.
Saini, R. S. (1964) Trans. ent. Soc. Lond., 116, 347–392.
Salkeld, E. H. (1967) Canad. J. Zool., 45, 967–973.

Sauer, J. R. and Mills, R. R. (1971) *J. Insect Physiol.*, 17, 1–8.

Sauer, J. R., Levy, J. J., Smith, D. W. and Mills, R. R. (1970) *Comp. Biochem. Physiol.*, 32, 601–614.

Schmidt, G., Liss, M. and Thannhauser, S. J. (1955) *Biochim. biophys. Acta*, 16, 533–535.

Schoffeniels, E. and Gilles, R. (1970) *Chem. Zool.*, 5A, 199–227.

Schöpf, C. and Becker, E. (1936) *Justis Liebigs Annln Chem.*, 524, 49–123.

Schöpf, G. and Reichert, R. (1941) *Justis Liebigs Annln Chem.*, 548, 82–94.

Schuler, W. and Reindel, W. (1935) *Hoppe-Seyler's Z. physiol. Chem.*, 234, 63–82.

Seegmiller, J. E. (1969) In: *Diseases of Metabolism* (P. K. Bondy, ed.), 516–599. Saunders, Philadelphia.

Self, L. S., Guthrie, F. E., and Hodgson, E. (1964) *J. Insect Physiol.*, 10, 907–914.

Seybold, W. D. (1973) *Experientia*, 29, 758.

Shaw, J. (1955) *J. exp. Biol.*, 32, 353–382.

Shinoda, T. and Glassman, E. (1968) *Biochim. biophys. Acta*, 160, 178–187.

Sidhu, H. S. and Patton, R. L. (1970) *J. Insect Physiol.*, 16, 1339–1348.

Simon, H., Wiygand, F., Walter, J., Wacker, H. and Schmidt, K. (1963) *Z. Naturforsch.*, 18b, 757–764.

Singh, A. (1964) *Current Sci.*, 33, 52.

Smissman, E. E. and Orme, J. P. R. (1969) *Ann. ent. Soc. Amer.*, 62, 246.

Smith, D. S. and Littau, V. C. (1960) *J. biophys. biochem. Cytol.*, 8, 103–133.

Smith, K. D., Ursprung, H. and Wright, T. R. E. (1963) *Science*, 142, 226–227.

Smith, J. N. (1962) *Ann. Rev. Entomol.*, 7, 465–480.

Smith, J. N. (1968) *Adv. comp. Physiol. Biochem.*, 3, 173–232.

Smith, J. N. and Turbert, H. B. (1964) *Biochem. J.*, 92, 127–131.

Smith, R. L. and Forrest, H. S. (1969) *J. Insect Physiol.*, 15, 953–957.

Snodgrass, R. E. (1935) *Principles of Insect Morphology*, McGraw-Hill, New York.

Sonne, J. C., Buchanan, J. M. and Delluva, A. M. (1948) *J. biol. Chem.*, 173, 69–79.

Sonobe, H. and Ohnishi, E. (1970) *Develop. Growth Different.*, 12, 41–52.

Speeg, K. V., Jr. and Campbell, J. W. (1968) *Amer. J. Physiol.*, 214, 1392–1402.

Spiegler, P. E. (1962) *J. Insect Physiol.*, 8, 127–132.

Srivastava, P. N. (1962) *J. Insect Physiol.*, 8, 223–232.

Srivastava, P. N. and Gupta, P. D. (1961) *J. Insect Physiol.*, 6, 163–167.

Srivastava, P. N. and Varshney, R. K. (1966) *Ent. exp. appl.*, 9, 209–212.

Staddon, B. W. (1955) *J. exp. Biol.*, 32, 84–94.

Staddon, B. W. (1959) *J. exp. Biol.*, 36, 566–574.

Stevens, T. H. (1961) *Comp. Biochem. Physiol.*, 3, 304–309.

Stobbart, R. H. (1968) *J. Insect Physiol.*, 14, 269–275.

Stobbart, R. H. (1971) *J. exp. Biol.*, 54, 29–66.

Stobbart, R. H. and Shaw, J. (1964) In: *Physiology of Insecta* (M. Rockstein, ed.), Vol. 3, 189–258. Academic Press, New York and London.

Strong, F. E. (1965) *Nature, Lond.*, 205, 1242.

Sugiura, K. and Goto, T. (1967) *Biochem. biophys. Res. Comm.*, 28, 687–691.

Szarkowska, L. and Porembska, Z. (1959) *Acta Biochim. Pol.*, 6, 273–276.

Taira, T. (1961) *Nature, Lond.*, 189, 231–232.

Takahashi, H. and Price, J. M. (1958) *J. biol. Chem.*, 233, 150–153.

Takahashi, S. Y., Suzaki, G. and Ohnishi, E. (1969) *J. Insect Physiol.*, 15, 403–407.

Tamada, T., Takamiya, M., Sasama, T., Mimuro, K. and Yatsum, S. (1967) *Nichidai Igaku Zasshi*, 26, 1262–1272.

Tamaki, Y. (1964a) *Jap. J. appl. Ent. Zool.*, 8, 159–163.

Tamaki, Y. (1964b) *Jap. J. appl. Ent. Zool.*, 8, 227–234.

Tartter, A. (1940) *Hoppe-Seyler's Z. physiol. Chem.*, 266, 130–134.

Tarver, R. U. and Pierre, L. L. (1967) *Nature, Lond.*, 213, 208–209.

Tatum, E. L. (1939) *Proc. Nat. Acad. Sci., USA,* 25, 486–497.

Taylor, H. H. (1971a) *Z. Zellforsch. Mikrosk. Anat.*, 118, 333–368.

Taylor, H. H. (1971b) *Z. Zellforsch. Mikrosk. Anat.*, 122, 411–424.

Taylor, R. L. (1969) *J. Invert. Pathol.*, 13, 167–187.

Taylor, R. L. and Freckleton, W. C., Jr. (1969) *J. Invert. Pathol.*, 13, 416–422.

Teigler, D. J. and Arnott, H. J. (1972) *Tissue Cell*, 4, 173–185.

Terranova, A. C., Leopold, R. A., Degrugillier, M. E. and Johnson, J. R. (1972) *J. Insect Physiol.*, 18, 1573–1591.

Terzian, L. A., Irreverre, F. and Stahler, N. (1957) *J. Insect Physiol.*, 1, 221–228.

Thayer, D. W. and Terzian, L. A. (1971) *J. Insect Physiol.*, 17, 1731–1734.

Thomas, K. K. and Nation, J. L. (1966) *Biol. Bull.*, 130, 442–449

Todd, A. C. (1944) *Trans. Amer. Microscop. Soc.*, 63, 54–67.

Tojo, S, (1971) *Insect Biochem*, 1, 249–263.

Tojo, S. and Hirano, C. (1966) *J. Insect Physiol.*, 12, 1467–1471.

Tojo, S. and Hirano, C. (1968) *J. Insect Physiol.*, 14, 1121–1133.

Tojo, S. and Yushima, T. (1972) *J. Insect Physiol.*, 18, 403–422.

Treherne, J. E. (1965) In: *Aspects of Insect Biochemistry* (T. W. Goodwin, ed.) 1–13. Academic Press, London and New York.

Trivellani, J. C. (1960) *Arch. Biochem. Biophys.*, 89, 149–150.

Truszkowski, R. and Chajkinowna, S. (1935) *Biochem. J.*, 29, 2361–2365.

Tsujita, M. and Sakurai, S. (1964) *Proc. Jap. Acad.*, 40, 461–465.

Tsujita, M. and Sakurai, S. (1965) *Proc. Jap. Acad.*, 41, 230–235.

Tsusue, M. (1967) *Experientia*, 23, 116–117.

Tsusue, M. (1971) *J. Biochem.*, 69, 781–788.

Umebachi, Y. (1962) *Sci. Rep. Kanazawa Univ.*, 8, 135–142.

Umebachi, Y. and Tsuchitani, K. (1955) *J. Biochem.*, 42, 817–824.

Umebachi, Y. and Takahashi, H. (1956) *J. Biochem.*, 43, 73–81.

Umebachi, Y. and Yamada, M. (1964) *Annot. Zool. Jap.*, 37, 51–57.

Umebachi, Y. and Katayama, M. (1966) *J. Insect Physiol.*, 12, 1539–1547.

Umebachi, Y. and Uchida, T. (1970) *J. Insect Physiol.*, 16, 1797–1812.

Unger, H. (1965) *Zool. Jbr. Physiol.*, 71, 710–717.

Ursprung, H. and Hadorn, E. (1961) *Experientia*, 17, 230–231.

Vietinghoff, U. (1967) *Vestn. Cesk. Spolecnosti Zool.*, 31, 376–382.

Wacker, H., Harvey, R. A., Winestock, C. H. and Plaut, G. W. E. (1964) *J. biol. Chem.*, 239, 3493–3497.

Wagner, R. P. and Mitchell, H. K. (1948) *Arch. Biochem.*, 17, 87–96.

Wall, B. J. (1965) *Zool. Jbr. Physiol.*, **71**, 702–709.

Wall, B. J. (1966) *Studies of Water Conservation in Cockroaches.* Doctoral Dissertation, Univ. of Pittsburg, Pennsylvania.

Wall, B. J. (1967) *J. Insect Physiol.*, **13**, 565–578.

Wall, B. J. (1970) *J. Insect Physiol.*, **16**, 1027–1042.

Wall, B. J. and Ralph, C. L. (1964) *Gen. comp. Endocrinol*, **4**, 452–456.

Wall, B. J. and Oschman, J. L. (1970) *Amer. J. Physiol.*, **218**, 1208–1215.

Wall, B. J., Oschman, J. L. and Schmidt-Nielsen, B. (1970) *Science*, **167**, 1497–1498.

Wang, C. M. and Patton, R. L. (1969) *J. Insect Physiol.*, **15**, 543–548.

Watanabe, H. and Kobara, R. (1968) *Jap. J. appl. Ent. Zool.*, **12**, 76–80.

Watt, W. B. (1967) *J. biol. Chem.*, **242**, 565–572.

Weber, J. and Roberts, C. W. (1967) *Canad. J. Gen. Cytol.*, **9**, 565–568.

Wehner, R., Gartenmann, G. and Jungi, T. (1969) *J. Insect Physiol.*, **15**, 815–823.

Weiner, H. (1902) *Ergebn. Physiol.*, **1**, 555–650.

Weinland, E. (1906) *Z. Biol.*, **47**, 232–247.

Wessing, A. (1965) In: *Sekretion und Exkretion.Wissenschaft. Konfer. Ges. Deut. Naturf. Arzte Func. Morph. Org. Zelle* 2, (1964), 228–268.

Wessing, A. (1966) *Naturwiss. Rundsch.*, **19**, 139–151.

Wessing, A. and Bonse, A. (1966) *Z. Naturforsch.*, **21b**, 1219–1223.

Wessing, A. and Eichelberg, D. (1968) *Z. Naturforsch.*, **23b**, 376–386.

Wessing, A. and Eichelberg, D. (1972) *Z. Zellforsch. Mikrosk. Anat.*, **125**, 132–142.

Weygand, F., Simon, H., Dahms, G., Waldschmidt, M., Schliep, H. J. and Wacker, H. (1961) *Angew. Chem.*, **73**, 402–407.

Wharton, D. R. A. and Wharton, M. L. (1961) *Rad. Res.*, **14**, 432–443.

Wharton, D. R. A. and Lola, J. E. (1969a) *J. Insect Physiol.*, **15**, 1647–1658.

Wharton, D. R. A. and Lola, J. E. (1969b) *J. Insect Physiol.*, **15**, 1877–1886.

Wharton, D. R. A. and Lola, J. E. (1970) *J. Insect Physiol.*, **16**, 199–209.

Whitmore, E. and Gilbert, L. I. (1972) *J. Insect Physiol.*, **18**, 1153–1167.

Whittaker, R. H. and Feeny, P. P. (1971) *Science*, **171**, 757–770.

Wieland, H. and Schöpf, C. (1925) *Ber. It. Chem. Ges.*, **58B**, 2178–2183.

Wieser, W. (1972) *Comp. Biochem. Physiol.*, **43A**, 859–868.

Wieser, W. and Schweizer, G. (1970) *J. exp. Biol.*, **52**, 267–274.

Wigglesworth, V. B. (1931) *J. exp. Biol.*, **8**, 411–451.

Wigglesworth, V. B. (1942) *J. exp. Biol.*, **19**, 56–77.

Wigglesworth, V. B. (1943) *Proc. Roy. Soc. B.* **131**, 313–339.

Wigglesworth, V. B. (1970) *J. reticuloendothel. Soc.*, **7**, 208–216.

Wigglesworth, V. B. (1972) *The Principles of Insect Physiology.* 7th edition. Chapman and Hall, London.

Wolf, J. P., III and Ewart, W. H. (1955) *Arch. Biochem. Biophys.*, **58**, 365–372.

Wolfram, R. (1949) *Z. Indukt. Abstamm. Vererbungsl.*, **83**, 254–298.

Wood, J. L., Farrand, P. S. and Harvey, W. R. (1969) *J. exp. Biol.*, **50**, 169–178.

Wyatt, G. R. (1961) *Ann. Rev. Entomol.*, **6**, 75–102.

Yang, Y. J. and Davies, D. M. (1971) *J. Insect Physiol.*, **17**, 2119–2123.

Yen, T. T. T. and Glassman, E. (1965) *Genetics*, **52**, 977–981.

Yen, T. T. T. and Glassman, E. (1967) *Biochim. biophys. Acta*, **146**, 35–44.

Yoshitake, N. and Aruga, H. (1950) *Nippon Sanshigaku Zasshi*, **19**, 536–537.

Yoshitake, N. and Aruga, H. (1952) *Nippon Sanshigaku Zasshi,* 21, 7–14.

Zayed, S. M. A. D., Hassan, A. and Fakhr, I. M. I. (1968), *Biochem. Pharmacol.,* 17, 1339–1347.

Zeutzschel, B. (1958) *Z. Vererbungslehre,* 89, 508–520.

Ziegler, I. (1961a) *Adv. Genet.,* 10, 349–403.

Ziegler, I. (1961b) *Z. Vererbungslehre,* 92, 239–245.

Ziegler, I. (1965) *Ergeb. Physiol.,* 56, 1–66.

Ziegler, I. and Harmsen, R. (1969) *Adv. Insect Physiol.,* 6, 139–203.

Zielinska, Z. M. (1957) *Acta Biol. Experiment.,* 17, 351–371.

Zielinska, Z. M. and Grzelakowska-Sztabert, B. (1968) *Acta Biochim. Pol.,* 15, 1–13.

ADDENDUM

Subsequent to completion of the literature search for this chapter (June 1973), an important review has appeared in the area of tryptophan metabolism in insects:

Linzen, B. (1974) *Adv. Insect Physiol.,* 10, 117–246.

4 Synaptic Transmission in Insects

G. G. LUNT

INTRODUCTION

Insects have never attracted the attentions of biochemists to the same extent as either microorganisms or vertebrates. In fact it is probably no great exaggeration to say that a major part of our current knowledge of biochemistry stems from observations using either tissue from the rat or *E. coli.*

The nervous system of insects has for a long time been the favoured material of many neurophysiologists and much basic knowledge of nerve function has arisen from studies on insects or other invertebrates. It is important to emphasize that, so far, the same basic mechanisms involving the same neurotransmitters have been found in all the systems studied be they vertebrate or invertebrate. However, biochemists or more particularly neuro-chemists seem largely to have ignored the nervous system of insects. Much of the information on the biochemistry of these systems has come from the work of neurophysiologists concerned with eluci-dating the molecular mechanisms underlying the physiological activities which they measured. It should be emphasized that our knowledge in this area is still sparse and is based on observations made on a very limited number of insect types. Thus it is not yet possible to present as comprehensive a picture as one can of the vertebrate nervous system. However, it is to be hoped that the very fact that such large gaps do exist in our knowledge of the biochemistry of insect nervous systems will encourage workers to leave their favoured 'rat cerebal cortex' and turn their attentions to the insect world.

GENERAL OUTLINE OF THE INSECT NERVOUS SYSTEM

A basic knowledge of the structure of the tissue is an essential prerequisite to any biochemical investigation, and this is probably more true of nervous tissue than of any other. It is however beyond the scope of this chapter to give a detailed description of the fine anatomical features of insect nervous tissue and the reader is referred to the excellent treatise on this subject by Bullock and Horridge (1965).

Briefly, the nervous system of insects may be considered to consist of three main units. A dorsal anterior brain, a ventral nerve cord often showing segmental ganglia, (frequently jointly referred to as the insect central nervous system) and a peripheral nervous system. The peripheral system may be further subdivided into three basic elements, the neuromuscular system, the sensory system and the stomatogastric system serving the anterior alimentary canal. A characteristic feature is the presence of a fibrous sheath or capsule surrounding the ganglia. Generally this takes the form of an outer layer of collagen termed the neural lamella which overlies a layer of rather flat cells — the perineurium. The presence of this sheath has frequently led to difficulties in experiments *in vitro* as it may present a relatively impermeable barrier to many solutes. Thus considerable differences may be seen in the concentration of compounds which produce a physiological response when they are perfused onto a 'de-sheathed', as opposed to an intact, ganglion (see Gerschenfeld, 1973). The general arrangement within a ganglion is a surface region of densely packed neuronal perikarya and glial cell bodies, within this is the neuropile or synaptic contact region containing fine processes from the neuronal perikarya, dendrites and their fine branches and some glial cells. The ultrastructure of the synapses is essentially the same as that found in vertebrate systems, (see Bloom and Aghajanian, 1968; Jones, 1970; Pappas and Purpura, 1972). One major difference from vertebrate systems is that to date no axo-somatic synapses have been described in insects.

The glial cells of the insect ganglia can be broadly divided into three main types. There are those found as a layer immediately beneath the perineurium which are termed capsular or perilemmal cells, those surrounding the neuronal perikarya — neuron satellite cells, and those surrounding the axons — nerve fibre satellite cells, (Trujillo-Cenoz, 1962; Smith, 1967). Whereas in vertebrates one of the major functions of glial cells is the formation of myelin, the classical myelin sheath has not been observed in insects. Wigglesworth (1960), however, has demonstrated that the glia have an important trophic function in *Periplaneta*. He showed that the glia contained large reserves of glycogen which were depleted during starvation, and rapidly restored after feeding. Wigglesworth (1960) suggested that the many invaginations which the glia make into the neuron enable nutrients to be transferred to the constituent neurons of a ganglion. It is important to remember here that unlike vertebrates there is no circulatory system for the circulation of haemolymph and its nutrients in the ganglia. However, the tracheae and tracheoles penetrate both central and peripheral nervous systems ensuring a direct supply of oxygen to the tissues.

The ultrastructural appearance of the neuromuscular junction in insects is very similar to that of vertebrates. There is a close apposition between the membranes of the muscle fibre and the axon, resulting in a synaptic cleft of about 20 nm. There are two basic types of nerve-muscle contact – the 'Doyere cone' (after Doyere, 1840) where the nerve terminal contacts with an elevated portion of the muscle fibre membrane, and the well known 'en passant' type of junction. Detailed descriptions of the anatomical and ultrastructural features of the junctions are given in Bullock and Horridge (1965). A feature that is quite different from the usual vertebrate system is that the innervation is multiterminal – the motor axons have numerous branches terminating at many points on a single fibre, as first described by Foettinger (1880) and since confirmed by many workers (see Gerschenfeld, 1973). The advantage of this for the biochemist is that insect muscle should be a richer source of presynaptic terminals and post-synaptic receptor material than is vertebrate muscle. This is partly borne out by the finding that higher yields of receptor material are obtained from *Schistocerca* muscle than from rat diaphragm (Lunt, 1973).

The whole process of synaptic transmission can be divided into a series of discrete steps. These are transmitter synthesis, storage and release, the interaction of the transmitter with a specific receptor in the post-synaptic membrane resulting in a change in membrane permeability, and finally the inactivation of the transmitter either by breakdown or re-uptake. In the case of vertebrates a number of these events are well characterised for a few transmitters. For example the synthesis, storage and breakdown of acetylcholine and the catecholamines has been studied in great detail. However, little is known of the mechanism by which these transmitters are released into the synaptic cleft, and neither are we able to define the precise nature of the interaction between transmitter and receptor. In the case of insects even less information is available and it is not yet possible to give a detailed account of the biochemistry of any one of the several transmitters found in the insect nervous system. In addition there is the problem that it is often necessary to compare work done on quite different insect types. For example, is it valid to present an overall picture of glutamate as an excitatory transmitter at the insect myoneural junction when some of the evidence comes from work on locusts, other evidence from houseflies and some from cockroaches? In spite of these problems however, the picture of synaptic transmission that emerges is very similar to that established in vertebrates. In this chapter an account will be given of the neurotransmitters whose presence is firmly established in the insect nervous system, and where possible, information of their respective receptor and associated enzyme systems.

CHEMICAL TRANSMITTERS IN THE CENTRAL NERVOUS SYSTEM

Acetylcholine

The first report of acetylcholine in the insect central nervous system was made by Gautrelet in 1938, in which he demonstrated the presence of the ester and of acetylcholinesterase activity in the heads of *Apis mellifera*. Shortly afterwards Corteggiani and Serfaty (1939) reported data for eight species of insect again indicating the presence of both acetylcholine and acetylcholinesterase in the central nervous system. Mikalonis and Brown (1941) showed that increased levels of acetylcholine occurred in ventral nerve cord preparations from *Periplaneta* treated with eserine and that these increases were correlated with electrical activity. Tobias, Kollros and Savit (1946) demonstrated also that *Periplaneta* nerve cord preparations could synthesize acetylcholine. It should be emphasized that many of the early measurements of acetylcholine levels are probably unreliable. For example, it is well known that tissue acetylcholine is rapidly destroyed by tissue acetylcholinesterase during the processes of grinding and homogenisation. However later workers (Colhoun, 1958a, b) using more reliable methods, confirmed in essence the early findings (see also Colhoun, 1963a). Thus, it is now well established that the three essentials of the cholinergic system – acetylcholine, acetylcholinesterase and choline acetylase occur in a wide variety of insect nervous systems. The acetylcholine content of the tissue although very variable may be as much as 100 times higher than that found in vertebrate central nervous tissue with values ranging from 100–500 μg/g tissue (Corteggiani and Serfaty, 1939; Lewis and Smallman, 1956; Smallman and Fischer, 1961). It is of interest to note that an even higher content is found in the royal jelly of *Apis mellifera* – 800 μg/g tissue (Henschler, 1956) though the function of acetylcholine in this material is not known.

It has been clearly demonstrated by standard electrophysiological techniques that acetylcholine acts as an excitatory transmitter on the neurons of insect ganglia (Kerkut, Pitman and Walker, 1969a, b). So far no reports of the subcellular distribution of this cholinergic system have been made. Thus it is not possible to establish whether the acetylcholine is localized in the synaptic vesicles as has been so clearly demonstrated in vertebrates. In spite of this however it does seem clear that acetylcholine is the main transmitter in the central ganglia (see Colhoun, 1963a; Gerschenfeld, 1973).

Acetylcholine receptors

An indispensable part of any neurotransmitter system is a specific receptor for the transmitter located in the post synaptic membrane. Until comparatively recently receptors had completely eluded neurochemists and consequently this aspect of neurotransmission was frequently glossed over in reviews on the subject. This situation has now changed and rapid advances in the isolation and characterization of synaptic receptor molecules have been made (De Robertis, 1971; Changeux, Kasai, Huchet and Meunier, 1971; Miledi, Molinoff and Potter, 1971). Most of this work has been concerned with the cholinergic receptor from the electroplax of *Torpedo* or *Electrophorus,* though De Robertis (1971) has also reported the isolation of cholinergic receptors from the mammalian central nervous system.

In view of the very high acetylcholine content of insect central nervous tissue it would seem to be an attractive source from which to isolate cholinergic receptors. In 1970 Eldefrawi and O'Brien made the first report of the attempted isolation of receptor material from the heads of *Musca domestica.* These workers prepared a 100 000 g aqueous supernatant from whole fly heads and then measured the binding of the potent cholinergic agonist muscarone by equilibrium dialysis. They reported a reversible binding to an unidentified constituent of the supernatant with a binding constant of 2.4×10^{-6} M. Competition binding experiments carried out with a very wide range of cholinergic agents suggested that the binding was of a mixed muscarinic – nicotinic nature. This certainly agrees with the little available data on the pharmacological characteristics of insect central nervous systems. Thus atropine, decamethonium, hexamethonium, tetraethylammonium, tetramethylammonium, nicotine and curare have all been shown to block various preparations of insect ganglia (Roeder and Roeder, 1939; Roeder, 1948; Twarog and Roeder, 1957; Suga and Katsuki, 1961; Callec and Boistel, 1967; Pitman and Kerkut, 1970; Shankland, Rose and Donninger, 1971). It would seem therefore that Eldefrawi and O'Brien (1970) were probably looking at binding to the physiological acetylcholine receptor, albeit in a very impure state. Further progress was made by Cattell and Donnellan (1972). Following the procedure of De Robertis and Fiszer (1970) these workers isolated a very hydrophobic protein – a proteolipid – from whole heads of *Musca domestica.* By chromatography of a chloroform/methanol extract of the fly heads on Sephadex LH20, three main peaks of proteolipid were isolated, one of which showed high affinity binding for acetylcholine and decamethonium. The amount of acetylcholine bound to the 'receptor' proteolipid was 10.2 nmol/mg of protein, a figure not far removed from the 13.4 nmol/g protein obtained by De Robertis, Lunt and La Torre (1971) for the isolated cholinergic receptor from the electroplax of *Torpedo*. An important point to emerge from the work of Cattell and Donnellan was that dimethyl tubocurarine did not affect the binding of acetylcholine to the receptor proteolipid. Eldefrawi and O'Brien (1970) showed that curare was much less effective at blocking the binding of muscarone to their fly-head supernatant fraction than to a particulate fraction from *Torpedo* electroplax. It has been known for some time that a number of insects are very resistant to injections of quite large amounts of curare (Hopf, 1952; Roeder, 1948), and Roeder, Kennedy and Samson (1949) showed that concentrations as high as 10^{-2} M were without effect on an abdominal ganglion of *Periplaneta*. It has been clearly demonstrated that curare binds very effectively to isolated nicotinic cholinergic receptors from vertebrate tissues. However, it has recently been found that Bungarotoxin, which is a highly specific antagonist for the nicotinic cholinergic receptor of vertebrate tissue (Chang and Lee, 1963; Miledi, Molinoff and Potter, 1971) does not interact with the proteolipid receptor from *Musca* heads (Donnellan, 1973 personal communication). Thus it seems clear that although the insect cholinergic receptor is a similar molecule to the vertebrate receptor, i.e. a very hydrophobic membrane protein or proteolipid, the topography of the binding site may be quite different.

Glutamic acid and γ-amino butyric acid (GABA)

Glutamate and GABA are both well established neurotransmitters at the myoneural junction (see below) but the situation with central nervous tissue is not quite so clear. Frontali (1961, 1965) has reported high levels of glutamate (5.33 µg/g tissue) and GABA (1.09 µg/g tissue) in *Apis mellifera* brain. Ray (1964) also found high levels of GABA in nerve cord of *Periplaneta*. Frontali (1965) has shown that both *Apis mellifera* and *Musca domestica* brain contain high glutamate decarboxylase (GAD) (EC 4 1.1.15) activity. However Bradford, Chain, Cory and Rose (1969) found that *Schistocerca* nerve preparations had very low GAD activity.

There have been a number of electrophysiological examinations of the effect of GABA on central nervous system neurons. Perfusion of

GABA onto intact ganglia is only effective when concentrations of the order of 10^{-2} to 10^{-3} M are employed (Suga and Katsuki, 1961). However, as with acetylcholine, if iontophoretic application is used then neurons may be found which are extremely sensitive. Thus Kerkut and co-workers (Kerkut, Pitman and Walker, 1969a, b; Pitman and Kerkut, 1970) found neurons from the abdominal ganglia of *Periplaneta* which were hyperpolarized by as little as 1.05×10^{-13} moles of GABA. All the electrophysiological data indicates GABA functioning as an inhibitory transmitter, i.e. it brings about a hyperpolarization of the neurones to which it is applied. It is interesting to note that Kerkut, Oliver, Rick and Walker (1970) reported that in *Periplaneta* the GABA level of a ganglion may be related to the behaviour of the insect. They found that during an avoidance learning procedure the GABA levels of the metathoracic ganglion fell and there was a concomitant decrease in acetyl-cholinesterase activity. Thus, assuming that acetylcholine is an excitatory and GABA an inhibitory transmitter, this would facilitate synaptic transmission in the ganglion and as such could constitute a simple memory system.

The status of glutamate as a central neurotransmitter is much less well defined. Although high levels are found in central nervous tissue it would seem not to function as a neurotransmitter. No electro-physiological studies of the effect of glutamate on central nervous system neurons have been published although Pitman (1971) refers to unpublished work of Pitman and Kerkut in which they found 'that glutamate had little or no action on electrically excitable cockroach central neurons'.

Recently Fox and Larsen (1972) have made one of the all too few attempts to carry out an extensive subcellular fractionation of central nervous tissue and to characterize the fractions enzymically. These workers measured the activities of GAD, GABA- ketoglutarate aminotransferase (GABAT) (EC 2.6.1.C) and succinate semi-aldehyde dehydrogenase (SSDH) (EC 1.2.1.b), in both total homogenates and subcellular fractions of *Apis mellifera* brains. They found that in general the kinetic properties of the GAD were quite similar to those of the enzyme from other sources, though they did find substrate inhibition by L-glutamate, also the Km of 0.06M was rather high. However, the authors point out that, based on Frontali's (1965) figures for the glutamate content of *Apis mellifera* brain, the local concentration of glutamate at a given subcellular site could be very much higher than this. Thus it is suggested that the high Km and substrate inhibition could be adaptations which would make the synthesis of GABA possible in an environment which was high in free glutamate and GABA. This would be in accordance with the view that a key role of the enzyme could be the production of GABA i.e. GAD should be considered as a synthetic enzyme rather than a means of removing glutamate from the synapse. Fox and Larsen (1972) did not find a marked concentration of GAD activity in any single subcellular fraction but about 75 per cent of the activity was membrane bound. One important observation was that GAD was not associated with membranes having high acetyl-cholinesterase activity which the authors suggested were probably synaptosomal in origin. This finding again lends support to the proposal that GAD should not be considered to be analogous to acetylcholinesterase and is not important for transmitter degrad-ation. This is also in accordance with the findings of Donnellan, Jenner (1974) and Ramsey (1974) on the myoneural junction, a discussion of which follows. GABAT and SSDH were both found to be associated primarily with mitochondria and this would be in agreement with the model for a GABA mediated synapse put forward by Salganicoff and De Robertis (1963, 1965), in which the released GABA is taken up and metabolized to succinate by mitochondrial GABAT and SSDH. However, no reports have yet been made of an active GABA uptake system in insect central nervous tissue though Iversen and Kravitz (1968) described such a system in lobster nerve-muscle preparations.

It would seem that complete elucidation of the glutamate — GABA system cannot be achieved until an extensive subcellular fractionation scheme for insect central nervous tissue is available. Telford and Matsumura (1970) described a procedure for cockroach (*Periplaneta americana* and *Blatella germanica*) heads and nerve cords which yielded six fractions. The authors identified these morpho-logically and judged only on this basis, the fractions would seem to be an excellent starting point for a detailed biochemical examination. Unfortunately, the authors reported only on the dieldrin binding capacity of the fractions and no biochemical characteristics were presented. Fox and Larsen (1972) on the other hand, give extensive biochemical data for their fractions from *Apis mellifera* brain but present no morphological data at all. One thing that is clear from these studies (see also Donnellan *et al.* 1974) is that it is not possible to take the conditions that have been so well established for mammalian cerebral cortex and apply them to insect nervous tissue in the hope of achieving analogous fractions.

Biogenic amines

Catecholamines may be detected histochemically by using the Hillarp – Falck fluorescence method (Falck, 1962). Basically the process involves the condensation of the amine group with formaldehyde thereby producing a new heterocyclic ring which has a characteristic fluorescence spectrum. Using this technique catecholamines have been shown to be present in the brains of *Locusta* (Plotnikova & Govyrin, 1966) and *Periplaneta* (Frontali, 1968; Frontali & Norberg, 1966). Frontali (1968) showed that in *Periplaneta* the fluorescence was reduced by the administration of reserpine and was increased by treatment with 3:4-dihydroxyphenylalanine (dopa) and the monoamine oxidase inhibitor nialamid. Catecholamines have also been visualized using the same fluorescence technique in ganglia of *Schistocerca* (Chanussot, Dando, Moulins and Laverack, 1969) and of the Trichopteran *Annabolia nervosa* (Björklund, Falck and Klemm, 1970). The latter workers showed that these ganglia contained dopamine and a smaller amount of noradrenaline.

A number of electrophysiological studies have been made and in most cases noradrenaline and dopamine have been shown to cause depolarization of central neurones (see Gerschenfeld, 1973; Pitman, 1971). However, considerably more work is required before the importance of these two potential transmitters can be assessed.

5-Hydroxytryptamine (5HT) has been shown to be present in the brains of *Periplaneta* (Colhoun, 1963b) and *Blaberus giganteus* (Welsh and Moorhead, 1960). Recently Hiripi and Rozsa (1973) have reported a quantitative determination of 5HT and catecholamines in *Locusta* tissues using a fluorimetric assay. They found that the brain contained $2.34~\mu g$ 5HT/g wet wt. Hiripi and Rozsa (1973) also reported that *Locusta* heart contained as much 5HT as the brain and that the dopamine concentrations was almost double – $2.42~\mu g/g$ tissue, as opposed to $1.31~\mu g/g$ tissue for brain. The authors suggest that these two compounds may act as transmitters in the heart as was suggested by an earlier electrophysiological investigation (Rozsa and Szöke, 1972). Colhoun (1963b) has shown that *Periplaneta* brain can decarboxylate exogenous 5-hydroxytryptophan to 5HT at a rate similar to that found in mouse brain. Ventral nerve cord preparations showed a lower rate of decarboxylase activity. Coulhoun (1967) has also reported the decarboxylation of L-dopa to dopamine by *Periplaneta* brain homogenates and in the same article he also recorded the lack of detectable monoamine oxidase activity in nervous tissue.

Pitman (1971) refers to unpublished work of Pitman and Kerkut in which they found no effect when large amounts of 5HT were applied to neurones in the abdominal ganglia of *Periplaneta*. Apart from this there are no reports of pharmacological studies on the 5HT system in central nervous tissue. Thus the status of 5HT as a central transmitter is even less well defined than that of the catecholamines.

CHEMICAL TRANSMITTERS AT THE NEUROMUSCULAR JUNCTION

Glutamic acid

As the early findings on the cholinergic system in central nervous tissue were confirmed and its presence firmly established, it became increasingly apparent that the situation at the neuromuscular junction was quite different. A number of workers observed that acetylcholine, curare, atropine, nicotine and a range of anti-cholinesterases had no effect on neuromuscular transmission in insects (Roeder and Roeder, 1939; Roeder, 1948; Harlow, 1958; Colhoun, 1958a, b; Faeder, O'Brien and Salpeter, 1970) and it is now generally accepted that neuromuscular transmission is definitely non-cholinergic. There are still many unanswered questions about transmission at the neuromuscular junction but there is a rapidly growing body of evidence that glutamate is the excitatory transmitter at this site.

Glutamate is an amino acid that plays a central role in metabolism and this fact has been largely responsible for considerable reluctance to accept it as a neurotransmitter. However, it is found in mammalian brain in a higher concentration than in any other organ and than any other amino acid (Waelsch, 1951) and as early as 1954 it was shown to be excitatory in both crustacean muscle and mammalian brain (Hayashi, 1954). A number of workers have subsequently shown that glutamate acts as an excitatory transmitter on crustacean neuromuscular junctions (Robbins, 1959; Van-Harreveld and Mendelson, 1959; Takeuchi and Takeuchi, 1964a, b). Thus it seems reasonable to suppose that in insects also, glutamate may act as a transmitter. In 1965 Kerkut's group (Kerkut, Shapira and Walker, 1965; Kerkut, Leake, Shapira, Cowan and Walker, 1965) made the important observations that low concentrations of glutamate caused contractions of *Periplaneta* muscles and that glutamate appeared in the perfusate from an electrically stimulated muscle. On the basis of these observations, glutamate could have been considered as a likely neurotransmitter since it satisfies the criteria put forward by Hebb and Krnjevic (1962) for identifying a neurotransmitter system.

Most of the early investigations were made on muscle preparations *in situ* and as such posed considerable technical problems. Usherwood and Machili (1966) made possible more rapid progress with the introduction of a preparation *in vitro* of the metathoracic retractor unguis muscle of locusts, the muscle being kept in a small perfusion chamber through which saline containing various drugs could be rapidly passed. They found that a contraction occurred when solutions of glutamate of concentrations greater than 6.8×10^{-6} M were passed through the bath. It was also found that both membrane potential and resistance were lowered by smaller concentrations of glutamate and that desensitization occurred with high concentrations. Similar studies were carried out using *Periplaneta* muscles by Kerkut and Walker (1966, 1967) and in *Sarcophaga* flight muscle by McCann and Reece (1967).

One drawback of all such experiments was that in each case the glutamate was applied in the organ bath fluid and as such it was impossible to localize precisely its site of action. Studies of the range of effective concentration, dose response curves etc. were difficult due to the necessity of replacing all the bath fluid and of washing the muscle between successive applications. The administration of the compound by the process of iontophoresis circumvents these difficulties and closely mimics the physiological transmitter-release situation (Del Castillo and Katz, 1955). Thus Beranek and Miller (1968) showed that the muscle fibres from *Schistocerca* and *Locusta* were sensitive to L-glutamate only at discrete sites on the membrane surface. These sites closely correspond to, and are probably identical with the neuromuscular junction. The authors further demonstrated that the induced depolarization was essentially the same as that produced naturally. Another point to emerge from their work was that D-glutamate was entirely without effect. Since then a vast amount of evidence has been collected, mainly by Usherwood and his co-workers, which supports the proposal that glutamate is the excitatory transmitter at the insect neuromuscular junction (see Usherwood 1972). Nevertheless the point is frequently made that in view of the high concentration of glutamate found in insect haemolymph (as high as 3.7×10^{-4} M; Usherwood and Machili, 1966) it is most unlikely to be a transmitter at the neuromuscular junction since such a concentration would cause depolarization of the muscle fibre and would rapidly lead to densitization.

Usherwood has always refuted this argument and as long ago as 1966 reported (Usherwood and Machili, 1966) that replacement of the saline perfusing a neuromuscular preparation with fresh haemolymph, resulted in no change in the response of the muscle to electrical stimulation. If saline containing an amount of glutamate equivalent to that in the haemolymph as determined chemically is used as the perfusing fluid then the muscle loses the ability to respond to neural stimulation. The conclusion of these important experiments is that the glutamate of haemolymph is not free in solution, and consequently under normal physiological conditions cannot effect the neuromuscular junction. Usherwood has recently reiterated that under his experimental conditions fresh haemolymph has no depolarizing effects unless glutamate is added to it, and that *in vivo* the glutamate of haemolymph is probably bound within the haematocytes (Usherwood 1971, 1972). These suggestions are supported by the findings of Murdock and Koidl (1972a, b) who examined the permeability of the gut wall of both *Schistocerca* and *Locusta* to a number of amino acids. It was found that the L-glutamate does not pass readily into the haemolymph from the lumen of the gut. Most of the glutamate was metabolized to glutamine and alanine neither of which would affect the myoneural junction except at extremely high concentrations which would never be reached under physiological conditions. Murdock and Koidl (1972a, b) suggest that this system is part of the mechanism for protecting the synapse from free glutamate and they make the point that under normal dietary conditions no free glutamate would enter the haemolymph. Tiegs (1955) showed that the neuromuscular junction is located in a deep invagination of the muscle membrane. Recently Clements and May (1974) have produced evidence that in the intact muscle of *Schistocerca* the nerve terminals are deeply seated between tightly packed muscle fibres and are separated from the haemolymph by a connective tissue sheath. They demonstrate very clearly that removal of the muscle causes considerable damage to this sheath and consequently muscles dissected out from the insect show much higher sensitivity to exogenous glutamate. Clements and May (1974) show also that calcium and magnesium ions can bind haemolymph amino acids, including glutamate, and they postulate that this could be a further protective mechanism.

It is clear then that there are a number of protective mechanisms which ensure that *in vivo* the synapses of the myoneural junction are not affected by haemolymph glutamate. Further evidence for glutamate's role as a transmitter has come from experiments on denervated muscle. It has been known for some time that denervation of vertebrate muscle leads to hypersensitivity of the muscle membrane to nicotine and acetylcholine (Langley, 1905; Brown, 1937). In normal muscle, acetylcholine sensitivity is confined to the myoneural junction and its immediately surrounding area. After denervation however, the sensitive region spreads and eventually the whole muscle surface becomes acetycholine-sensitive (Ginetzinsky and Shamarina, 1942; Axelsson and Thesleff, 1959; Miledi, 1960a, b). The first account of analogous studies on insect muscles was made by Usherwood (1969). He unilaterally denervated the retractor unguis muscle of *Schistocerca* by sectioning the motor nerve, the contralateral muscle serving as the control. The sensitivity of the muscles to iontophoretically applied glutamate was then measured and in the denervated muscle there was a time-dependent spread of glutamate sensitivity away from the myoneural junction. In some cases the sensitivity of the non-synaptic membrane approached that of the junction, a situation analogous to that found in vertebrate muscle. Thus again the conclusion to be drawn from these findings is that glutamate is the natural transmitter at the insect myoneural junction. One major difficulty in experiments concerned with the glutaminergic system is the lack of substances known to have well characterized pharmacological activities at the glutamate-sensitive synapse. A very wide range of compounds have been examined (Usherwood and Machili, 1968; Holman and Cook, 1970), but so far, without success. Recently there has been a very brief report that kainic acid potentiates the action of glutamate at the neuromuscular junction of the crayfish (Shinozaki and Shibuya, 1972) but a detailed report of the effects of this compound in insect systems has not yet appeared. Thus at the moment it would appear that the excitatory receptor at the insect myoneural junction shows a specificity for its transmitter which is considerably higher than in any other transmitter-receptor system so far studied.

The glutamate receptor

Lunt (1973) has recently applied the techniques of receptor isolation, developed for cholinergic systems, to the neuromuscular system of *Schistocerca*. It was found that hydrophobic proteins or proteolipids with glutamate binding properties could be extracted from the whole muscles. Using a detergent-extraction procedure based on that of Changeux, Kasai, Huchet and Meunier (1970), a soluble protein extract was obtained that showed a specific binding for glutamate as measured by equilibrium dialysis. Although the dissociation constant was low, 5×10^{-7} M, the amount of glutamate bound was rather high − 1 mol/60 000 g protein. However, these experiments were carried out on crude detergent extracts and it is hoped that future work will lead to the isolation of purified proteins showing glutamate binding properties. The proteolipid isolation procedure of De Robertis and his group (De Robertis, Fiszer and Soto, 1967) when applied to *Schistocerca* muscle yielded two hydrophobic proteins (or proteolipids) showing specific high affinity glutamate binding. Fiszer de Plazas and De Robertis (1973) have recently reported the isolation of a glutamate binding proteolipid from the muscle of the crustacean *Artemia longinaris,* and in the same work the isolation of a proteolipid which bound γ-amino butyric acid.

Recently, preliminary binding studies have been carried out on the glutamate binding proteolipids from *Schistocerca* using the phase-partition method described by Weber *et al.* (1971). It has been found that under these conditions kainic acid at concentrations between 10^{-5} and 10^{-3} M causes an increased binding of glutamate to the receptor proteolipids (Lunt unpublished findings). It is emphasized that these are preliminary findings and much further work is required to fully characterize the system. Nevertheless, they are in accordance with the observations of Shinozaki and Shibuya (1972) referred to above. The amount of glutamate binding proteolipid from the muscle was of the order of 10 μg mg total muscle protein. This is about ten times higher than the amount of cholinergic receptor proteolipid found in rat diaphragm (Lunt, Stefani and De Robertis, 1971) and may well be a reflection of the fact that insect muscle is multiply innervated, the receptor proteins therefore representing a higher proportion of the total muscle proteins than in vertebrate tissue.

Enymes associated with the glutamate system

It is well established that glutamate occupies a central role in metabolism in both vertebrates and invertebrates and the details of its involvement in metabolic processes are to be found in any general biochemistry textbook.

Barron and Tahmisian (1948) reported that homogenates of cockroach muscle showed increased oxygen uptake on the addition of glutamate and also reported an alanine/glutamate transamination activity. Kilby and Neville (1956, 1957) made a comprehensive study of general amino acid metabolism in a number of locust tissues but did not report specifically on nervous tissue. Consideration of the metabolism of glutamate as a neurotransmitter must involve the examination of four main stages, synthesis, storage, release and inactivation. Some information is available on the first and last of these but virtually no work has been published on storage and release, though the assumption by analogy with other systems, is that the synaptic vesicles contain glutamate.

Donnellan, Jenner and Ramsey (1974) have recently attempted the first detailed investigation of the subcellular localization of a number of glutamate metabolising enzymes in *Sarcophaga* flight muscle, with a view to determining their role in transmission. They applied the techniques developed for the isolation of pinched-off nerve terminals (synaptosomes) from mammalian brain (De Robertis, De Iraldi, Rodriguez de Lores Arnaiz and Gomez, 1961; Gray and Whittaker, 1962) to flight muscle, but failed to obtain a fraction enriched in intact nerve terminals. They were able to show by electron microscope examination of the material, that the initial homogenate contained large numbers of intact terminals but these were apparently disrupted during the subsequent fractionation procedures. Donnellan *et al.* (1974) went on to assay the activities of glutamate dehydrogenase (GDH) (EC 1.4.1.3), glutamic-oxaloacetic transaminase (GOT) (EC 2.6.1.1), glutamic-pyruvic transaminase (GPT) (EC 2.6.1.2), glutamine synthetase (GS) (EC 6.3.1.2) and glutamate decarboxylase (GAD) (EC 4.1.1.15) in each of the subcellular fractions. GDH, GAD and GPT were all found to be mitochondrial enzymes, GOT had a bimodal distribution appearing in both the mitochondrial and in the soluble fractions and GS was almost entirely in the soluble fraction. Unfortunately the authors were unable to show a specific association of any of the glutamate enzyme systems with membranous fractions that could have arisen from the nerve terminals. It would appear that the homogenisation and centrifugation conditions which were developed primarily for mammalian cortex are not suitable for insect muscle. Donnellan *et al.* (1974) make the important observation that no occluded glutamic acid was detected in any membranous fraction, which would suggest that considerable damage had been done to the synaptic vesicles during the isolation procedure and that their transmitter content had been lost. It should be pointed out however that as yet there is no experimental evidence to show that the synaptic vesicles seen at the insect myoneural junction contain glutamic acid.

By analogy with the cholinergic system it has been suggested that the released glutamate could be enzymically degraded at the synapse (Usherwood and Machili, 1968). However, the observation that in *Sarcophaga* at least, GAD is exclusively mitochondrial (Donnellan *et al.*, 1974), suggests that this is not the case, and as described above, GAD does not have this role in the central nervous system either. However degradation is by no means the only way in which inactivation of a transmitter can occur. It is well established that in adrenergic systems re-uptake by the nerve terminal or adjacent cells is the major mechanism for removing transmitter from the synapse (see Iversen, 1967). Faeder and Salpeter (1970) examined the uptake of glutamate by an isolated cockroach (*Gromphadorhina portentosa*) nerve-muscle preparation. They found that glutamate was readily taken up by their preparations and stimulation of the nerve led to increased uptake. Using electron microscope radio-autography they were able to show that the main site of uptake was the sheath cells, particularly those in the region of the neuromuscular junction. The authors suggest that the sheath cells could have an important role in regulating the supply of glutamate to the axon and nerve terminal in addition to removing released glutamate from the synapse.

γ-Amino butyric acid (GABA)

Friedrich (1933) made the first report of an inhibitory innervation in the muscles of *Dixipus morosus*. This was confirmed by several other workers (see Gerschenfeld, 1973). However, it was not until 1964 that Usherwood and Grundfest (1964, 1965) showed that the inhibitory, post-synaptic potentials produced by electrical stimulation of an inhibitory nerve could be produced by the application of $10^{-8} - 10^{-7}$ M GABA. They were also able to show that in both cases the hyperpolarization was the result of an increased chloride permeability of the post-synaptic muscle membrane. Usherwood and Grundfest (1965) showed that the effects of GABA were blocked by the administration of picrotoxin. Virtually no biochemical information is available on the inhibitory system of insect muscles. Huggins, Rick and Kerkut (1967) showed that *Periplaneta* muscle could convert labelled glutamate to labelled GABA suggesting the presence of GAD. However, the experiments could not distinguish between GABA produced in the muscles and that produced in the inhibitory pre-synaptic terminals.

Certainly the electrophysiological data that are available suggest

GABA is an inhibitory transmitter at the myoneural junction and if one considers the data relating to crustaceans (see Gerschenfeld, 1973) the case is considerably strengthened. However, much work needs to be done before the biochemical characteristics of the amino acid transmission system can be described.

Biogenic amines

It seems quite certain that neither the catecholamines nor 5HT have any transmitter role at the insect myoneural junction. Adrenaline was shown to be without action on a *Locusta* muscle preparation (Harlow, 1958) and O'Conner, O'Brien and Salpeter (1965) showed that noradrenaline was also inactive. Usherwood (1963) and Hill and Usherwood (1961) reported some effects of 5HT and tryptamine on locust muscle and suggested that these compounds may block neuromuscular transmission. However, the concentrations required are very high – of the order of 10^{-3} M – and it is unlikely that the findings have any physiological significance.

SUMMARY

Acetylcholine

Acetylcholine is firmly established as the major excitatory transmitter in the central nervous system, and acetylcholinesterase and choline acetylase are both present. Pharmacologically the cholinergic system does not fall into either the nicotinic or muscarinic classes found in vertebrates. This fact is substantiated by the characteristics of the isolated cholinergic receptor. Acetylcholine does not have a role as a neurotransmitter at the myoneural junction.

Glutamic acid

Glutamic acid is the excitatory transmitter at the myoneural junction. The enzymes associated with glutamate synthesis and breakdown are all found in muscle tissue, although no evidence has been presented to show that they are located in the synaptic membranes. A specific glutamate receptor protein has been isolated from muscle tissue. The presence of an active uptake system for glutamate in the sheath cells has been demonstrated. Glutamate does not seem to be a major transmitter in the central nervous system.

GABA

GABA is an inhibitory transmitter at the myoneural junction and GABA synthesis at this site has been demonstrated. In the central nervous system GABA is an important inhibitory transmitter and a number of associated enzymes have been shown to be present.

Biogenic amines

Both noradrenaline and dopamine are present in the central nervous system and may function as minor transmitters. 5HT occurs in central tissues and has been suggested as a transmitter in the heart. None of the amines seem to have any role as transmitters at the myoneural junction.

ACKNOWLEDGEMENTS

The Author thanks Shell Research Ltd. for supporting the original work referred to. Thanks are also due to Dr. J. Donnellan for making copies of his papers available before publication.

REFERENCES

Axelsson. J. and **Thesleff**, S. (1959) *J. Physiol.* (Lond.), **149**, 178–193.
Barron, E. S. G. and **Tahmisian**, T. N. (1948) *J. Cell. Comp. Physiol.*, **32**, 57–76.
Beranek, R. and **Miller**, P. L. (1968) *J. Exptl Biol.*, **49**, 83–93.
Björklund, A., **Falck**, B. and **Klemm**, N. (1970) *J. Insect Physiol.*, **16**, 1147–1154.
Bloom, F. E. and **Agahajanian**, G. K. (1968) *J. Ultrastruct. Res.*, **22**, 361–375.
Bradford, H. F., **Chain**, E. B., **Cory**, H. T. and **Rose**, S. O. R. (1969) *J. Neurochem.*, **16**, 969–978.
Brown, G. L. (1937) *J. Physiol.* (Lond.), **89**, 438–461.
Bullock, T. H. and **Horridge**, G. A. (1965) *Structure and function in the nervous system of invertebrates.* Freeman, San Francisco.
Callec, J. J. and **Boistel**, J. (1967) *Compt, Rendu. Soc. Biol.*, **161**, 442–446.
Cattell, K. J. and **Donnellan**, J. F. (1972) *Biochem. J.*, **128**, 187–189.
Chang, C. C. and **Lee**, C. Y. (1963) *Arch. Int. Pharmacodyn. Ther.*, **144**, 241–257.
Changeux, J. P., **Kasai**, M., **Huchet**, M. and **Meunier**, J. C. (1970) *C. R. Acad. Sci.* Paris, **270**, 2864–2867.
Changeux, J. P., **Kasai**, M., **Huchet**, M. and **Meunier**, J. C. (1971) In: *Cholinergic ligand interactions.* (Triggle, J., Moran, J. F. & Barnard, E. A. eds.), pp. 33–47, Academic Press, New York.
Chanussot, B., **Dando**, J., **Moulins**, M. and **Laverack**, M. S. (1969) *Compt. Rend.*, **268**, 2101–2104.
Clements, A. N. and **May**, T. E. (1974) *J. Expt: Biol.*, **60**, 673–705.
Colhoun, E. H. (1958a) *J. Insect Physiol.*, **2**, 108–116.
Colhoun, E. H. (1958b) *J. Insect Physiol.*, **2**, 117–127.
Colhoun, E. H. (1963a) *Advan. Insect Physiol.*, **1**, 1–46.

Colhoun, E. H., (1963b) *Experientia,* **19**, 9–10.

Colhoun, E. H. (1967) In: *Insects and Physiology* (Beament, J. W. and Treherne, J. E. eds.), pp. 201–213, Oliver and Boyd, Edinburgh.

Corteggiani, E. and Serfaty, A. (1939) *Compt. Rend. Soc. Biol.,* **131**, 1124–1126.

Del Castillo, J. and Katz, B. (1955) *J. Physiol.* (Lond.), **128**, 157–181.

De Robertis, E. (1971) *Science,* **171**, 963–971.

De Robertis, E, De Iraldi, A. P., Rodriguez de Lores Arnaiz, G. and Gomez, J. (1961) *J. Biophys. Biochem. Cytol.,* **9**, 229–235.

De Robertis, E. and Fiszer, S. (1970) *Biochim, Biophys. Acta,* **219**, 388–397.

De Robertis, E. and Fiszer, S. and Soto, E. (1967) *Science,* **158**, 928–929.

De Robertis, E., Lunt, G. and La Torre, J. L. (1971) *Mol. Pharmacol.,* **7**, 97–103.

Donellan, J. F., Jenner, D. W. and Ramsey, A. (1974) *J. Insect Biochem.,* **4**, 243–265.

Doyere, M. (1840) *Ann. Sci. Nat. Ser.* II, **14**, 269–361.

Eldefrawi, A. T. and O'Brien, R. D., (1970) *J. Neurochem.,* **17**, 1287–1293.

Faeder, I. R., O'Brien, R. D. and Salpeter, M. M. (1970) *J. Exp. Zool.,* **173**, 203–214.

Faeder, I. R. and Salpeter, M. M. (1970) *J. Cell. Biol.,* **46**, 300–307.

Falck, B. (1962) *Acta Physiol. Scand.,* **56**, Suppls 197.

Fiszer De Plazas, S. and De Robertis, E. (1970) *FEBS Letts.,* **33**, 45–48.

Foettinger, A. (1880) *Arch. Biol.,* **1**, 279–304.

Fox, P. M. and Larsen, J. R. (1972) *J. Insect Physiol.,* **18**, 439–457.

Friedrich, F. (1933) *Z. Vergleich. Physiol.,* **18**, 536–561.

Frontali, N. (1961) *Nature,* **191**, 178–179.

Frontali, N. (1965) In: *Comparative Neurochemistry.* (Richter, D., ed.) pp. 185–192, Pergamon Press, Oxford.

Frontali, N. (1968) *J. Insect Physiol.,* **14**, 881–886.

Frontali, N. and Norberg, K. A. (1966) *Acta Physiol. Scand.,* **66**, 243–244.

Gautrelet, J. (1938) *Bull. Acad. Nat. Med.,* **120**, 285–291.

Gerschenfeld, H. M. (1973) *Physiol. Revs.,* **53**, 1–119.

Ginetzinsky, A. G and Shamarina, N. M. (1942) *Adv. Mod. Biol (USSR) Usp. Sovrem. Biol.,* **15**, 283–294.

Gray, E. G. and Whittaker, V. P. (1962) *J. Anat.,* **96**, 79–88.

Harlow, P. A. (1958) *Ann. Appl. Biol.,* **46**, 55–73.

Hayashi, T. (1954) *Chemical Physiology of excitation in muscle and nerve,* Natayama Shaten, Tokyo.

Hebb, C. O. and Krnjevic (1962) In: *Neurochemistry* (Elliot, K. A. C., Page, I. H. and Quastel, J. H. eds.), Thomas Springfield.

Henschler, D. (1965) *Hoppe-Seylers Z.,* **305**, 34–41.

Hill, R. B. and Usherwood, P. N. R. (1961) *J. Physiol.* (Lond.), **157**, 393–401.

Hiripi, L. and Rozsa, K. S. (1973) *J. Insect Physiol.,* **19**, 1481–1485.

Holman, G. M. and Cook, B. J. (1970) *J. Insect Physiol.,* **16**, 1891–1907.

Hopf, H. S. (1952) *Ann. Appl. Biol.,* **39**, 193–202.

Huggins, A. K., Rick, J. T. and Kerkut, G. A. (1967) *Comp. Biochem. Physiol.,* **21**, 23–30.

Iversen, L. L. (1967) *The uptake and storage of noradrenalin in sympathetic nerves,* Cambridge University Press, Cambridge.

Iversen, L. L. and Kravitz, E. A. (1968) *J. Neurochem.*, 15, 609–620.
Jones, D. G. (1970) *Z. Zellforsch. Mikrosk. Anat.*, 103, 48–60.
Kerkut, G. A., Leake, L. D., Shapira, A., Cowan, S. and Walker, R. J. (1965) *Comp. Biochem. Physiol.*, 15, 485–502.
Kerkut, G. A., Oliver, G., Rick, J. T. and Walker, R. J. (1970) *Nature*, 227, 722–723.
Kerkut, G. A., Pitman, R. M. and Walker, R. J. (1969a) *Comp. Biochem. and Physiol.*, 31, 611–633.
Kerkut, G. A., Pitman, R. M. and Walker, R. J. (1969b) *Nature*, 222, 1075–1076.
Kerkut, G. A., Shapira, A. and Walker, R. J. (1965) *Comp. Biochem. Physiol.*, 16, 37–48.
Kerkut, G. A. and Walker, R. J. (1966) *Comp. Biochem. Physiol.*, 17, 435–454.
Kerkut, G. A. and Walker, R. J. (1967) *Brit. J. Pharmacol.*, 30, 644–654.
Kilby, B. A. and Neville, E. (1956) *Biochim. Biophys. Acta*, 19, 389–90.
Kilby, B. A. and Neville, E. (1957) *J. Exptl. Biol.*, 34, 276–289.
Langley, J. N. (1905) *J. Physiol.* (Lond.), 33, 374–398.
Lewis, S. E. and Smallman, B. N. (1956) *J. Physiol.* (Lond.), 134, 241–256.
Lunt, G. G. (1973) *Comp. Gen. Pharmacol.*, 4, 75–79.
Lunt, G. G., Stefani, E. and De Robertis, E. (1971) *J. Neurochem.*, 18, 1545–1553.
McCann, F. V. and Reece, R. W. (1967) *Comp. Biochem. Physiol.*, 21, 115–124.
Mikelonis, S. J. and Brown, R. H. (1941) *J. Cellular Comp. Physiol.*, 18, 401–403.
Miledi, R. (1960a) *J. Physiol.*, (Lond.), 151, 1–23.
Miledi, R. (1960b) *J. Physiol.*, (Lond.), 151, 24–30.
Miledi, R., Molinoff, P. and Potter, L. T. (1971) *Nature*, 229, 554–557.
Murdock, L. L. and Koidl, B. (1972a) *J. Exptl. Biol.*, 56, 781–794.
Murdock, L. L. and Koidl, B. (1972b) *J. Exptl. Biol.*, 56, 795–808.
O'Conner, A. K., O'Brien, R. D. and Salpeter, M. M. (1965) *J. Insect Physiol.*, 11, 1351–1358.
Pappas, G. D. and Purpura, D. P. (1972) *Structure and function of synapses*, Raven, New York.
Pitman, R. M. (1971) *Comp. Gen. Pharmacol.*, 2, 347–371.
Pitman, R. M. and Kerkut, G. A. (1970) *Comp. Gen. Pharmacol.*, 1, 221–230.
Plotnikova, S. N. and Govyrin, W. A. (1966) *Archs. Anat. Histol. Embryol.*, 50, 79–87.
Ray, J. W. (1964) *J. Insect Physiol.*, 10, 587–597.
Robbins, J. (1959) *J. Physiol.* (Lond.), 148, 39–50.
Roeder, K. D. (1948) *Bull. Johns Hopkins Hosp.*, 83, 587–600.
Roeder, K. D., Kennedy, N. K. and Samson, E. A. (1947) *J. Neurophysiol.*, 10, 1–10.
Roeder, K. D. and Roeder, S. (1939) *J. Cellular Comp. Physiol.*, 14, 1–12.
Rozsa, S. K. and Szöke, V. (1972) *Acta Physiol. Acad. Sci. Hung.*, 41, 27–36.
Salganicoff, L. and De Robertis, E. (1963) *Life Sci.*, 2, 85–91.
Salganicoff, L. and De Robertis, E. (1965) *J. Neurochem.*, 12, 287–309.
Shankland, D. J., Rose, J. A. and Donninger, C. (1971) *J. Neurobiol.*, 2 247–262.
Shinozaki, H. and Shibuya, I. (1972) *Jap. J. Pharmacol.*, 22, (Suppl). 100.

Smallman, B. N. and Fischer, R. W. (1961) *Can. J. Biochem. Physiol.*, **36**, 575–586.

Smith, D. S. (1967) In: *Insects and Physiology* (Beament J. W. and Treherne, J. E., eds.) pp. 189–198, Oliver and Boyd, Edinburgh.

Suga, N. and Katsuki, Y. (1961) *J. Exptl. Biol.*, **38**, 759–770.

Takeuchi, A. and Takeuchi, N. (1964b) *Nature*, **203**, 1074–1075.

Takeuchi, A. and Takeuchi, N. (1964a) *J. Physiol.* (Lond.), **170**, 296–317.

Telford, J. N. and Matsumura, F. (1970) *J. Econom. Entomol.*, **63**, 795–800.

Tiegs, O. W. (1955) *Phil. Trans. Roy. Soc.* London Series B, **238**, 221–348.

Tobias, J M., Kollros, J. J. and Savit, J. (1946) *J. Cellular Comp. Physiol.*, **28**, 159–182.

Trujillo-Cenoz (1962) *Z. Zellforsch*, **56**, 649–682.

Twarog, B. M. and Roeder, K. D. (1957) *Ann. Entomol. Soc. Am.*, **50**, 231–237.

Usherwood, P. N. R. (1963) *J. Physiol. Lond.*, **169**, 149–160.

Usherwood, P. N. R. (1969) *Nature*, **223**, 411–413.

Usherwood, P. N. R. (1971) *Proc. XXV Int. Congr. Union Physiol. Sci.*, **8**, 251–252.

Usherwood, P. N. R. (1972) *Neurosci. Res. Prog. Bull.*, **10**, 136–143.

Usherwood, P. N. R. and Grundfest, H. (1964) *Science*, **143**, 817–818.

Usherwood, P. N. R. and Grundfest, H. (1965) *J. Neurophysiol.*, **28**, 497–518.

Usherwood, P. N. R. and Machili, P. (1966) *Nature*, **210**, 643–636.

Usherwood, P. N. R. and Machili, P. (1968) *J. Exptl. Biol.*, **49**, 341–361.

Van-Harreveld, A. and Mendelson, M. (1959) *J. Cellular Comp. Physiol.*, **54**, 85–94.

Waelsch, H. (1951). In: *Advances in Protein Chemistry* (Anson, M. L., Edsall, J. T. and Bailey, K. eds.) vol. 6, pp. 299–341, Academic Press, New York.

Weber, G., Borris, D. P., De Robertis, E., Barrantes, F. J., La Torre, J. L. and Llorente De Carlin, M. C. (1971) *Mol. Pharmacol.*, **7**, 530–570.

Welsh, J. H. and Moorhead, M. (1960) *J. Neurochem.*, **6**, 146–169.

Wigglesworth, V. B. (1960) *J. Exptl. Biol.*, **37**, 500–512.

ADDENDUM

It has recently been shown that 2-amino, 4-phosphono butyric acid (a phosphonate derivative of glutamic acid) competes with glutamate for the binding site on the isolated proteolipid receptor. Binding of 10^{-5} M glutamate to the receptor from *Schistocerca gregaria* was reduced 65% by 1.4×10^{-4} M phosphonate. A similar concentration of phosphonate decreased glutamate binding to a proteolipid from *Musca domestica* muscle by up to 70% (James, Lunt and Donnellan, 1974).

A recent report by Roberts (1974) suggests that it may be possible to distinguish between glutamate synaptic receptors and glutamate re-uptake sites. The data reported were obtained from rat cerebral cortex and revealed that DL-homocysteic acid and glutamic diethyl ester both produced a very significant inhibition of the binding of glutamate to the synaptic receptor. The effect of homocysteic acid on the uptake sites was less pronounced while glutamic acid diethyl ester was without effect. It is to be hoped that similar studies on the insect myoneural junction will add greatly to our limited knowledge of the pharmacology and biochemistry of the glutamate transmitter system.

James, R. W., Lunt, G. G. and Donnellan, J. F. (1974) *Proc. 9th. FEBS Meeting*, p. 258. Hung. Biochem. Soc. Budapest.
Roberts, P. J. (1974) *Nature Lond.*, 252, 399—400.

Index

Acetate conversion to fatty acids, 125, 128, 133
Acetate thiokinase, 125
Acetoacetate metabolism, 113, 114
Acetylcholine, 287-290, 301
Acetylcholinesterase, 288, 301
Acetylcholine receptors, 289, 290
Acetyl CoA carboxylase, 118, 119, 124
Acetyl CoA oxidation, 47-49
Aconitase, 55, 127
Actomyosin ATPase, 17, 61, 74-76
Adenine nucleotide phosphorylation state, 26, 70
Adenine deaminase (adenase), 203, 204, 206, 219
Adenosine deaminase, 202-204, 219
Adenylate kinase, 54, 55
Adipokinetic hormone, 143-145
ADP
  Concentration in muscle, 26
  Isocitrate dehydrogenase regulation, 67, 68, 80, 81
  α-Ketoglutarate dehydrogenase regulation, 69, 70
  Oxidative phosphorylation, 57-60, 63, 67, 77, 79, 80
  Phosphofructokinase regulation, 25, 130
  Proline dehydrogenase regulation, 42, 63-65
  Pyruvate kinase regulation, 28, 29
Alanine from glutamate, 296
Alanine from proline, 41-43
Aldolase, 24, 27, 43
Allantoic acid excretion, 191, 193
Allantoicase, 208, 209
Allantoin excretion, 191, 193, 195, 207
Allantoinase, 208, 225
AMP
  Concentration in muscle, 26
  Phosphofructokinase regulation, 25, 26, 130-132
  Phosphorylase regulation, 8-14, 159-163
AMP, cyclic, 18
  Effect on lipase, 46
  Effect on lipid transport, 143
  Phosphofructokinase regulation, 25, 130
  Phosphorylase regulation, 159, 160
Amino acids
  Active transport, 232

Conversion to carbohydrate, 157, 158
Conversion to lipid, 118, 120, 126
Excretion, 190, 228-231
Fuel for flight, 92, 93, 164-166
In haemolymph, 164, 165
Oxidases, 218, 221
Oxidation in muscle, 40
γ-Aminobutyric acid, *see* GABA
2-Amino, 4-hydroxypteridine (pterine), 243-246, 252
Ammonia
  Excretion, 215-223
  Formation, 217-221
  Toxicity, 179, 216
  Transport, 222
Anal papillae in excretion, 264
Anthranilic acid metabolism, 235, 240, 241
Antidiuretic hormone, 186, 191
Arginase, 225-227
Arginine phosphate, 25, 26
Aspartate, 36, 44
Asynchronous muscle, 17, 74
ATP-citrate lyase, 127, 128, 136
ATP
  Concentration in muscle, 26
  Glycogen synthetase regulation, 19, 24
  Isocitrate dehydrogenase regulation, 67, 68
  α-Ketoglutarate dehydrogenase regulation, 69, 70
  Oxidative phosphorylation, 57-60
  Phosphofructokinase regulation, 25, 26, 130-132
  Phosphorylase regulation, 11, 12, 14
  Proline dehydrogenase regulation, 65
  Pyruvate kinase regulation, 29, 132
  Synthetase, 53-55, 74

Biogenic amines, 293, 301, 302
Biopterin, 245, 247
Brain neurosecretory cells effects on metabolism, 116, 117, 123, 161, 163, 185
Bungarotoxin, 290

Calcium carbonate excretion, 255
Calcium ions
  Control of muscle metabolism, 71, 74-76, 80, 81

Fructose-1,6-diphosphatase regulation, 34
α-Glycerol phosphate dehydrogenase regulation, 61-63, 75, 76, 79
Isocitrate dehydrogenase regulation, 68, 69, 75, 76
Mitochondrial release, 73, 74
Mitochondrial uptake, 71-74
Phosphorylase *b* kinase regulation, 15-17
Pyruvate oxidation regulation, 68, 69
Trehalase regulation, 22
Calcium oxalate excretion, 255
Carbamyl phosphate synthetase, 227
Carbohydrates
Biosynthesis, 157, 158
Conversion to lipid, 118, 120-128, 130, 133
Excretion, 254, 255
Fuel for flight, 5, 92, 93, 146-150
Gut content, 147, 149, 150
Carbon dioxide excretion, 179
Carcinogenic tryptophan derivatives, 242, 243
Carnitine acyl transferase, 47, 48, 54, 55
Carnitine, 46-48
Catecholamines, 293, 301
Chemi-osmotic coupling, 56, 70
Chloride ions in excretion, 188, 189
Choline acetylase, 288, 301
Citrate, 36, 37
Role in fatty acid synthesis, 127-129
Citrate lyase, 127, 128, 136
Citrate synthetase, 54, 55
Citric acid cycle, *see* tricarboxylic acid cycle
Citrulline, 226, 227
Conjugation reactions, 257
Constant proportion groups of enzymes, 28, 47
Corpora allata metabolic effects, 116, 123, 134-136, 161, 164
Corpora cardiaca hormones, 113, 116, 117, 138, 143-145, 161-163, 185, 191
Creatine excretion, 253
Creatinine excretion, 253
Curare, 290
Cyclic AMP, *see* AMP cyclic
Cystine excretion, 231
Cytochrome oxidase, 53
Cytochromes, 76-81

Denervation of muscle, 297
Detoxification mechanisms, 257
Diglycerides
Fat body, 100-102, 108-110, 145
Haemolymph, 103-105, 108, 109, 137, 138, 141, 142
Hydrolysis, 46
Release from fat body, 137-139, 141-144
Role in lipid digestion, 108
Synthesis, 123, 133, 134, 137
Uptake by fat body, 108-110, 145
3,4-Dihydroxyphenylalanine, *see* DOPA
Diuretic hormones, 185, 186, 191
DOPA, 240
DOPA decarboxylation, 293
Dopamine, 293, 302
Dye excretion, 256

Electron transport, 38, 53, 55, 56, 76
Energy charge, 70
Enolase, 24
Erythopterin, 245, 247, 248, 250, 251
Excretion definition, 179-181

Fat body
Lipid degradation, 110, 111, 112
Lipase, 110, 114
Role in excretion, 205, 241, 242
Symbionts, 212
Fat content of muscle, 45
Fat oxidation, 45-48
Fat stores, 45
Fatty acids
Desaturation, 121, 122, 126
Essential, 107, 120, 121
Fat body, 95, 100-102
Haemolymph, 103-105, 108, 138, 139
Oxidation in fat body, 110-112, 114
Oxidation in muscle, 45-47, 49
Release from fat body, 137-139, 143
Synthesis, 117-133
Uptake by fat body, 109
Fatty acyl CoA synthetase, 48, 49
Fluorescein excretion, 263
Folic acid, 244, 245, 250
Formed body excretion, 187, 258
Fructose-1,6-diphosphatase, 24, 26, 33, 34, 157-159
Fructose-1,6-diphosphate, 25
Pyruvate kinase regulation, 29

Fructose
  Haemolymph, 147
  Metabolism, 150-152
Fuel reserves, 5, 45, 93, 94
Fumarase, 55
Fumarate, 36

GABA, 290-292, 300-302
GABA-aminoglutarate
  aminotransferase, 291, 292
Galactose metabolism, 150-152
Ganglia structure, 286
Glial cells, 286
Gluconeogenesis, 33, 157, 158
Glucose
  Fat body, 146
  Haemolymph, 147-149
Glucose-1-phosphate, 9, 10, 12
Glucose-6-phosphatase, 24, 33, 156
Glucose-6-phosphate, 19, 27
Glucose-6-phosphate dehydrogenase,
  23, 127, 129, 136
Glutamate, 36, 37, 40, 42-44
  Haemolymph, 296, 297
  Metabolism, 298-300
  Neurotransmitter, 290-301
  Proline dehydrogenase regulation,
  65
Glutamate decarboxylase, 290-292,
  299
Glutamate dehydrogenase, 43, 44,
  217, 218, 220, 299
Glutamate-oxaloacetate
  aminotransferase, 299
Glutamate-pyruvate aminotransferase,
  299
Glutamate receptor, 297, 298
Glutamine from glutamate, 296
Glutamine synthetase, 299
Glyceraldehyde-3-phosphate
  dehydrogenase, 24, 28, 29
Glycerol, 34, 35
  Conversion to carbohydrate, 157,
  158
Glycerol kinase, 35, 133, 157, 158
α-Glycerol phosphate cycle, 24, 29-32,
  35, 57
α-Glycerol phosphate dehydrogenase
  of cytosol, 24, 29-31, 35, 157,
  158
α-Glycerol phosphate dehydrogenase
  of mitochondria, 24, 29, 30, 38,
  39, 53-55, 61-63, 75

Glycerol phosphate oxidation, 36,
  38-40, 43, 57-63, 73
α-Glycerol phosphate phosphatase, 35
Glycine, 40
Glycogen
  Haemolymph, 147
  Muscle, 5-8, 18, 19, 43, 93
  Metabolic control by hormones, 161,
  162
  Molecular weight, 8
  Reserves, 5-8, 93, 146-150
  Synthesis, 18, 19, 123, 152-154,
  156
Glycolysis
  Fat body, 127-133, 150, 158
  Muscle, 22-30, 57
  Intermediates, 24-30
Glyoxalate cycle, 159
Guanine deaminase (guanase), 219
Guanine excretion, 191, 193
Guanosine conversion to pteridines,
  249, 251
Gut
  Fuel supplies, 94, 95
  Lipids, 95, 106, 107
  Role in ammonia excretion, 222,
  223

Haematin excretion, 253
Haemoglobin digestion, 253, 260
Hexokinase
  Muscle, 22-24
  Specificity, 27, 150, 151
  Regulation, 27
Hexose monophosphate pathway, 23
Hexose phosphate isomerase, 24, 27,
  43, 151
Heat generation in flight, 34
Hindgut role in excretion, 263, 264
Histidine excretion, 231
Honeydew, 230, 254, 255
Hormonal regulation
  Carbohydrate metabolism, 161, 162,
  164
  Lipid synthesis, 134-136
  Lipid transport, 138, 143-145
  Urine excretion, 185, 186, 191
Hydrocarbons of fat body, 100, 101
3-Hydroxyacyl CoA dehydrogenase,
  47, 49, 54, 55, 111-113
3-Hydroxyanthranilic acid metabolism,
  235, 240-242
β-Hydroxybutyrate metabolism, 113,
  114

3-Hydroxykynurenine formation and excretion, 235-240, 242
Hydroxymethyl glutaryl CoA, 113, 114
8-Hydroxyquinaldic acid, 235, 237, 242
5-Hydroxytryptamine
    Decarboxylation, 293
    Formation, 233
    Physiological activities, 233, 234
    Transmitter action, 293, 294, 301, 302
    Urine excretion effects, 185
Hyperglycaemic hormone, 18, 145, 161, 163
Hypoxanthine excretion, 191-195, 198 202, 203

Iodoacetate, 28
Ion excretion, 182
Insecticide excretion, 256-258
Isocitrate, 36, 37
Isocitrate dehydrogenases, 32, 43, 54, 55, 67, 127, 129, 136
    Regulation, 67, 68, 80, 81
Isosepiapterin, 245
Isoxanthopterin, 245-248, 252

Juvenile hormone
    Effect on glycogen metabolism, 164
    Effect on lipid synthesis, 135, 136

3-Ketoacyl CoA thiolase, 49, 111, 112
α-Ketoglutarate, 36, 37, 44
α-Ketoglutarate dehydrogenase, 54, 55, 69
    Regulation, 69, 70
Ketone body metabolism, 113, 114
Krebs cycle, *see* tricarboxylic acid cycle
Kynurenic acid formation and excretion, 235-237, 240
Kynurenine formation and excretion, 235-239
Kynurenine foramidase, 238, 239, 241
Kynurenine 3-hydroxylase, 239-241
Kynurenine transaminase, 240, 241

Labial glands structure and function, 259, 260
Lactate dehydrogenase, 24, 29-31
Leucopterin, 245, 246, 248, 251, 252
Lipases
    Fat body, 110, 139, 140, 145
    Gut, 107

Haemolymph, 138-141
Muscle, 45, 46
Lipids
    Absorption, 107, 108
    Fat body, 98-102, 106, 111, 135
    Fuel for flight, 5, 92-99
    Haemolymph, 103-106
    Honeydew, 255
    Muscle, 93, 137
    Release from fat body, 137-144
    Reserves in whole insect, 96, 97
    Synthesis, 92, 117-120, 137
    Transport, 45, 108, 137-145
    Uptake by fat body, 145
Lipoproteins of haemolymph, 108, 141-143

Magnesium ions and phosphorylase *b* kinase, 17
Malate, 36, 37
Malate dehydrogenase, 32, 54, 55, 127, 128
Malic enzyme, 127, 128, 136
Malpighian tubules
    Amino acid excretion, 229
    Ammonia excretion, 222
    Secretion, 183-186, 194
    Structure, 183, 185, 190, 194
Malpighian tubule-rectum complex, 179, 181
Mannose metabolism, 150-152
Midgut role in excretion, 262, 263
Migration, 96, 97, 99
Mitochondria
    Cristae, 50, 52-54
    Enzyme localization, 53-55
    Morphology, 50-54
    Permeability, 29, 32, 35-38, 41, 42, 47-49, 53, 70-74, 127
    Respiratory rates, 57-59
    Substrate oxidation, 36-41
Mixed function oxidases, 257
Monoamine oxidase, 219, 221
Monoglycerides
    Fat body, 100-102
    Haemolymph, 103-105
    Hydrolysis, 46
Muscarone, 289
Muscle morphology, 50, 51, 73
Mutants and tryptophan metabolism, 238-240

NADH oxidation, 29-33, 36, 54
NADP reduction, 32, 33

NADPH in fatty acid synthesis, 124, 127, 128
Nephrocytes, 260
Nervous system structure, 285-287
Neural lamellae, 286
Neuromuscular junction, 286, 287, 294-301
Neurosecretions, 116, 117, 123, 161, 163, 185
Noradrenaline, 293, 301
Nucleic acid degradation, 198-201

Ommochrome pigments, 232, 234-243
Ornithine cycle, 224-227
Ornithine transaminase, 226
Osmoregulation, 188, 222, 263
Ovary symbionts, 212
Oxalic acid excretion, 255
Oxaloacetate decarboxylase, 32, 42
Oxaloacetate from proline, 41, 42
β-Oxidation, 49, 55, 111, 112
Oxidative phosphorylation, 56-60, 76, 115, 117
Oxygen uptake during flight, 3, 4, 49

Pentose phosphate pathway, 23, 127-130, 150
Peptide excretion, 253, 254, 260
Pericardial cells, 260, 261
Phosphate inorganic, 9-15, 17, 28
    Concentration in muscle, 26
    Mitochondria accumulation, 70
    Oxidative phosphorylation, 57-60, 66, 67
    Phosphofructokinase regulation, 25
    Proline dehydrogenase regulation, 65
    Tricarboxylic acid cycle regulation, 67
    Triose phosphate isomerase regulation, 28
Phosphatidic acid in glyceride synthesis, 133, 134
Phosphoenolpyruvate carboxykinase, 33, 157-159
Phosphofructokinase, 22-27, 33, 34, 128, 130-133, 136, 158
Phosphoglucomutase, 24, 43
Phosphoglucose isomerase, 24, 27, 43, 151
Phosphoglycerate kinase, 24, 28
Phosphoglyceromutase, 24
Phospholipids
    Digestion, 108

Fat body, 99-102
Haemolymph, 103-105, 142, 143
Membrane, 120
Mitochondria, 74
Synthesis, 122, 123, 125
Phosphorylase, 8-15, 23, 24, 43, 153, 154, 159-163
Phosphorylase *a* phosphatase, 18
Phosphorylase *b* kinase, 15-19, 61, 74, 75
Potassium ions
    Malpighian tubules, 183, 184, 186, 194, 210, 211
    Midgut, 263
    Rectum, 188, 189
Proline
    Fuel for flight, 92, 93, 164-166
    Oxidation, 40-43, 58, 63-65, 164, 165, 228
    Synthesis, 165, 166, 226-228
Proline dehydrogenase, 41-43, 53-55, 63, 65
Protein excretion, 253, 254, 260
Pteridines
    Effect on tryptophan pyrrolase, 239, 249
    Formation and excretion, 243-252
    Pigments, 180, 243-247
Pterines, 243-252
Purines
    Conversion to pteridines, 249
    Synthesis, 197, 198
Pyrimidines excretion, 201
Pyrroline-5-carboxylate dehydrogenase, 41, 42
Pyrroline-5-carboxylate reductase, 226
Pyruvate carboxylase, 33, 66, 128, 157-159
Pyruvate dehydrogenase, 38, 54, 55, 66, 75, 127
Pyruvate kinase, 24, 28, 128, 130-133, 136
Pyruvate oxidation, 36, 38-43, 45, 56-60, 63, 65-69

Receptors, 289, 290, 297, 298
Rectum in excretion, 187-190
Respiration
    During flight, 3, 4, 91
    Hormonal control in fat body, 115, 116
Respiratory chain, 115-117
    Oxidation state of components, 76-81

Respiratory control, 49, 57-61
Respiratory quotients, 5, 92, 110, 111
Respiratory system in ammonia
    excretion, 223
Riboflavin, 244, 245, 248, 250

Sarcoplasmic reticulum, 17, 50, 51,
    61, 71, 73
Scyllo-inositol, 147
Secretion definition, 179, 180
Sepiapterin, 245, 247, 248, 250
Serine dehydrase, 219
Sodium ions
    Malpighian tubules, 183, 184
    Rectum, 188
Sorbitol metabolism, 150-152
Sterols and sterol esters
    Fat body, 100, 101
    Haemolymph, 103-105
Storage excretion, 180, 195, 196, 210,
    211, 244
Subcellular fractionation of nervous
    tissue, 292
Substrate cycling, 33, 34
Sugar excretion, 255
Succinate, 36, 37
Succinate dehydrogenase, 43, 54, 55
Succinate semialdehyde
    dehydrogenase, 291, 292
Symbionts, 212-214
Synchronous muscle, 17, 74
Synapses, 286
Synaptic transmission, 287-301

Transaminases, 32, 41-44, 54, 55, 217,
    218, 220
Transmitters, 287-301
Trehalase, 20-22, 24, 43, 53-55, 156
Trehalose, 20, 43
    Fat body, 146
    Haemolymph, 146-150, 152, 156
    Muscle, 93
    Regulation óf synthesis, 155, 156
    161, 162
    Synthesis, 154-156
    Utilization during flight, 20
Trehalose-6-phosphate, 155, 156
Tricarboxylic acid cycle, 36, 37, 41,
    42, 44, 49, 56, 57, 65, 66, 68,
    112, 114, 115, 127

Triglycerides
    Fat body, 95, 100-102
    Fatty acid composition, 97, 98,
        102, 103
    Haemolymph, 103-105, 109, 138
    Hydrolysis, 46
    Release from fat body, 137-139
    Synthesis, 122, 123, 125, 133, 134,
        137
Triose phosphate isomerase, 24, 28
Tryptophan derivatives, 232-243
    Carcinogenic properties, 242, 243
Tryptophan metabolism mutants,
    238-240
Tryptophan pyrrolase, 235, 238, 239,
    241

UDP glucose, 18, 19
UDP glucose pyrophosphorylase, 152,
    154
Urate ion excretion, 186, 194, 210-212
Urea, 209
    Excretion, 223-227
    Formation from purines, 224, 225
    Formation via ornithine cycle,
        224-227
Urease, 209, 217
Uric acid
    Excretion, 180, 190-215, 261, 262
    Oxidation, 209, 213, 214
    Transport, 214, 215
    Utilization as nitrogen store, 212,
        261, 262, 265
Uricase, 207, 208, 225
Uricose glands, 180
Urine
    Composition, 183
    Definition, 182
Utriculi majores, 261, 262

Violopterin, 247, 248

Water excretion, 182, 184, 188, 191

Xanthine dehydrogenase, 191,
    203-207, 250-252
Xanthine excretion, 192, 193, 203
Xanthine oxidase, 252
Xanthopterin, 245, 247, 248, 250-252
Xanthurenic acid, 235-237, 240, 242
Xenobiotics excretion, 255